BARRON'S

AP®
Human Geography
PREMIUM

WITH 4 PRACTICE TESTS

NINTH EDITION

Meredith Marsh, Ph.D.

Peter S. Alagona, Ph.D.

D0608261

ACKNOWLEDGMENTS

It's hard to believe that this project has been going on for over ten years now. The best part of its success is that it allows me to maintain my friendship with Pete Alagona; I couldn't ask for a more delightful colleague! I also am grateful to Jennifer Giammusso, our editor, for her patient provision of resources, extended deadlines, and much encouragement. Finally, my colleagues, students, and, most important, my family (Lucy especially), consistently keep me smiling and excited about what I do, which makes projects like this much more enjoyable.

Meri Marsh, Ph.D.

This book has become something of an old friend. It has appeared and reappeared in my life many times over the past several years. Each time it reappears it brings back old memories and leaves me with new ones. Many people have made this volume possible. My family, friends, mentors, and students have all played important roles. Most of all, however, this book owes its existence to my friend and colleague, Meri Marsh, without whom it would never have reached a first draft, much less a ninth edition.

Peter S. Alagona

ABOUT THE AUTHORS

Meredith J. Marsh received her bachelor's degree in geography from Calvin College in 1999, her master's degree in geography from the University of California, Santa Barbara (UCSB) in 2003, and her Ph.D. (also in geography) from UCSB in 2009. Meredith has worked as a reader for the APHG exam in 2007, 2009, 2010, 2011, 2013, and 2014, and she is currently an associate professor of geography at Lindenwood University in St. Charles, Missouri.

Peter S. Alagona is an assistant professor of history and environmental studies at the University of California, Santa Barbara. He holds an M.A. degree in geography from the University of California, Santa Barbara, and M.A. and Ph.D. degrees in history from the University of California, Los Angeles. He has also held academic positions at both Harvard and Stanford universities.

© Copyright 2020, 2018, 2016, 2014, 2012, 2010, 2008 by Kaplan, Inc.,
d/b/a Barron's Educational Series

Previous edition © copyright 2003 by Kaplan, Inc., d/b/a Barron's Educational Series under the title *How to Prepare for the AP Human Geography Exam*

Photo Credits: *www.chartsbin.com*, page 122

Published by Kaplan, Inc., d/b/a Barron's Educational Series
750 Third Avenue
New York, NY 10017
www.barronseduc.com

ISBN: 978-1-5062-5884-3

10 9 8 7 6 5 4 3 2 1

Kaplan, Inc., d/b/a Barron's Educational Series print books are available at special quantity discounts to use for sales promotions, employee premiums, or educational purposes. For more information or to purchase books, please call the Simon & Schuster special sales department at 866-506-1949.

Contents

As you review the content of this book to work toward earning a **5** on your AP HUMAN GEOGRAPHY exam, you **MUST** focus on these five essential points:

Barron's
Essential
5

1 **Know the course goals and how the subdisciplines of human geography meet them.** This may seem fairly obvious. However, you should make sure that as you study each subdiscipline in the course, you can clearly describe geography within that subdiscipline. For example, what's unique about political organization as compared with politics? Cities and urban land use as compared with urban studies? Or cultural patterns and processes as compared with cultural studies?

2 **Think spatially.** Think in multiple ways! You need to read and understand any spatial data in the text and test (e.g., maps) carefully. You must also be able to think across scales. How might spatial patterns and processes (or representations of them) differ when looking at large-scale spaces (e.g., the world) compared with small-scale spaces (e.g., your neighborhood)?

3 **Know current examples.** As you will quickly notice when completing the practice tests, free-response questions often ask for examples. As you learn the concepts in each chapter, keep lists of current events that accurately illustrate concepts, processes, and models. Watch and read the news. Think about how the news connects to human geography!

4 **Look for interconnections**. Pay special attention to the different subdisciplines. Can your free-response question on urban models bring in some of your knowledge about economic development? How does culture influence economic development? What are the connections between population dynamics and urbanization? Do not memorize concepts in isolation. Instead, constantly try to connect across topic areas.

5 **Focus on figures.** Figures (maps, graphics, charts, and tables) frequently appear on the test in both multiple-choice and free-response questions. **Do not skip over the figures in the text!** Instead, know them by name, make sure you know how they work, and be able to describe what they represent.

Introduction

- → **WELCOME TO THE NINTH EDITION**
- → **ABOUT THE BOOK**
- → **ABOUT THE TEST**
- → **STRATEGIES FOR PREPARING FOR THE AP HUMAN GEOGRAPHY TEST**
- → **COMMONLY ASKED QUESTIONS ABOUT THE AP HUMAN GEOGRAPHY TEST**
- → **ADDITIONAL RESOURCES**

WELCOME TO THE NINTH EDITION

The ninth edition of Barron's *AP Human Geography* contains some substantial updates to account for recent changes in the course outline and exam structure. This chapter, the Introduction, has been extensively revised to reflect more accurately the topics and skills covered on the new exam (beginning May 2020). In addition, both practice exams have been completely rewritten; every question now contains "metadata" that show the specific unit, skill, and "essential knowledge" covered by the question. Inclusion of these data ensures alignment between the practice tests and the actual exam while also providing students with the information needed to target their studies.

ABOUT THE BOOK

The AP Human Geography Exam covers a range of material that would normally be included in a semester-long college-level course in Introductory Human Geography. These courses typically cover a wide variety of topics, with the basic goal of understanding patterns and processes that have shaped human relationships on Earth over space and time. This overarching goal can be further divided into a set of goals derived from *Geography for Life: National Geography Standards, Second Edition*, introduced in 2012.

> The multiple-choice questions in each chapter should only be reviewed as content/concept review. These questions don't necessarily reflect the same level of difficulty as the exam questions.

The APHG Exam covers seven major content areas. Each content area, described later in this chapter, will make up a certain percentage of the exam. This book has been designed to correspond with the College Board's objectives, and one chapter is dedicated to each of the Board's seven major content areas. Each chapter contains content knowledge about the topic under discussion, tips on preparing for the exam, and practice multiple-choice and free-response questions with answers explained.

ABOUT THE TEST

The AP Human Geography Exam contains two sections and lasts for two hours and 15 minutes. The first section includes 60 multiple-choice questions; students are given 60 minutes to complete this portion of the exam. In the remaining 75 minutes, students answer three free-response essay questions.

Beginning in 2019, the AP Human Geography course and exam now contains an explicit emphasis on geographic skills in addition to the traditional content areas. Every question—multiple-choice and free-response—assesses one of the following five skills or skill components.

Skill Category 1: Concepts and Processes	**1.A:** Describe geographic concepts, processes, models, and theories. **1.B:** Explain geographic concepts, processes, models, and theories. **1.C:** Compare geographic concepts, processes, models, and theories. **1.D:** Describe a relevant geographic concept, process, model, or theory in a specified context. **1.E:** Explain the strengths, weaknesses, and limitations of different geographic models and theories in a specified context.
Skill Category 2: Spatial Relationships	**2.A:** Describe spatial patterns, networks, and relationships. **2.B:** Explain spatial relationships in a specified context or region of the world using geographic concepts, processes, models, and theories. **2.C:** Explain a likely outcome in a geographic scenario using geographic concepts, processes, models, and theories. **2.D:** Explain the significance of geographic similarities and differences among different locations and/or at different times. **2.E:** Explain the degree to which a geographic concept, process, model, or theory effectively explains geographic effects in different contexts and regions of the world.
Skill Category 3: Data Analysis	**3.A:** Identify the different types of data presented in maps and in quantitative and geospatial data. **3.B:** Describe spatial patterns in maps and in quantitative and geospatial data. **3.C:** Explain patterns and trends in maps and in quantitative and geospatial data to draw conclusions. **3.D:** Compare patterns and trends in maps and in quantitative and geospatial data to draw conclusions. **3.E:** Explain what maps or data imply or illustrate about geographic principles, processes, and outcomes. **3.F:** Explain possible limitations of the data provided.

Skill Category 4: Source Analysis	**4.A:** Identify the different types of information presented in visual sources. **4.B:** Describe the spatial patterns presented in visual sources. **4.C:** Explain patterns and trends in visual sources to draw conclusions. **4.D:** Compare patterns and trends in sources to draw conclusions. **4.E:** Explain how maps, images, and landscapes illustrate or relate to geographic principles, processes, and outcomes. **4.F:** Explain possible limitations of visual sources provided.
Skill Category 5: Scale Analysis	**5.A:** Identify the scales of analysis presented by maps, quantitative and geospatial data, images, and landscapes. **5.B:** Explain spatial relationships across various geographic scales using geographic concepts, processes, models, and theories. **5.C:** Compare geographic characteristics and processes at various scales. **5.D:** Explain the degree to which a geographic concept, process, model, or theory effectively explains geographic effects across various geographic scales.

Content Areas

The test covers seven units related to different fundamental components of Human Geography. Each of the seven units focuses on three different "big ideas":

1. Patterns and Spatial Organization (PSO)
2. Impacts and Interactions (IMP)
3. Spatial Processes and Societal Change (SPS)

These big ideas are further broken down into learning objectives specific to the unit's content, which are then broken down into more specific essential knowledge items. Every practice test question, in the answer explanation, includes coded data on the unit, essential knowledge, and the specific skill addressed by the question. As you are practicing, you can use the data to discern areas of strength and areas that require more studying.

Unit 1: Thinking Geographically (8–10% of exam)

Big Ideas:

1. *Patterns and Spatial Organization (PSO):* Why do geographers study relationships and patterns among and between places?
2. *Impacts and Interactions (IMP):* How do geographers use maps to help them discover patterns and relationships in the world?
3. *Spatial Processes and Societal Change (SPS):* How do geographers use a spatial perspective to analyze complex issues and relationships?

Topic 1.1: Introduction to Maps

IMP-1: Geographers use maps and data to depict relationships of time, space, and scale.

   A.   IMP-1.A: Identify types of maps, the types of information presented in maps, and different kinds of spatial patterns and relationships portrayed in maps.

      a.   IMP-1.A.1: Types of maps include reference maps and thematic maps.
      b.   IMP-1.A.2: Types of spatial patterns represented on maps include absolute and relative distance and direction, clustering, dispersal, and elevation.
      c.   IMP-1.A.3: All maps are selective in information; map projections inevitably distort spatial relationships in shape, area, distance, and direction.

Topic 1.2: Geographic Data

IMP-1: Geographers use maps and data to depict relationships of time, space, and scale.

   A.   IMP-1.B: Identify different methods of geographic data collection.

      a.   IMP-1.B.1: Data may be gathered in the field by organizations or by individuals.
      b.   IMP-1.B.2: Geospatial technologies include geographic information systems (GIS), satellite navigation systems, remote sensing, and online mapping and visualization.
      c.   IMP-1.B.3: Spatial information can come from written accounts in the form of field observations, media reports, travel narratives, policy documents, personal interviews, landscape analysis, and photographic interpretation.

Topic 1.3: The Power of Geographic Data

IMP-1: Geographers use maps and data to depict relationships of time, space, and scale.

   A.   IMP-1.C: Explain the geographical effects of decisions made using geographical information.

      a.   IMP-1.C.1: Geospatial and geographical data, including census data and satellite imagery, are used at all scales for personal, business and organizational, and governmental decision-making purposes.

Topic 1.4: Spatial Concepts

PSO-1: Geographers analyze relationships among and between places to reveal important spatial patterns.

   A.   PSO-1.A: Define major geographic concepts that illustrate spatial relationships.

      a.   PSO-1.A.1: Spatial concepts include absolute and relative location, space, place, flows, distance decay, time-space compression, and pattern.

Topic 1.5: Human-Environmental Interaction

PSO-1: Geographers analyze relationships among and between places to reveal important spatial patterns.

   A.   PSO-1.B: Explain how major geographic concepts illustrate spatial relationships.

      a.   PSO-1.B.1: Concepts of nature and society include sustainability, natural resources, and land use.

b. PSO-1.B.2: Theories regarding the interaction of the natural environment with human societies have evolved from environmental determinism to possibilism.

Topic 1.6: Scales of Analysis

PSO-1: Geographers analyze relationships among and between places to reveal important spatial patterns.

A. PSO-1.C: Define scales of analysis used by geographers.

a. PSO-1.C.1: Scales of analysis include global, regional, national, and local.

B. PSO-1.D: Explain what scales of analysis reveal.

a. PSO-1.D.1: Patterns and processes at different scales reveal variations in and different interpretations of data.

Topic 1.7: Regional Analysis

SPS-1: Geographers analyze complex issues and relationships with a distinctively spatial perspective.

A. SPS-1.A: Describe different ways that geographers define regions.

a. SPS-1.A.1: Regions are defined on the basis of one or more unifying characteristics or on patterns of activity.
b. SPS-1.A.2: Types of regions include formal, functional, and perceptual/vernacular.
c. SPS-1.A.3: Regional boundaries are transitional and often contested and overlapping.
d. SPS-1.A.4: Geographers apply regional analysis at local, national, and global scales.

Unit 2: Population and Migration Patterns and Processes (12–17% of exam)

Big Ideas:

1. *Patterns and Spatial Organization (PSO):* How does where and how people live impact global cultural, political, and economic patterns?
2. *Impacts and Interactions (IMP):* How does the interplay of environmental, economic, cultural, and political factors influence change in the population?
3. *Spatial Patterns and Societal Change (SPS):* How do changes in population affect a place's economy, culture, and politics?

Topic 2.1: Population Distribution

PSO-2: Understanding where and how people live is essential to understanding global, cultural, and economic patterns.

A. PSO-2.A: Identify the factors that influence the distribution of human populations at different scales.

a. PSO-2.A.1: Physical factors (e.g., climate, landforms, water bodies) and human factors (e.g., culture, economics, history, politics) influence the distribution of the population.

b. PSO-2.A.2: Factors that illustrate patterns of population distribution vary according to the scale of analysis.

B. PSO-2.B: Define methods geographers use to calculate population density.

a. PSO-2.B.1: The three methods for calculating population density are arithmetic, physiological, and agricultural.

C. PSO-2.C: Explain the differences between and the impact of methods used to calculate population density.

a. PSO-2.C.1: The method used to calculate population density reveals different information about the pressures the population exerts on the land.

Topic 2.2: Consequences of Population Distribution

PSO-2: Understanding where and how people live is essential to understanding global, cultural, and economic patterns.

A. PSO-2.D: Explain how population distribution and density affect society and the environment.

a. PSO-2.D.1: Population distribution and density affect political, economic, and social processes, including the provision of such services as medical care.
b. PSO-2.D.2: Population distribution and density affect the environment and natural resources; this is known as carrying capacity.

Topic 2.3: Population Composition

PSO-2: Understanding where and how people live is essential to understanding global, cultural, and economic patterns.

A. PSO-2.E: Describe elements of population composition used by geographers.

a. PSO-2.E.1: Patterns of age structure and sex ratio vary across different regions and may be mapped and analyzed at different scales.
b. PSO-2.E.2: Population pyramids are used to assess population growth and decline and to predict markets for goods and services.

Topic 2.4: Population Dynamics

IMP-2: Changes in population are due to mortality, fertility, and migration, which are influenced by the interplay of environmental, economic, cultural, and political factors.

A. IMP-2.A: Explain factors that account for contemporary and historical trends in population growth and decline.

a. IMP-2.A.1: Demographic factors that determine a population's growth and decline are fertility, mortality, and migration.
b. IMP-2.A.2: Geographers use the rate of natural increase and population-doubling time to explain population growth and decline.
c. IMP-2.A.3: Social, cultural, political, and economic factors influence fertility, mortality, and migration rates.

Topic 2.5: The Demographic Transition Model

IMP-2: Changes in population are due to mortality, fertility, and migration, which are influenced by the interplay of environmental, economic, cultural, and political factors.

A. IMP-2.B: Explain theories of population growth and decline.

    a. IMP-2.B.1: The demographic transition model can be used to explain population change over time.

    b. IMP-2.B.2: The epidemiological transition explains causes of changing death rates.

Topic 2.6: Malthusian Theory

IMP-2: Changes in population are due to mortality, fertility, and migration, which are influenced by the interplay of environmental, economic, cultural, and political factors.

A. IMP-2.C: Explain theories of population growth and decline.

    a. IMP-2.C.1: Malthusian theory and its critiques are used to analyze population change and its consequences.

Topic 2.7: Population Policies

SPS-2: Changes in population have long- and short-term effects on a place's economy, culture, and politics.

A. SPS-2.A: Explain the intent and effects of various population and immigration policies on population and composition.

    a. SPS-2.A.1: Types of population policies include those that promote or discourage population growth, such as pronatalist, antinatalist, and immigration policies.

Topic 2.8: Women and Demographic Change

SPS-2: Changes in population have long- and short-term effects on a place's economy, culture, and politics.

A. SPS-2.B: Explain how the changing role of women has demographic consequences in different parts of the world.

    a. SPS-2.B.1: Changing social values and access to education, employment, health care, and contraception have reduced fertility rates in most parts of the world.

    b. SPS-2.B.2: Changing social, economic, and political roles for women have influenced patterns of fertility, mortality, and migration, as illustrated by Ravenstein's laws of migration.

Topic 2.9: Aging Populations

SPS-2: Changes in population have long- and short-term effects on a place's economy, culture, and politics.

A. SPS-2.C: Explain the causes and consequences of an aging population.

   a. SPS-2.C.1: Population aging is determined by birth and death rates and life expectancy.

   b. SPS-2.C.2: An aging population has political, social, and economic consequences, including the dependency ratio.

Topic 2.10: Causes of Migration

IMP-2: Changes in population are due to mortality, fertility, and migration, which are influenced by the interplay of environmental, economic, cultural, and political factors.

A. IMP-2.A: Explain how different causal factors encourage migration.

   a. IMP-2.A.1: Migration is commonly divided into push factors and pull factors.

   b. IMP-2.A.2: Push/pull factors and intervening opportunities/obstacles can be cultural, demographic, economic, environmental, or political.

Topic 2.11: Forced and Voluntary Migration

IMP-2: Changes in population are due to mortality, fertility, and migration, which are influenced by the interplay of environmental, economic, cultural, and political factors.

A. IMP-2.B: Describe types of forced and voluntary migration.

   a. IMP-2.B.1: Forced migrations include slavery and events that produce refugees, internally displaced persons, and asylum seekers.

   b. IMP-2.B.2: Types of voluntary migration include transhumance, internal, chain, step, guest worker, and rural-to-urban.

Topic 2.12: Effects of Migration

IMP-2: Changes in population are due to mortality, fertility, and migration, which are influenced by the interplay of environmental, economic, cultural, and political factors.

A. IMP-2.C: Explain historical and contemporary geographic effects of migration.

   a. IMP-2.C.1: Migration has political, economic, and cultural effects.

Unit 3: Cultural Patterns and Processes (12–17% of exam)

Big Ideas:

1. *Patterns and Spatial Organization (PSO):* How do where people live and what resources they have access to impact their cultural practices?
2. *Impacts and Interactions (IMP):* How does the interaction of people contribute to the spread of cultural practices?
3. *Spatial Patterns and Societal Change (SPS):* How and why do cultural ideas, practices, and innovations change or disappear over time?

Topic 3.1: Introduction to Culture

PSO-3: Cultural practices vary across geographical locations because of physical geography and available resources.

A. PSO-3.A: Define the characteristics, attitudes, and traits that influence geographers when they study culture.

a. PSO-3.A.1: Culture comprises the shared practices, technologies, attitudes, and behaviors transmitted by a society.

b. PSO-3.A.2: Cultural traits include such things as food preferences, architecture, and land use.

c. PSO-3.A.3: Cultural relativism and ethnocentrism are different attitudes toward cultural difference.

Topic 3.2: Cultural Landscapes

PSO-3: Cultural practices vary across geographical locations because of physical geography and available resources.

A. PSO-3.B: Describe the characteristics of cultural landscapes.

a. PSO-3.B.1: Cultural landscapes are combinations of physical features, agricultural and industrial practices, religious and linguistic characteristics, evidence of sequent occupancy, and other expressions of culture, including traditional and postmodern architecture and land-use patterns.

B. PSO-3.C: Explain how landscape features and land and resource use reflect cultural beliefs and identities.

a. PSO-3.C.1: Attitudes toward ethnicity and gender, including the role of women in the workforce; ethnic neighborhoods; and indigenous communities and lands help shape the use of space in a given society.

Topic 3.3: Cultural Patterns

PSO-3: Cultural practices vary across geographical locations because of physical geography and available resources.

A. PSO-3.D: Explain patterns and landscapes of language, religion, ethnicity, and gender.

a. PSO-3.D.1: Regional patterns of language, religion, and ethnicity contribute to a sense of place, enhance placemaking, and shape the global cultural landscape.

b. PSO-3.D.2: Language, ethnicity, and religion are factors in creating centripetal and centrifugal forces.

Topic 3.4: Types of Diffusion

IMP-3: The interaction of people contributes to the spread of cultural practices.

A. IMP-3.A: Define the types of diffusion.

a. IMP-3.A.1: Relocation and expansion—including contagious, hierarchical, and stimulus expansion—are types of diffusion.

Topic 3.5: Historical Causes of Diffusion

SPS-3: Cultural ideas, practices, and innovations change or disappear over time.

A. SPS-3.A: Explain how historical processes impact current cultural patterns.

    a. SPS-3.A.1: Interactions between and among cultural traits and larger global forces can lead to new forms of cultural expression; for example, creolization and lingua franca.

    b. SPS-3.A.2: Colonialism, imperialism, and trade helped to shape patterns and practices of culture.

Topic 3.6: Contemporary Causes of Diffusion

SPS-3: Cultural ideas, practices, and innovations change or disappear over time.

A. SPS-3.B: Explain how historical processes impact current cultural patterns.

    a. SPS-3.B.1: Cultural ideas and practices are socially constructed and change through both small-scale and large-scale processes such as urbanization and globalization. These processes come to bear on culture through media, technological change, politics, economics, and social relationships.

    b. SPS-3.B.2: Communication technologies, such as the internet and the time-space convergence, are reshaping and accelerating interactions among people; changing cultural practices, as in the increasing use of English and the loss of indigenous languages; and creating cultural convergence and divergence.

Topic 3.7: Diffusion of Religion and Language

IMP-3: The interaction of people contributes to the spread of cultural processes.

A. IMP-3.A: Explain what factors lead to the diffusion of universalizing and ethnic religions.

    a. IMP-3.A.1: Language families, languages, dialects, world religions, ethnic cultures, and gender roles diffuse from cultural hearths.

    b. IMP-3.A.2: Diffusion of language families, including Indo-European, and religious patterns and distributions can be visually represented on maps, in charts and toponyms, and in other representations.

    c. IMP-3.A.3: Religions have distinct places of origin from which they diffused to other locations through different processes. Practices and belief systems impacted how widespread the religion diffused.

    d. IMP-3.A.4: Universalizing religions, including Christianity, Islam, Buddhism, and Sikhism, are spread through expansion and relocation diffusion.

    e. IMP-3.A.5: Ethnic religions, including Hinduism and Judaism, are generally found near the hearth or spread through relocation diffusion.

Topic 3.8: Effects of Diffusion

SPS-3: Cultural ideas, practices, and innovations change or disappear over time.

A. SPS-3.A: Explain how the process of diffusion results in changes to the cultural landscape.

    a. SPS-3.A.1: Acculturation, assimilation, syncretism, and multiculturalism are effects of the diffusion of culture.

Unit 4: Political Patterns and Processes (12–17% of exam)

Big Ideas:

1. *Patterns and Spatial Organization (PSO):* How do historical and current events influence political structures around the world?
2. *Impacts and Interactions (IMP):* How are balances of power reflected in political boundaries and government power structures?
3. *Spatial Patterns and Societal Change (SPS):* How can political, economic, cultural, or technological changes challenge state sovereignty?

Topic 4.1: Introduction to Political Geography

PSO-4: The political organization of space results from historical and current processes, events, and ideas.

A. PSO-4.A: For world political maps:

   a. Define the different types of political entities.

     i. PSO-4.A.1: Independent states are the primary building blocks of the world political map.

   b. Identify a contemporary example of political entities.

     i. PSO-4.A.2: Types of entities include nations, nation-states, stateless nations, multinational states, multistate nations, and autonomous and semiautonomous regions, such as American Indian reservations.

Topic 4.2: Political Processes

PSO-4: The political organization of space results from historical and current processes, events, and ideas.

A. PSO-4.B: Explain the processes that have shaped contemporary political geography.

   a. PSO-4.B.1: The concepts of sovreignty, nation-states, and self-determination shape the contemporary world.
   b. PSO-4.B.2: Colonialism, imperialism, independence movements, and devolution along national lines have influenced contemporary political boundaries.

Topic 4.3: Political Power and Territoriality

PSO-4: The political organization of space results from historical and current processes, events, and ideas.

A. PSO-4.C: Describe the concepts of political power and territoriality as used by geographers.

   a. PSO-4.C.1: Political power is expressed geographically as control over people, land, and resources, as illustrated by neocolonialism, shatterbelts, and choke points.
   b. PSO-4.C.2: Territoriality is the connection of people, their culture, and their economic systems to the land.

Topic 4.4: Defining Political Boundaries

IMP-4: Political boundaries and divisions of governance between states and within them reflect balances of power that have been negotiated or imposed.

    A.   IMP-4.A: Define the types of political boundaries used by geographers.

        a.   IMP-4.A.1: Types of political boundaries include relic, superimposed, subsequent, antecedent, geometric, and consequent boundaries.

Topic 4.5: The Function of Political Boundaries

IMP-4: Political boundaries and divisions of governance between states and within them reflect balances of power that have been negotiated or imposed.

    A.   IMP-4.B: Explain the nature and function of international and internal boundaries.

        a.   IMP-4.B.1: Boundaries are defined, delimited, demarcated, and administered to establish limits of sovereignty, but they are often contested.

        b.   IMP-4.B.2: Political boundaries often coincide with cultural, national, or economic divisions. However, some boundaries may be created by demilitarized zones or a policy, such as the one established at the Berlin Conference.

        c.   IMP-4.B.3: Land and maritime boundaries and international agreements can influence national and regional identity and encourage or discourage international or internal interactions or disputes over resources.

        d.   IMP-4.B.4: The United Nations Convention on the Law of the Sea defines the rights and responsibilities of nations in the use of international waters, established territorial seas, and exclusive economic zones.

Topic 4.6: Internal Boundaries

IMP-4: Political boundaries and divisions of governance between states and within them reflect balances of power that have been negotiated or imposed.

    A.   IMP-4.C: Explain the nature and function of international and internal boundaries.

        a.   IMP-4.C.1: Voting districts, redistricting, and gerrymandering affect election results at various scales.

Topic 4.7: Forms of Governance

IMP-4: Political boundaries and divisions of governance between states and within them reflect balances of power that have been negotiated or imposed.

    A.   IMP-4.D: Define federal and unitary states.

        a.   IMP-4.D.1: Forms of governance include unitary states and federal states.

    B.   IMP-4.E: Explain how federal and unitary states affect spatial organization.

        a.   IMP-4.E.1: Unitary states tend to have a more top-down, centralized form of governance, while federal states have more locally based, dispersed power centers.

Topic 4.8: Defining Devolutionary Factors

SPS-4: Political, economic, cultural, or technological changes can challenge state sovereignty.

A.   SPS-4.A: Define factors that lead to the devolution of states.

    a.   SPS-4.A.1: Factors that can lead to the devolution of states include the division of groups by physical geography, ethnic separatism, ethnic cleansing, terrorism, economic and social problems, and irredentism.

Topic 4.9: Challenges to State Sovereignty

SPS-4: Political, economic, cultural, or technological changes can challenge state sovereignty.

A.   SPS-4.B: Explain how political, economic, cultural, and technological changes challenge state sovereignty.

    a.   SPS-4.B.1: Devolution occurs when states fragment into autonomous regions or into subnational political-territorial units, such as those within Spain, Belgium, Canada, and Nigeria; or when states disintegrate, as happened in Eritrea, South Sudan, East Timor, and states that were a part of the former Soviet Union.

    b.   SPS-4.B.2: Advances in communication technology have facilitated devolution, supranationalism, and democratization.

    c.   SPS-4.B.3: Global efforts to address transnational and environmental challenges and to create economies of scale, trade agreements, and military alliances help to further supranationalism.

    d.   SPS-4.B.4: Supranational organizations—including the United Nations (UN), North Atlantic Treaty Organization (NATO), European Union (EU), Association of Southeast Asian Nations (ASEAN), Arctic Council, and African Union—can challenge state sovereignty by limiting the economic or political actions of member states.

Topic 4.10: Consequences of Centrifugal and Centripetal Forces

SPS-4: Political, economic, cultural, or technological changes can challenge state sovereignty.

A.   SPS-4.C: Explain how the concepts of centrifugal and centripetal forces apply at the state scale.

    a.   SPS-4.C.1: Centrifugal forces may lead to failed states, uneven development, stateless nations, and ethnic nationalist movements.

    b.   SPS-4.C.2: Centripetal forces can lead to ethnonationalism, more equitable infrastructure development, and increased cultural cohesion.

Unit 5: Agriculture and Rural Land-Use Patterns and Processes (12–17% of exam)

Big Ideas:

1. *Patterns and Spatial Organization (PSO):* How do a people's culture and the resources available to them influence how they grow food?
2. *Impacts and Interactions (IMP):* How does what people produce and consume vary in different locations?
3. *Spatial Patterns and Societal Change (SPS):* What kind of cultural changes and technological advances have impacted the way people grow and consume food?

Topic 5.1: Introduction to Agriculture

PSO-5: Availability of resources and cultural practices influence agricultural practices and land-use patterns.

A. PSO-5.A: Explain the connection between physical geography and agricultural practices.

   a. PSO-5.A.1: Agricultural practices are influenced by the physical environment and climatic conditions, such as those associated with a Mediterranean or tropical climate.
   b. PSO-5.A.2: Intensive farming practices include market gardening, plantation agriculture, and mixed crop/livestock systems.
   c. PSO-5.A.3: Extensive farming practices include shifting cultivation, nomadic herding, and ranching.

Topic 5.2: Settlement Patterns and Survey Methods

PSO-5: Availability of resources and cultural practices influence agricultural practices and land-use patterns.

A. PSO-5.B: Identify different rural settlement patterns and methods of surveying settlements.

   a. PSO-5.B.1: Specific agricultural practices shape different land-use patterns.
   b. PSO-5.B.2: Rural settlement patterns are classified as clustered, dispersed, or linear.
   c. PSO-5.B.3: Rural survey methods include metes and bounds, township and range, and long lot.

Topic 5.3: Agricultural Origins and Diffusions

SPS-5: Agriculture has changed over time because of cultural diffusion and advances in technology.

A. SPS-5.A: Identify major centers of domestication of plants and animals.

   a. SPS-5.A.1: Early hearths of domestication of plants and animals arose in the Fertile Crescent and several other regions of the world, including the Indus River Valley, Southeast Asia, and Central America.

B. SPS-5.B: Explain how plants and animals diffused globally.

    a. SPS-5.B.1: Patterns of diffusion, such as the Columbian Exchange and the agricultural revolutions, resulted in the global spread of various plants and animals.

Topic 5.4: The Second Agricultural Revolution

SPS-5: Agriculture has changed over time because of cultural diffusion and advances in technology.

A. SPS-5.C: Explain the advances and impacts of the second agricultural revolution.

    a. SPS-5.C.1: New technology and increased food production in the second agricultural revolution led to better diets and longer life expectancies, and thus more people were available to work in factories.

Topic 5.5: The Green Revolution

SPS-5: Agriculture has changed over time because of cultural diffusion and advances in technology.

A. SPS-5.D: Explain the consequences of the Green Revolution on food supply and the environment in the developing world.

    a. SPS-5.D.1: The Green Revolution was characterized in agriculture by the use of high-yield seeds, increased use of chemicals, and mechanized farming.

    b. SPS-5.D.2: The Green Revolution had positive and negative impacts for both human populations and the environment.

Topic 5.6: Agricultural Production Regions

PSO-5: Availability of resources and cultural practices influence agricultural practices and land-use patterns.

A. PSO-5.A: Explain how economic forces influence agricultural practices.

    a. PSO-5.A.1: Agricultural production regions are defined by the extent to which they reflect subsistence or commercial practices (monocropping or monoculture).

    b. PSO-5.A.2: Intensive and extensive farming practices are determined in part by land costs (bid-rent theory).

Topic 5.7: Spatial Organization of Agriculture

PSO-5: Availability of resources and cultural practices influence agricultural practices and land-use patterns.

A. PSO-5.B: Explain how economic forces influence agricultural practices.

    a. PSO-5.B.1: Large-scale commercial agricultural operations are replacing small family farms.

    b. PSO-5.B.2: Complex commodity chains link production and consumption of agricultural products.

    c. PSO-5.B.3: Technology has increased economies of scale in the agricultural sector and the carrying capacity of the land.

Topic 5.8: Von Thunen Model

PSO-5: Availability of resources and cultural practices influence agricultural practices and land-use patterns.

A. PSO-5.C: Describe how the von Thunen model is used to explain patterns of agricultural production at various scales.

a. PSO-5.C.1: Von Thunen's model helps to explain rural land use by emphasizing the importance of transportation costs associated with distance from the market; however, regions of specialty farming do not always conform to von Thunen's concentric rings.

Topic 5.9: The Global System of Agriculture

PSO-5: Availability of resources and cultural practices influence agricultural practices and land-use patterns.

A. PSO-5.D: Explain the interdependence among regions of agricultural production and consumption.

a. PSO-5.D.1: Food and other agricultural products are part of a global supply chain.
b. PSO-5.D.2: Some countries have become highly dependent on one or more export commodities.
c. PSO-5.D.3: The main elements of global food distribution networks are affected by political relationships, infrastructure, and patterns of world trade.

Topic 5.10: Consequences of Agricultural Practices

IMP-5: Agricultural production and consumption patterns vary in different locations, presenting different environmental, social, economic, and cultural opportunities and challenges.

A. IMP-5.A: Explain how agricultural practices have environmental and societal consequences.

a. IMP-5.A.1: Environmental effects of agricultural land use include pollution, land-cover change, desertification, soil salinization, and conservation efforts.
b. IMP-5.A.2: Agricultural practices—including slash and burn, terraces, irrigation, deforestation, draining wetlands, shifting cultivation, and pastoral nomadism—alter the landscape.
c. IMP-5.A.3: Societal effects of agricultural practices include changing diets, the role of women in agricultural production, and economic purpose.

Topic 5.11: Challenges of Contemporary Agriculture

IMP-5: Agricultural production and consumption patterns vary in different locations, presenting different environmental, social, economic, and cultural opportunities and challenges.

A. IMP-5.B: Explain challenges and debates related to the changing nature of contemporary agriculture and food-production practices.

    a. IMP-5.B.1: Agricultural innovations such as biotechnology, genetically modified foods, and aquaculture have been accompanied by debates over sustainability, soil and water usage, reductions in biodiversity, and extensive fertilizer and pesticide use.

    b. IMP-5.B.2: Patterns of food production and consumption are influenced by movements related to individual food choice, such as urban farming, community-supported agriculture (CSA), organic farming, value-added specialty crops, fair trade, local-food movements, and dietary shifts.

    c. IMP-5.B.3: Challenges of feeding a global population include lack of food access, as in cases of food insecurity and food deserts; problems with distribution systems; adverse weather; and land use lost to suburbanization.

    d. IMP-5.B.4: The location of food-processing facilities and markets, economies of scale, distribution systems, and government policies all have economic effects on food-production practices.

Topic 5.12: Women in Agriculture

IMP-5: Agricultural production and consumption patterns vary in different locations, presenting different environmental, social, economic, and cultural opportunities and challenges.

A. IMP-5.C: Explain geographic variations in women's roles in production and consumption.

    a. IMP-5.C.1: The role of women in food production, distribution, and consumption varies in many places depending on the type of production involved.

Unit 6: Cities and Urban Land-Use Patterns and Processes (12–17% of exam)

Big Ideas:

1. *Patterns and Spatial Organization (PSO):* How do physical geography and resources impact the presence and growth of cities?
2. *Impacts and Interactions (IMP):* How are the attitudes, values, and balance of power of a population reflected in the built landscape?
3. *Spatial Patterns and Societal Change (SPS):* How are urban areas affected by unique economic, political, cultural, and environmental challenges?

Topic 6.1: The Origin and Influences of Urbanization

PSO-6: The presence and growth of cities vary across geographical locations because of physical geography and natural resources.

A. PSO-6.A: Explain the processes that initiate and drive urbanization and suburbanization.

    a. PSO-6.A.1: Site and situation influence the origin, function, and growth of cities.

b. PSO-6.A.2: Changes in transportation and communication, population growth, migration, economic development, and government policies influence urbanization.

Topic 6.2: Cities Across the World

PSO-6: The presence and growth of cities vary across geographical locations because of physical geography and natural resources.

A. PSO-6.B: Explain the processes that initiate and drive urbanization and suburbanization.

   a. PSO-6.B.1: Megacities and metacities are distinct spatial outcomes of urbanization increasingly located in countries on the periphery and semiperiphery.

   b. PSO-6.B.2: Suburbanization, sprawl, and decentralization have created new land-use forms—including edge cities, exurbs, and boomburbs—and new challenges.

Topic 6.3: Cities and Globalization

PSO-6: The presence and growth of cities vary across geographical locations because of physical geography and resources.

A. PSO-6.C: Explain how cities embody processes of globalization.

   a. PSO-6.C.1: World cities function at the top of the world's urban hierarchy and drive globalization.

   b. PSO-6.C.2: Cities are connected globally by networks and linkages and mediate global processes.

Topic 6.4: The Size and Distribution of Cities

PSO-6: The presence and growth of cities vary across geographical locations because of physical geography and resources.

A. PSO-6.D: Identify the different urban concepts such as hierarchy, interdependence, relative size, and spacing that are useful for explaining the distribution, size, and interaction of cities.

   a. PSO-6.D.1: Principles that are useful for explaining the distribution and size of cities include rank-size rule, the primate city, gravity, and Christaller's central place theory.

Topic 6.5: The Internal Structure of Cities

PSO-6: The presence and growth of cities vary across geographical locations because of physical geography and resources.

A. PSO-6.E: Explain the internal structure of cities using various models and theories.

   a. PSO-6.E.1: Models and theories that are useful for explaining internal structure of cities include the Burgess concentric-zone model, the Hoyt sector model, the Harris and Ullman multiple-nuclei model, the galactic city model, bid-rent theory, and urban models drawn from Latin America, Southeast Asia, and Africa.

Topic 6.6: Density and Land Use

IMP-6: The attitudes and values of a population, as well as the balance of power within that population, are reflected in the built landscape.

A. IMP-6.A: Explain how low-, medium-, and high-density housing characteristics represent different patterns of residential land use.

a. IMP-6.A.1: Residential buildings and patterns of land use reflect and shape the city's culture, technological capabilities, cycles of development, and infilling.

Topic 6.7: Infrastructure

IMP-6: The attitudes and values of a population, as well as the balance of power within that population, are reflected in the built landscape.

A. IMP-6.B: Explain how a city's infrastructure relates to local politics, society, and the environment.

a. IMP-6.B.1: The location and quality of a city's infrastructure directly affects its spatial patterns of economic and social development.

Topic 6.8: Urban Sustainability

IMP-6: The attitudes and values of a population, as well as the balance of power within that population, are reflected in the built landscape.

A. IMP-6.C: Identify the different urban design initiatives and practices.

a. IMP-6.C.1: Sustainable design initiatives and zoning practices include mixed land use, walkability, transportation-oriented development, and smart-growth policies, including New Urbanism, greenbelts, and slow-growth cities.

B. IMP-6.D: Explain the effects of different urban design initiatives and practices.

a. IMP-6.D.1: Praise for urban design initiatives includes the reduction of sprawl, improved walkability and transportation, improved and diverse housing options, improved livability, and promotion of sustainable options. Criticisms include increased housing costs, possible de facto segregation, and the potential loss of historical or place character.

Topic 6.9: Urban Data

IMP-6: The attitudes and values of a population, as well as the balance of power within that population, are reflected in the built landscape.

A. IMP-6.E: Explain how qualitative and quantitative data are used to show the causes and effects of geographic change within urban areas.

a. IMP-6.E.1: Quantitative data from census and survey data provide information about changes in population composition and size in urban areas.

b. IMP-6.E.2: Qualitative data from field studies and narratives provide information about individual attitudes toward urban change.

Topic 6.10: Challenges of Urban Changes

SPS-6: Urban areas face unique economic, political, cultural, and environmental challenges.

A. SPS-6.A: Explain causes and effects of geographic change within urban areas.

    a. SPS-6.A.1: As urban populations move within a city, economic and social challenges result, including issues related to housing and housing discrimination, such as redlining, blockbusting, and affordability; access to services; rising crime; environmental injustice; and the growth of disamenity zones or zones of abandonment.

    b. SPS-6.A.2: Squatter settlements and conflicts over land tenure within large cities have increased.

    c. SPS-6.A.3: Responses to economic and social challenges in urban areas can include inclusionary zoning and local-food movements.

    d. SPS-6.A.4: Urban renewal and gentrification have both positive and negative consequences.

    e. SPS-6.A.5: Functional and geographic fragmentation of governments—the way government agencies and institutions are dispersed between state, county, city, and neighborhood levels—presents challenges in addressing urban issues.

Topic 6.11: Challenges of Urban Sustainability

SPS-6: Urban areas face unique economic, political, cultural, and environmental challenges.

A. SPS-6.B: Describe the effectiveness of different attempts to address urban sustainability challenges.

    a. SPS-6.B.1: Challenges to urban sustainability include suburban sprawl, sanitation, climate change, air and water quality, the large ecological footprint of cities, and energy use.

    b. SPS-6.B.2: Responses to urban sustainability challenges can include regional planning efforts, remediation and redevelopment of brownfields, establishment of urban growth boundaries, and farmland protection policies.

Unit 7: Industrial and Economic Development Patterns and Processes (12–17% of exam)

Big Ideas:

1. *Patterns and Spatial Organization (PSO):* Why does economic and social development happen at different times and at different rates in different places?
2. *Impacts and Interactions (IMP):* How might environmental problems stemming from industrialization be remedied through sustainable development strategies?
3. *Spatial Patterns and Societal Change (SPS):* Why has industrialization helped improve standards of living while also contributing to geographically uneven development?

Topic 7.1: The Industrial Revolution

SPS-7: Industrialization, past and present, has facilitated improvements in standards of living, but it has also contributed to geographically uneven development.

A. SPS-7.A: Explain how the Industrial Revolution facilitated the growth and diffusion of industrialization.

a. SPS-7.A.1: Industrialization began as a result of new technologies and was facilitated by the availability of natural resources.

b. SPS-7.A.2: As industrialization spread, it caused food supplies to increase and populations to grow; it allowed workers to seek new industrial jobs in the cities; and it changed class structures.

c. SPS-7.A.3: Investors in industry sought out more raw materials and new markets, a factor that contributed to the rise of colonialism and imperialism.

Topic 7.2: Economic Sectors and Patterns

SPS-7: Industrialization, past and present, has facilitated improvements in standards of living, but it has also contributed to geographically uneven development.

A. SPS-7.B: Explain the spatial patterns of industrial production and development.

a. SPS-7.B.1: The different economic sectors—including primary, secondary, tertiary, quaternary, and quinary—are characterized by distinct development patterns.

b. SPS-7.B.2: Labor, transportation (including shipping containers), the break-of-bulk point, the least-cost theory, markets, and resources influence manufacturing locations (i.e., core, semiperiphery, and periphery locations).

Topic 7.3: Measures of Development

SPS-7: Industrialization, past and present, has facilitated improvements in standards of living, but it has also contributed to geographically uneven development.

A. SPS-7.C: Describe social and economic measures of development.

a. SPS-7.C.1: Measures of social and economic development include gross domestic product (GDP), gross national product (GNP), and gross national income (GNI) per capita; sectoral structure of an economy, both formal and informal; income distribution; fertility rates; infant mortality rates; access to health care; use of fossil fuels and renewable energy; and literacy rates.

b. SPS-7.C.2: Measures of gender inequality, such as the gender inequality index (GII), include reproductive health, indices of empowerment, and labor-market participation.

c. SPS-7.C.3: The human development index (HDI) is a composite measure used to show spatial variation among states in levels of development.

Topic 7.4: Women and Economic Development

SPS-7: Industrialization, past and present, has facilitated improvements in standards of living, but it has also contributed to geographically uneven development.

A. SPS-7.D: Explain how and to what extent changes in economic development have contributed to gender parity.

a. SPS-7.D.1: The roles of women change as countries develop economically.

b. SPS-7.D.2: Although there are more women in the workforce, they do not have equity in wages or employment opportunities.

c. SPS-7.D.3: Microloans have provided opportunities for women to create small local businesses, which have improved standards of living.

Topic 7.5: Theories of Development

SPS-7: Industrialization, past and present, has facilitated improvements in standards of living, but it has also contributed to geographically uneven development.

    A.   SPS-7.E: Explain different theories of economic and social development.

        a.   SPS-7.E.1: Rostow's stages of economic growth, Wallerstein's world system theory, dependency theory, and commodity-dependence theory are some of the theories that help explain variations in development.

Topic 7.6: Trade and the World Economy

PSO-7: Economic and social development happen at different times and rates in different places.

    A.   PSO-7.A: Explain causes and geographic consequences of recent economic changes, such as the increase in international trade, deindustrialization, and growing interdependence in the world economy.

        a.   PSO-7.A.1: Complementarity and comparative advantage establish the basis for trade.

        b.   PSO-7.A.2: Neoliberal policies, including free-trade agreements, have created new organizations, spatial connections, and trade relationships, such as the European Union (EU), World Trade Organization (WTO), Mercosur, and OPEC, that foster greater globalization.

        c.   PSO-7.A.3: Government initiatives at all scales, including tariffs, may affect economic development.

        d.   PSO-7.A.4: Global financial crises (e.g., debt crises), international lending agencies (e.g., the International Monetary Fund), and strategies of development (e.g., microlending) demonstrate how different economies have become more closely connected, even interdependent.

Topic 7.7: Changes as a Result of the World Economy

PSO-7: Economic and social development happen at different times and rates in different places.

    A.   PSO-7.B: Explain causes and geographic consequences of recent economic changes, such as the increase in international trade, deindustrialization, and growing interdependence in the world economy.

        a.   PSO-7.B.1: Outsourcing and economic restructuring have led to a decline in jobs in core regions and an increase in jobs in newly industrializing countries.

        b.   PSO-7.B.2: In countries outside the core, the growth of industry has resulted in the creation of new manufacturing zones—including special economic zones, free-trade zones, and export-processing zones—and the emergence of an international division of labor in which developing countries have lower-paying jobs.

        c.   PSO-7.B.3: The contemporary economic landscape has been transformed by post-Fordist methods of production, multiplier effects, economies of scale, agglomeration, just-in-time delivery, the emergence of service sectors, high-technology industries, and growth poles.

Topic 7.8: Sustainable Development

IMP-7: Environmental problems stemming from industrialization may be remedied through sustainable development strategies.

A. IMP-7.A: Explain how sustainability principles relate to and impact industrialization and spatial development.

    a. IMP-7.A.1: Sustainable development policies attempt to remedy problems stemming from natural-resource depletion, mass consumption, the effects of pollution, and the impact of climate change.

    b. IMP-7.A.2: Ecotourism, tourism in natural environments—often environments that are threatened by looming industrialization or development—frequently helps to protect the environment in question while also providing jobs for the local population.

    c. IMP-7.A.3: The UN's Sustainable Development Goals help measure progress in development of such projects as small-scale finance and public transportation projects.

STRATEGIES FOR PREPARING FOR THE AP HUMAN GEOGRAPHY EXAM

Outside of thoroughly studying the content described in the APHG course, the best way to prepare for the exam is by actually answering questions similar to those that you will see on test day. The following are strategies to help you become an effective test-taker, and we would advise you to practice these strategies while taking the model tests provided in this study guide, such that these strategies essentially become second-nature for test day.

Multiple-Choice Questions

When taking the multiple-choice section of the exam, do not leave any blanks.

- Go through the entire set of multiple-choice questions and answer ONLY the ones you know for sure.
- When you reach a question you are not quite sure of, but think you can answer given more time, mark that question (devise a strategy that works best for you; perhaps you can circle the question number).
- After you have answered all the questions you know for sure, go back to the questions you circled. If you can narrow down the choices to two possible responses, go ahead and make your best guess.
- If time remains, scan through all the questions you marked with a circle and answer them.

DIFFERENT TYPES OF MULTIPLE-CHOICE QUESTIONS

The following are different types of multiple-choice questions with examples of each type. Beginning with the 2020 exam, stimulus-based questions will make up close to half of all multiple-choice questions. A stimulus question, like the one below, involves interpretation of a map, graph, cartoon, chart, or other bit of data. The remaining questions will be a mix of the other types. Preparing yourself for different ways of being tested on similar content will enhance your ability to effectively answer questions on test day.

STIMULUS-RESPONSE QUESTIONS: In these types of questions you are given a graph, cartoon, figure, map, or other bit of information that you must interpret before being able to correctly answer the question.

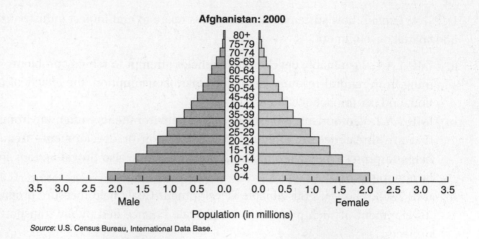

Afghanistan: 2000

Population (in millions)

Source: U.S. Census Bureau, International Data Base.

1. Based on this population pyramid, which of the following statements accurately describes Afghanistan's most likely population pyramid in twenty years?

   (A) While growth is currently very rapid, in twenty years, it will have begun to stabilize, causing the base to narrow.

   (B) Most likely, the population pyramid will resemble the current figure, as the current large base will begin reproducing, causing an even larger base.

   (C) The top of the pyramid will begin to widen as the death rate begins decreasing.

   (D) The pyramid will look less like a pyramid and more like a rectangle as the larger base move up and fills in the middle.

   (E) The bottom of the pyramid will be narrower than the type showing the successful implementation of strict population policy.

DEFINITIONAL: These types of questions tend to be factual and test you on your ability to define particular concepts, processes, models, or theories.

2. Which of the following best defines a functional region?

   (A) The boundary including all areas within the circulation of a particular newspaper

   (B) The boundary including all people who speak Creole in Louisiana

   (C) The boundary around gerrymandered voting districts

   (D) The boundary that includes the American "Deep South"

   (E) The boundary encompassing the United States of America

CAUSE-AND-EFFECT RELATIONSHIPS: With these types of questions, you will be given a cause and asked to determine the correct effect from the possible responses given.

3. As a country becomes increasingly developed, economic activities become dominant in which sector?

   (A) Primary sector
   (B) Tertiary sector
   (C) Non-basic sector
   (D) Secondary sector
   (E) Basic sector

GENERALIZATION: In a generalization question, you will be given a specific event, process, model, or theory, and you must be able to identify the general principle that came about as a result of that event, process, model, or theory.

4. The von Thunen model describes agricultural activity as it takes place in relation to the market. Which of the following statements generally represents the agricultural landscape according to the model?

   (A) Agricultural activity is solely determined by the longevity of the agricultural product.
   (B) Goods that are expensive to transport and spoil quickly must be located closer to the market.
   (C) Smaller agricultural goods like beans, herbs, and berries will be grown closer to the market than bigger goods like pumpkins.
   (D) Animals will be located closer to the market, like grazing cattle and hens, because they are difficult to move.
   (E) Agricultural goods that require more inputs, such as fertilizers and pesticides, will be grown closer to the market.

SOLUTION TO A PROBLEM: Solution problems tend to be combination questions; in this case you must first know the definition of a particular concept before being able to effectively solve the problem.

5. Which of the following economic enterprises is the best example of a footloose industry?

   (A) Shoe store
   (B) Jewelry store
   (C) Dance company
   (D) Cheese factory
   (E) Big box store

HYPOTHETICAL SITUATION: In these types of questions, you put yourself in a hypothetical scenario and use your knowledge of certain principles to determine an appropriate response.

6. You decide to eat dinner at the local diner and get there right during the dinner rush when the restaurant is full of people. What is the most likely makeup of the group of diners?

   (A) You don't recognize many of them so they must be mainly tourists, with a couple of familiar faces from the local area.

   (B) You don't recognize too many faces, but they seem very neighborly so most of them probably come from the neighboring town because they don't have a local diner, and the rest are from around town.

   (C) It's probably about an equal mix of tourists, locals, and people from the neighboring towns.

   (D) You recognize most faces, but there are a few you don't know, that are probably from the neighboring towns, and possibly a couple of tourists stopping through.

   (E) You probably don't know most of the people since the diner is located at the intersection of two major freeways.

TIP

To get more practice answering multiple-choice questions, search the online resources that accompany major introductory Human Geography textbooks. These sites almost always contain test banks with practice questions. Also be sure to check out AP Central (*www.apcentral. collegeboard.com*), where example test questions are provided.

CHRONOLOGICAL PROBLEM: Chronological problems are simply those that draw on your memory of the time period of certain notable events.

7. The Industrial Revolution took hold in England when?

   (A) Late 20th century
   (B) Late 19th century
   (C) Late 18th century
   (D) Early 20th century
   (E) Early 18th century

COMPARING/CONTRASTING CONCEPTS AND EVENTS: In these types of questions you are asked to compare two events, processes, models, or concepts. As you answer these types of questions, define the two terms, processes, or events being compared before even looking at your options. This will spare you a lot of confusion.

8. India's population policy differs from China's in that

   (A) they aren't different; they both strictly enforce a one-child-per-couple policy.

   (B) India is much more stringent, forcefully implementing their population policy.

   (C) China strictly adheres to a national population policy with benefits for those who conform and punishments for those who don't, while India does not really have a policy.

   (D) India encourages rather than forces couples to limit the number of children they have.

   (E) China allows provinces to determine local strategies for implementing national policy, while India has one strict policy that the entire nation must abide by.

NEGATIVE QUESTIONS: When answering negative questions, which are easily identified through the terms "not" or "except," it is usually easier to determine all the true statements that have to be incorrect, leaving you with the one false statement.

NOTE

Negative questions rarely appear on the AP Human Geography exam.

9. All of the following are characteristics of agribusiness EXCEPT

   (A) it, in large part, has led to the demise of the American family farm.

   (B) it has incorporated production, consumption, and marketing into an integrated whole.

   (C) it has allowed for increased family market gardening.

   (D) it has transformed agricultural productivity such that agricultural activities yield much more than they have historically.

   (E) it often leads to monoculture forms of farming.

Free-Response Questions

Beginning with the May 2020 exam, the three free-response questions will each contain nine parts (a–g). One question will not have a stimulus, one question will have one stimulus (a graph, map, chart, picture, or reading), and one question will contain two stimuli. Each question will be worth 7 points (1 point/part, regardless of the difficulty of the part). Each of the nine parts will likely use one of five following *task* verbs:

COMPARE: Provide an explanation or description of the similarities and/or differences between two things.

For example: Compare a country that conforms to the rank-size rule of urban areas to a country that has a primate city pattern.

DEFINE: Provide a specific meaning for a given word, concept, or idea.

For example: Define the concept of distance decay.

DESCRIBE: Provide the relevant characteristics of a specified process or trend.

For example: Describe how the opening of China's economy led to demographic shifts within China.

EXPLAIN: Provide information on how or why a pattern, process, or relationship occurs (use evidence or reasoning to support your explanation).

For example: Explain how a strict population policy, such as China's former one-child-per-couple law, affects a country's population pyramid.

IDENTIFY: Provide information on a specific pattern, process, or concept without elaboration or explanation.

For example: Identify one specific example of an effect of agribusiness on food production patterns in the United States.

When constructing your response, it is not necessary to provide the typical five-paragraph essay used in other humanities courses (introduction, paragraphs supporting the thesis statement, conclusion). Instead, you should focus on answering each of the three questions as directly and comprehensively as possible.

As you practice answering free-response questions, focus first on answering the question as accurately, clearly, and succinctly as possible. The readers are not evaluating you on your ability to construct an essay; they are simply looking for correct responses to each portion of the question! This means that you can receive full points for a response that is NOT a constructed essay as long as it contains correct responses/examples to each of the question's elements. Make sure to provide only the information necessary to answer the question. For example, if the question asks you to define and provide an example of a particular concept or feature, provide only **one** definition and **one** example, as the reader will look only at the first definition/example provided. Your response should be written neatly with complete sentences and correct grammar (this makes it much easier for your reader to understand your communication).

When preparing for this portion of the exam, it is always good practice to take time to construct a brief outline that directly corresponds to the question. This will ensure that you address all parts of the question, and it will likely help you maintain organization as you construct your response.

Taking and Scoring Practice Tests

The two practice tests in the book and the three online tests each consist of 60 multiple-choice questions and three free-response questions, just like the real exam. In order to utilize these tests effectively, simulate the test experience as closely as possible when taking each one. In other words, be sure to take them in a quiet place where you can be disruption-free for a full 2 hours and 15 minutes. Once you feel you've adequately studied the material you needed to work on, you can take one of the practice tests at the end of the book to assess your improvements and areas that might still require greater review. You might consider saving the online tests (if applicable) until the end of your preparation time when you feel completely ready (i.e., how you should feel on test day). These practice tests are a good final study opportunity before the actual exam to brush up on any remaining concepts or theories that need greater studying. In addition to using the tests to expose you to areas that need greater attention, also use them as an opportunity to apply the test-taking strategies described above and to learn how to pace yourself on both sections.

When scoring the tests, be sure to carefully read over the explanations for the multiple-choice questions you missed. Use the rubrics provided for each free-response question to score your responses more accurately. Then carefully compare your responses to the samples provided to understand areas where you need to improve. Finally, in order to get an idea of your overall score, you'll have to crunch some numbers. Generally, if you feel you answered the free-response questions well, and you consistently answer 50–60% of the multiple-choice questions correctly, you can expect a minimum score of a 3; if you answer 65–75% correctly, you can expect a minimum score of a 4; and if you correctly answer 80–100% of the questions, you should be able to score a 5 on the exam. Remember, these are only score *approximations*. You might consider asking a teacher for feedback.

COMMONLY ASKED QUESTIONS ABOUT THE AP HUMAN GEOGRAPHY EXAM

WHY SHOULD I TAKE THIS TEST?

First, in preparing for this test, you will be introduced to interesting material that you probably have not yet encountered during your high school experience. Preparation for the test exposes you to an exciting discipline that explains many of the processes occurring on the earth that you encounter on a daily basis. Secondly, preparing for the test provides you with an opportunity to improve your analytical and writing skills. Finally, if you do well on the test, you will be given Advanced Placement credit that may exempt you from having to take this course at a college or university, meaning you can get jump-started on your university career.

HOW MUCH DOES IT COST TO TAKE THE TEST?

The current fee for every AP test is $94.00. Many states and schools pay for students to take these tests. Be sure to check with your AP coordinator or high school counselor before paying the fee yourself. Fee reductions are available to students who require financial assistance. Check the following College Board web page *https://parents.collegeboard.org/faq/is-there-financial-assistance-available-low-income-students-want-take-ap-exam* for details on how to qualify.

DO ALL COLLEGES ACCEPT THE SCORES?

Advanced Placement is awarded to you by colleges or universities, not by the College Board. The best way to find out about a college's policy is to look in that institution's catalog or on its website, or to contact the registrar's office at the particular college or university you are interested in attending. Additionally, you can use the College Board website (*www.collegeboard.com*) to investigate each university's policies. The College Board performs an annual survey and reports each college's policies on its website.

WHAT MATERIALS SHOULD I BRING WITH ME ON TEST DAY?

> **NOTE**
> Do NOT bring any electronics or communication devices, such as cell phones.

1. Several sharpened #2 pencils (with erasers) for the multiple-choice answer sheet.
2. Black or blue ballpoint pens for the free-response questions.
3. Your school code—ask your AP coordinator if you are unsure. If you are home schooled, you will be given a code at the test.
4. A watch to keep track of time.
5. Your Social Security number for identification purposes.
6. A photo I.D. if you do not attend the school where the test is being administered.

HOW IS THE TEST GRADED?

Each portion of the test is worth about ½ of the student's overall grade. The multiple-choice section is graded electronically by tallying the number of correct responses. To grade the free-response questions, each year in June a reading is held that convenes college faculty and secondary school AP teachers to score the responses. These readings are led by a Chief Reader (a college professor), who ensures a rigorous system for maintaining the reliability and validity of scoring across readers.

Once scores have been obtained for both the multiple-choice and free-response sections of the test, they are combined, and the total raw scores are converted to a composite score on AP's 5-point scale:

AP Grade	Qualification
5	Extremely well qualified
4	Well qualified
3	Qualified
2	Possibly qualified
1	No recommendation

Grade distribution charts are available at AP Central (*http://apcentral.collegeboard.com*), as is information on how the grade boundaries for each AP grade are established. Most colleges or universities will accept a score of 4 or 5 for credit and placement. Scores of 1 and 2 are not accepted, and many universities will accept a 3 for credit and/or placement.

WHEN WILL I GET MY SCORE?

AP Grade Reports are sent in July to the college(s) you designated on your answer sheet, to you, and to your high school. You are able to view your scores online and send them to additional institutions (for a small fee) at *https://apscore.collegeboard.org*.

IF I FEEL THAT I PERFORMED POORLY, CAN I CANCEL THE TEST SCORE?

In order to have a grade canceled, AP services must receive a signed letter requesting cancellation by June 15. The letter must be sent to:

AP Services
P.O. Box 6671
Princeton, NJ 08541-6671

Grade cancellation deletes an AP test grade permanently from your records. Grades may be canceled at any time. There is no fee for this service; however, your test fee is not refunded. The grade report that you and your school receive will indicate that the grade has been canceled.

HOW DO I REGISTER TO TAKE THE TEST?

The AP coordinator at your school is your best source for registering for the test and getting all the specific details of test date and location. If you do not have this resource, contact the College Board (*http://apcentral.collegeboard.com*), and they will guide you to the nearest location to register for the test.

WHEN IS THE TEST OFFERED?

All AP tests, including the AP Human Geography test, are offered in May. Check the College Board website (*http://apcentral.collegeboard.com*) to find out the exact date of the test.

HOW MANY TIMES CAN I TAKE THE TEST?

You can take the test every year it is offered, but your grade report will include grades for every AP test you have taken, including scores for any repeated tests.

ARE COPIES OF OLD TESTS AVAILABLE?

The College Board does release copies of past AP Human Geography tests. The tests cost $25.00 each and can be purchased by phone (1-800-323-7155) or from the website (*http://store.collegeboard.com*). Additionally, if you go to the College Board website and navigate to the Human Geography Test page, sample multiple-choice questions along with the actual free-response questions from past years are available for free.

SHOULD I GUESS ON THE TEST?

You should guess whenever you are not sure about any question. As of May 2011, the College Board no longer penalizes you for incorrect answers. You gain points for every correct answer.

WHAT IF I HAVE MORE QUESTIONS?

If you still have questions about the test, the best source is the College Board website (*www.collegeboard.com*). If you cannot find the information you are looking for within the site, it is easy to contact the Board with any questions or concerns you might have about the test.

ADDITIONAL RESOURCES

Text

Any of the major introductory Human Geography textbooks on the market can help provide additional content information to that contained in this book. Some popular ones include:

- Bjelland, Mark, Daniel R. Montello, Jerome Fellman, Arthur Getis, and Judith Getis. *Human Geography: Landscapes of Human Activities.* New York: McGraw-Hill, 2013.
- Domosh, Mona, Roderick P. Neumann, Patricia L. Price, and Terry G. Jordan-Bychkov. *The Human Mosaic: The Cultural Approach to Human Geography.* New York: W.H. Freeman and Company, 2011.
- Fouberg, Erin H., Alexander B. Murphy, and H.J. de Blij. *Human Geography: People, Place, and Culture.* Hoboken, NJ: John Wiley & Sons, 2015.
- Knox, Paul L., and Sallie Marston. *Human Geography: Places and Regions in Global Context.* Upper Saddle River, NJ: Prentice Hall, 2015.
- Malinowski, Jon C., and David Kaplan. *Human Geography.* New York: McGraw-Hill, 2012.
- Norton, William. *Human Geography.* New York: Oxford University Press, 2014.
- Rubenstein, James M. *The Cultural Landscape: An Introduction to Human Geography.* Upper Saddle River, NJ: Prentice Hall, 2016.

Additionally, it would be helpful to have a good atlas on hand. The more you familiarize yourself with the geographic locations of specific places and their relative locations, the better your ability to effectively answer multiple-choice and free-response questions.

In addition to these suggestions, at the end of each chapter we have included a list of text materials relevant to the content area discussed within that chapter.

Internet

The following set of websites provides information relevant either to the test specifically (the College Board's site) or to information that will help you strengthen your general knowledge of human geography.

www.apcentral.collegeboard.com The website for the test; as already mentioned throughout this chapter, you should definitely visit this site. Here you will find answers to any questions you might have about the test; additionally, you will find example multiple-choice questions AND the free-response questions and example answers for the last five years.

One great way to assure a better score on free-response questions is through inclusion of relevant examples or current events. By reading a major newspaper or news magazine on a weekly basis for a couple of months before the exam, and taking brief notes on the relevant articles, you will build up a substantial database of possible examples to highlight in your responses—you'll be surprised at how often the concepts, processes, and theories of human geography surface in everyday events!

www.cia.gov/library/publications/the-world-factbook This website is fantastic for obtaining general information on any country in the world.

www.census.gov The factfinder data provided on this website allows you to see spatial variation (at the county level) in nearly every variable measured by the census.

www.economist.com The website that accompanies a good global news magazine. Visiting this website can provide you with some great examples of how human geography concepts and processes apply in current events across the globe.

https://www.nationalgeographic.org/education/ap-human-geography/ This website provides a variety of helpful materials from maps to brief interesting stories, to lesson plans and projects that can help you strengthen your knowledge of different areas of human geography.

PRACTICE, PRACTICE, PRACTICE

Now that you have all the information and strategies for the test in AP Human Geography, be sure to use them whenever you do the practice questions at the end of each chapter, the practice tests in the book, and, of course, when you take the actual test.

Thinking Geographically

1

IN THIS CHAPTER

→ GEOGRAPHY AS A FIELD OF INQUIRY
→ GEOGRAPHY BASICS
→ DESCRIBING LOCATION
→ SPACE AND SPATIAL PROCESSES
→ MAP FUNDAMENTALS
→ APPLICATIONS OF GEOGRAPHY

Key Terms

Absolute distance
Absolute location
Accessibility
Aggregation
Anthropogenic
Azimuthal projection
Breaking point
Cartograms
Cartography
Choropleth map
Cognitive map
Complementarity
Connectivity
Contagious diffusion
Coordinate system
Cultural ecology
Cultural landscape
Distance decay effect
Dot maps
Environmental geography
Expansion diffusion
Formal region

Friction of distance
Fuller projection
Functional region
Geographic Information System (GIS)
Geographic scale
Geoid
Global Positioning System (GPS)
Gravity model
Hierarchical diffusion
Human geography
International Date Line
Intervening opportunity
Isolinc
Large scale
Latitude
Law of retail gravitation
Location charts
Longitude
Map projection
Map scale
Mercator projection

Meridian
Natural landscape
Nature-society
Parallel
W. D. Pattison
Perceptual region
Peters projection
Physical geography
Preference map
Prime Meridian
Projection
Proportional symbols map
Ptolemy
Qualitative data
Quantitative data
Reference map
Region
Regional geography
Relative distance
Relative location
Relocation diffusion
Remote sensing
Resolution

GEOGRAPHY AS A FIELD OF INQUIRY

During the last 3,000 years, geography has evolved from a speculative philosophical endeavor into a vigorous area of academic research and applied science. The first geographers studied places and regions for practical purposes. They were interested in learning about geography primarily for developing trade routes to distant, and often dangerous, lands. Without knowledge of geography, travelers were lost. During the early centuries of geographic research, Chinese, Greek, and North African scholars took the lead. Their amazingly precise measurements and detailed maps laid the foundations for what later became the art and science of mapmaking, or **cartography**. Cartography has been central to the study of geography ever since. Today, cartography remains an important aspect of geography, though the field has grown to encompass much more.

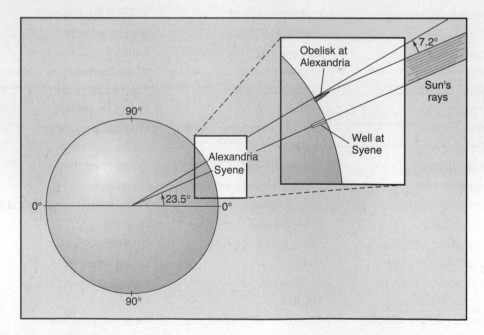

Figure 1.1 In 247 B.C., Eratosthenes calculated the circumference of the earth by measuring the sun's angles on June 21st at Alexandria and Syene in modern-day Egypt. He measured the distance between the two cities and then, using basic geometry, computed Earth's circumference to be 46,250 km, only about 175 km too long.

Eratosthenes, who served as the head librarian at Alexandria during the third century B.C.E., was one of these early cartographers. One of his greatest accomplishments was a remarkably accurate computation of Earth's circumference, which he based on the sun's angle at the summer solstice and the distance between the two Egyptian cities of Alexandria and Syene. He is also credited with coining the term *geography*, which literally means "Earth writing." About 500 years later, in the second century A.D., **Ptolemy** published his *Guide to Geography*, which included rough maps of the landmasses, as he understood them at the time, and a global grid system. Eratosthenes's and Ptolemy's efforts represent significant early contributions, both to the technical aspects of cartography and to our general understanding of geography.

Beginning in about 1400 A.D., scholars from Western Europe arrived at the cutting edge of geographic thought. During this period, explorers traveled the globe mapping landforms, climates, indigenous cultures, and the distribution of plants and animals. Names such as Bartholomeu Dias, Christopher Columbus, and Ferdinand Magellan are representative of this period of European exploration and colonization. Later, explorers such as Alexander von Humboldt and the members of the Lewis and Clark expedition became famous for their adventures traveling the globe in the name of geography and natural history.

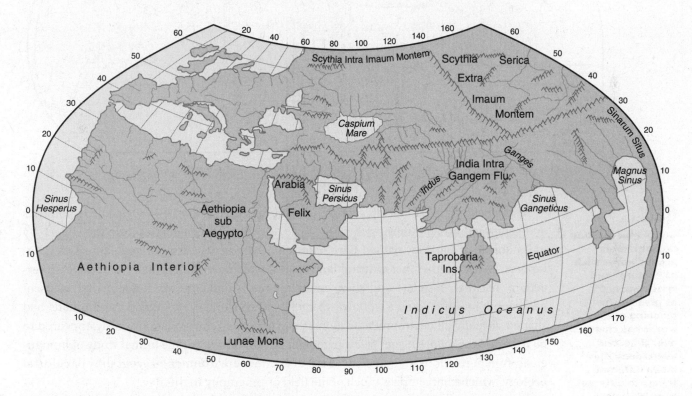

Figure 1.2 This early world map was created by Ptolemy in the second century A.D., more than 1,200 years before Europeans reached the New World.

Although these people were certainly not the first humans to see the places they "discovered"—European explorers encountered native peoples in almost every corner of the globe—their efforts did contribute much to the integration of geographic knowledge and to our understanding of spatial relationships. During the 18th, 19th, and early 20th centuries, scholars began to use the geographic information that was pouring into European museums and universities to synthesize theories about people and nature. This period saw the development of many modern academic disciplines, such as anthropology, geology, and ecology. Many key theories that rely heavily on geographic information were also proposed during this period, such as Charles Darwin's theory of evolution through natural selection and Alfred Wegner's theory of continental drift.

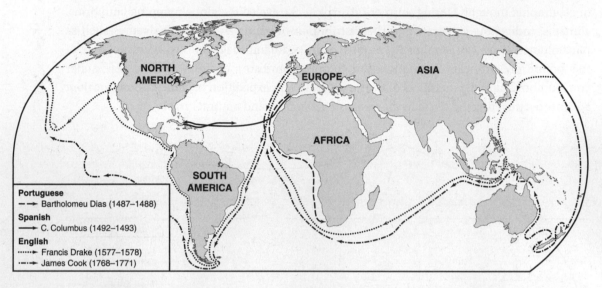

Figure 1.3 Routes of four European sailors during the Age of Exploration

TIP

Geographers today do not merely accept words such as "nature" and "environment" as givens. The meaning of these words has changed over time, and even today they mean different things to different people. In your essays, always be sure to define the terms you use and maintain a critical distance from culturally constructed concepts.

By the beginning of the 20th century, geographers had proposed few testable hypotheses and generated no real unifying theories. In 1925, though, a geographer from the University of California at Berkeley named **Carl Sauer** charted a new course for generations of geographers. Sauer argued that **cultural landscapes**, which are the products of complex interactions between humans and their environments, should be the fundamental focus of geographic inquiry. Sauer's work was, largely, part historical. He argued that humans had shaped virtually all environments around the world, even those that outwardly appeared to be **natural landscapes**. Sauer's new paradigm paved the way for the formal study of human-environment relations, frequently referred to as either **environmental geography** or **cultural ecology**, which characterizes much of the field of geography to this day.

Some 40 years after Carl Sauer's groundbreaking work, geographers at the University of Washington and elsewhere once again altered the direction of professional geography by joining a greater movement throughout the social sciences known as the quantitative revolution. The quantitative revolution stressed the use of empirical measurements, the testing of hypotheses, the development of mathematical models, and the use of computer programs to explain geographic patterns.

Since the 1970s, geography has undergone a new revolution. Geographers now use high-tech tools such as remote sensing, the Global Positioning System (GPS), and Geographic Information Systems (GIS) to collect and analyze spatial data. Since spatial data include any type of information associated with a particular location on Earth's surface, these technologies have radically changed the methodologies many geographers use.

Remote sensing is the process of capturing images of Earth's surface from airborne platforms such as satellites or airplanes. Researchers use remotely sensed images in several ways. One important method uses colors of visible light, often called *bands*, from the ultraviolet to the infrared, to distinguish features of the physical environment. Different kinds of soil, rock, water, vegetation, and built structures reflect different wavelengths of light and can thus be used to map and analyze geographic patterns.

The **Global Positioning System (GPS)** is an integrated network of satellites that orbit Earth, broadcasting location information to handheld receivers on Earth's surface. With a handheld GPS receiver, a person standing at any point on Earth can obtain highly accurate information about his or her geographic location in terms of latitude and longitude. This accuracy allows geographers to determine precise distances between two points, making GPS a valuable tool for navigational purposes.

Geographic Information Systems (GIS) are a family of software programs that enables geographers to map, analyze, and model spatial data. Most Geographic Information Systems use **thematic layers**. Each thematic layer consists of an individual map that contains specific features, such as roads, stream networks, or elevation contours. Multiple thematic layers may be united into one comprehensive map, combining many useful features that help geographers understand spatial relationships among different phenomena. For example, if geographers wanted to understand why a particular area of a city has higher property values than another area, they might overlap thematic layers. Each layer would have a different type of information, such as the locations of certain amenities (beaches, schools, or recreation areas), income levels, and employment opportunities. This method allows geographers to examine the role of multiple variables in shaping geographic phenomena, including property values. With the development of increasingly powerful computers and sophisticated software packages, GIS has become a tremendously powerful tool for understanding spatial processes on Earth's surface.

Modern geography is an extremely diverse discipline covering several major areas of study and involving researchers with different backgrounds from all over the world. Because of this diversity, geography has been divided into a handful of broad categories: human geography, physical geography, environmental geography, and geographical techniques. **Human geography** can be broadly defined as the study of human activities on Earth's surface—a broad definition! However, people who study human geography generally focus their work on more specific subdisciplines. The major subdisciplines of human geography include population geography, cultural geography, political geography, economic geography, agricultural and rural geography, and urban geography. Each of these fields is the subject of a chapter in this book. Each has its own theoretical frameworks, methodologies, and applications.

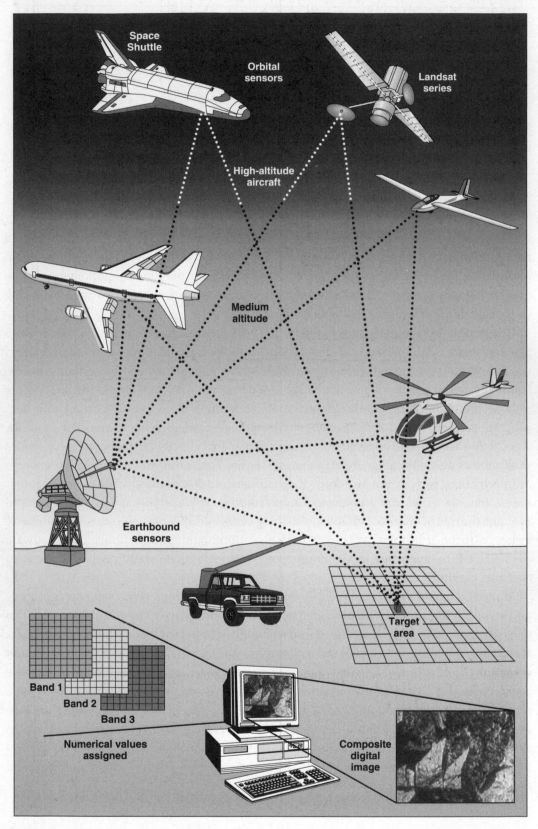

Figure 1.4 Remote sensing includes the use of both airplanes and satellites to capture images of the earth's surface.

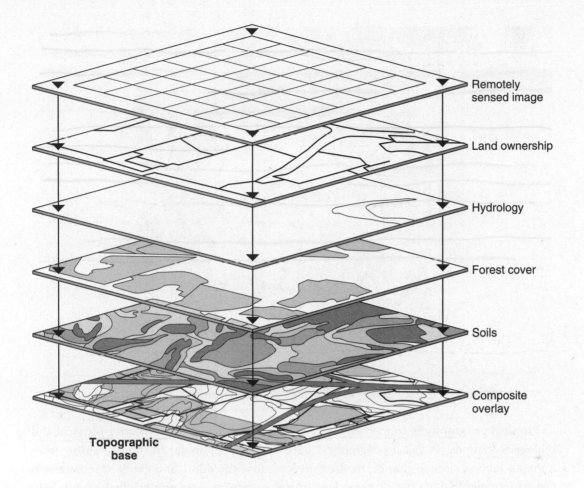

Remotely sensed image

Land ownership

Hydrology

Forest cover

Soils

Composite overlay

Topographic base

Figure 1.5 Geographical Information Systems unite multiple thematic layers to create one data-rich map.

Many human geographers combine two or more subfields in their research to promote a better understanding of the spatial dimensions of complex, interlinked social systems. A medical geographer studying the spread of disease might combine epidemiological data with information about socio-economic factors to understand why people in certain places are more susceptible to particular maladies than others. The Amish people, belonging to an old-order Protestant sect that lives in farming areas and does not participate in technology, popular music, and other common American innovations, such as automobiles, concentrated in rural Pennsylvania, for example, often abstain from vaccinations based on their religious beliefs, leaving them with low levels of "herd" immunity. In 1987 and 1988, a large measles epidemic struck Amish communities in Pennsylvania and the surrounding eastern states. As this example demonstrates, human geographers cannot fully understand patterns in culture and society without examining a variety of potential variables and relationships.

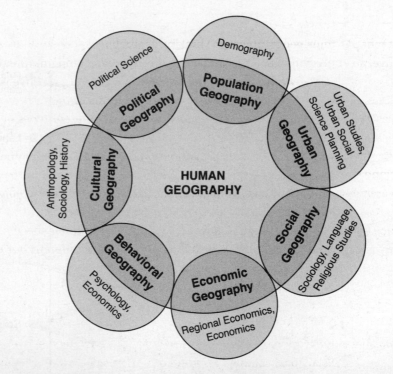

Figure 1.6 The discipline of human geography combines many subfields.

TIP

Geography is a synthetic, or interdisciplinary, field that brings together many scholars with different interests. In your essays, remember that geography provides a wide range of perspectives for you to use when considering how to answer any given question.

Physical geography is concerned with the spatial characteristics of Earth's physical and biological systems. Physical geographers have been instrumental in understanding phenomena such as climate change, biodiversity loss, desertification, and the El Niño Southern Oscillation, all of which have tremendous importance for people around the world. Where physical and human geography meet—and they often do—one enters the realm of environmental geography. Environmental geographers come from almost every academic discipline. **Sustainability** means many things to many people. It thus pops up at several different points in this book. In 1987 the World Commission on Environment and Development (WCED), or Brundtland Commission, defined sustainable development as "development that meets the needs of the present without compromising the ability of future generations to meet their own needs." As you will see later, however, even such apparently straightforward definitions of sustainability have proven highly contentious.

In 1964, **W. D. Pattison** introduced a slightly different way of thinking about the structure of the discipline. According to Pattison, geography drew from four distinct traditions: the Earth-science tradition, the culture-environment tradition, the locational tradition, and the area-analysis tradition. Pattison's Earth-science tradition is another name for physical geography. His culture-environment tradition is what we now call environmental geography. His locational tradition relates to the analysis of spatial data through cartography. Finally, the area-analysis tradition refers to regional geography, which is covered later in the chapter.

GEOGRAPHY BASICS

What does it mean to think geographically? Thinking geographically means developing a spatial perspective, an appreciation of scale, and the ability to analyze and interpret varied forms of geographic data. Let's begin with **spatial perspective**. A spatial perspective is an intellectual framework that allows geographers to look at Earth in terms of the relationships among various places. Geographers look at the spatial distribution of different types of phenomena and ask why and how certain patterns or processes occur in certain places. For example, a cultural geographer might look at the spatial distribution of McDonald's restaurants by where they are. Beyond simply noting the fast-food conglomerate's locations, a geographer asks *why* McDonald's are both located and successful in various parts of the world and *how* they spread to so many areas. This is the spatial perspective: observing the spatial location of things on Earth's surface and determining why and how those things occupy their specific locations.

Geography is based on the simple premise that all places are different but many places have important similarities. All places on Earth are related to each other, but some places are more related than others. Geographers seek to understand not only spatial patterns but also spatial interrelations. How do two places interact economically, socially, and culturally? Why do some places have more in common than others? How are social phenomena connected over time and space? Answering these questions requires a spatial perspective.

Thinking geographically also requires an appreciation of scale. Map scale is simply the ratio of the distance between two objects depicted on a map and the distance of the same two objects where they actually exist on Earth's surface. **Geographic scale**, however, is a broader concept than map scale; it refers to a conceptual hierarchy of spaces, from small to large, that reflects levels of spatial organization in the world. Examples of characteristic scales in human geography include the neighborhood, the urban area, the metropolitan area, and the region. Other frequently cited geographic scales include the watershed, ecosystem, landscape, and biome. A geographer seeks to understand how processes occurring at one scale may affect activities on other scales. For example, much of the pollution that causes acid rain in North America is emitted from a small number of industrial facilities, such as power plants. Yet the pollution emitted from these point sources has environmental consequences that may be national, continental, or even global in scope.

One of the most important and hotly contested concepts in all of geography is the **region**. No two places on Earth are the same. But some places share enough characteristics that geographers—as well as just regular people—tend to group these places together in larger areas with coherent properties or identities. A region is generally defined as an area larger than a city that contains some unifying social or physical characteristics. Regions do not exist as well-defined units in the landscape. Instead, they are conceptual constructions that geographers use for convenience and comparison. Every region can be described by its unique area, location, and set of boundaries. A region's area may be small or large, such as Little Italy in New York City or all of Western Europe. Its location may be based on physical characteristics such as mountain ranges, or on cultural characteristics such as the predominant language or religion. Its boundary may be a strictly demarcated political border, or a fuzzy imagined boundary such as the area that contains the American Midwest. Despite much debate about exactly what makes a region, the concept itself remains important because regions often form the basic units of geographic research. **Regional geography**, or Pattison's area-analysis tradition, is the study of regions. Courses such as the Geography of the United States take a regional approach and explore the common characteristics unique to those particular regions.

The 2003 AP Human Geography exam included a free-response question on the ways that tourism has both increased and decreased regional landscape distinctiveness. Could you provide examples of each?

TIP

It is useful to think of regions not only as places but also as processes. Regions change through time—and they even move! In the 18th century, for example, European colonists thought of the "American West" as including upstate New York and Kentucky. In the early 19th century, the "American West" was located in the area we now call the "Midwest." By the late 19th century, the term "American West" referred to the Rocky Mountains, Great Basin, and Pacific Coast.

Geographers have described several different types of regions. **Functional regions** have special identities because of the social and economic relationships that tie them together. They are often also referred to as nodal regions because they are defined by the connections and interactions that occur between them and surrounding areas. For example, at the heart of Northern California's sprawling Bay area is the city of San Francisco. The boundary around the Bay area is loosely defined, but it certainly includes the major population and economic centers, such as Oakland and San Jose.

Formal regions have specific characteristics that are relatively uniform from one place to another within the designated region. These properties may include physical features, such as rolling hills and redwood trees, or cultural traits, such as religion and ethnicity. Tibet, for example, is a region of high mountains and plateaus where most people have a common culture and observe the religious practices of Buddhism. This example might lead you to think that any country can be categorized as a formal region. Yet most countries contain significant natural and cultural diversity. China, for example, is a vast and extremely diverse country, and Tibet is only one region within it. Many Tibetans hope someday to achieve independence from China, but the Chinese government, based more than 2,000 miles away in Beijing, has other plans. For the time being, therefore, Tibet provides a good example of a region whose areal extent is determined by cultural and physical properties rather than national boundaries.

Perceptual regions exist in the minds of people. A common example used to illustrate this concept is that of the American Deep South. When individuals from other regions are asked to draw a boundary around this region, they tend to consider physical and cultural characteristics, and historical legacies of slavery and war, as well as stereotypes about this region today. These might include a muggy climate, magnolia trees, a distinct accent (or cluster of accents), culinary traditions, and the Southern Baptist Protestant religion. The boundaries of this region, as with all perceptual regions, tend to be fuzzy at their borders, but they frequently involve important issues of identity. People's attachment to the region that they perceive as their home is what geographers call **a sense of place**.

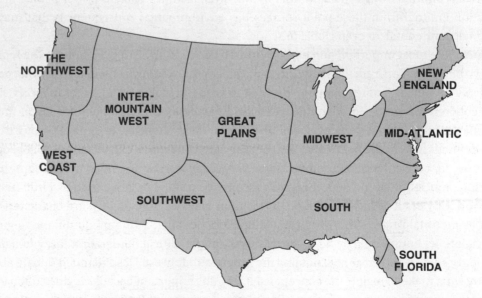

Figure 1.7 The contiguous 48 states are frequently divided into about 10 major regions.

One example of a distinctive region is South Florida. South Florida is bounded by the Gulf of Mexico to the west, the Florida Strait to the south, the Atlantic Ocean to the east, and by Interstate 4 to the north. It is distinctive because of its humid subtropical, monsoonal climate; its sprawling urban centers; its high proportion of Caribbean immigrants and retirees; and its unique and expansive wetland system known as the Everglades. Miami is an important city that spawns numerous social and economic interactions throughout the region. Thus, South Florida is a unique place tied together by both social and natural factors.

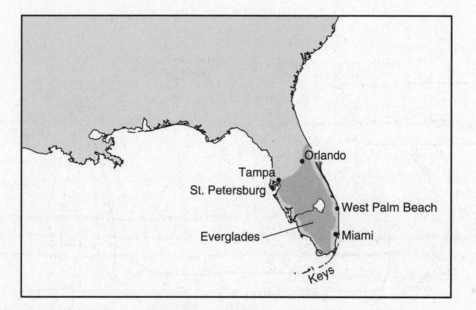

Figure 1.8 South Florida is a relatively well-defined region, with recognizable borders and cohesive geographical characteristics.

Thinking geographically also requires the ability to understand and synthesize various types of data. Human geographers work with two main forms of data: qualitative and quantitative. **Qualitative approaches** are often associated with cultural or regional geography because they tend to be more unique to and descriptive of particular places and processes. **Qualitative data** are not well suited to statistical analyses and modeling. It is often collected through interviews, empirical observations, or the interpretation of primary source mateirals, such as texts, artwork, old maps, and other archives. **Quantitative approaches**, by contrast, use rigorous mathematical techniques and are particularly important in economic, political, and population geography, where hard, numerical data abound. Most physical geography is also based on quantitative methods.

DESCRIBING LOCATION

One of the basic tasks of human geography is to describe places in terms of both their location and their relationships to other places. Some features on Earth's surface are commonly represented as points, some as lines, and some as areas. These representations make the concepts of location, distance, and scale critical for describing Earth's features.

The precise location of any object on Earth's surface can be pinpointed on a standard grid, or **coordinate system**, on which you can designate and describe a place's **absolute location**.

Coordinates are made up of lines of **longitude** and **latitude**. Lines of longitude, or **meridians**, originate at the **Prime Meridian** (0°), which passes through Greenwich, England, and end at the **International Date Line** (180°) in the Pacific Ocean. All lines of longitude also meet at the poles. Unlike meridians, lines of latitude never intersect. As a result, lines of latitude are often referred to as **parallels**. They originate at the equator (0°) and terminate at the poles (90°). An example of a well-known parallel is 49 degrees north latitude, which marks the boundary between the western United States and Canada.

TIP

Free-response questions often provide graphs or maps, and ask students to interpret those figures using specific terminology from the AP Human Geography course. You should always strive to use and define specific terms from this book when answering such essay questions.

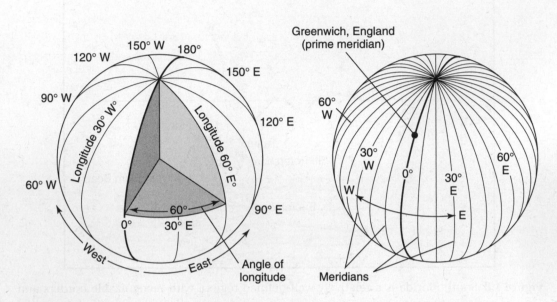

Figure 1.9 The Cartesian coordinate system uses both parallels and meridians. Meridians, or lines of longitude, originate at the Prime Meridian centered on Greenwich, England.

There are other ways to describe location than simply using coordinates. A place's **site** refers to its physical and cultural features, independent of its relationship to other places around it. For instance, the city of San Francisco is located at about 37° N latitude and 123° W longitude on a windswept peninsula separating the San Francisco Bay from the Pacific Ocean. San Francisco is characterized by diverse ethnic neighborhoods; a large harbor; fine Victorian-era architecture; a cool, foggy climate; and hilly topography. **Situation**, or **relative location**, describes a place's relationship to other places around it. San Francisco is the economic hub of northern California and the center of a large metropolitan area containing more than 6 million people. To its east lie the San Francisco Bay and the cities of Oakland and Berkeley. To its west is the Pacific Ocean. To its north is Marin County. To its south are the sprawling peninsula South Bay areas, including the city of San Jose.

Distance between places can be described in several ways. **Absolute distance** is a measurement of linear space, in standard or metric units, between two places. Bangor, Maine, for example, is 130 miles northeast of Portland, Maine. Although knowing absolute distances is important, **relative distance** measures are often much more meaningful. One way to think about relative distance is through the concept of **connectivity**. Connectivity is an important concept because the absolute distance between places is often not an accurate characterization of their social, cultural, political, or economic connectivity. Some places that are close together in absolute distance are actually less connected than other places that are farther apart. For example, Honolulu, Hawaii, is thousands of miles away from the American mainland. Yet it is very closely connected, culturally and economically, to cities like Los Angeles and San Diego and to the larger American society in general. Havana, Cuba, however, is less than two hundred miles from Key West, Florida, but beginning in 1962, the US embargo of Cuba made these two places seem quite remote. This may now be changing with policies meant to reopen relations between the US and Cuba introduced under the Obama administration. Two other distance measures, time and money, are related to this notion of connectivity. Often, the distances between two places are described in terms of the amount of time or money needed to travel from one to the other rather than by a standard unit of absolute distance. For example, when asked how far it is from your home to your school, you might be more likely to give a time unit such as 10 minutes rather than a distance unit such as 1.5 miles.

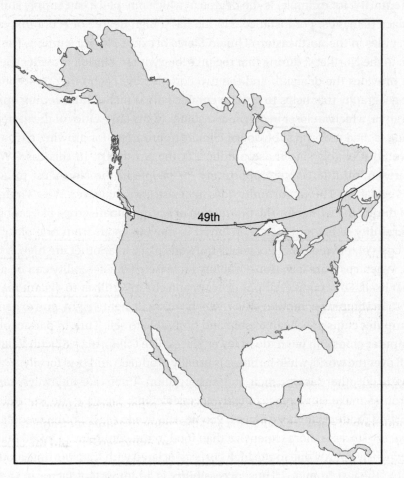

Figure 1.10 The 49th parallel forms a political boundary between the western United States and Canada, though it has no relationship to natural features.

Another related concept is **time-space convergence**. Due to improved transportation and communication technologies, the relative distance between some places is, in effect, shrinking. Crossing the Atlantic, from London to New York, used to take days or even weeks by boat, but it now takes only half a day by plane. A handwritten letter used to take weeks to arrive in Paris from Los Angeles, but residents in these cities can now instantly communicate by telephone or any of several forms of electronic communication. In these ways, many places now seem much closer to each other than they once were. It is important to recognize, however, that this process of shrinking relative distance has not occurred equally throughout the world. Remote areas of some developing countries remain distant, or isolated, from the developed world in the sense that absolute distance must still be overcome in order for interaction to occur. For example, packages originating in North America still take weeks to arrive in certain parts of Africa. In parts of Africa and Asia, traveling a couple hundred miles can be a true adventure, sometimes taking days to complete.

SPACE AND SPATIAL PROCESSES

When thinking about connectivity, geographers often borrow concepts from economics. **Complementarity**, for example, is the degree to which one place can supply something that another place demands. For instance, Florida has a high degree of economic complementarity with cities in the northeastern United States because Florida supplies fresh fruits and vegetables to the Northeast during that region's long winter. Florida provides the supply, the Northeast provides the demand, and the two complement each other. A second idea from economic geography that helps to explain connectivity is that of **intervening opportunities**. If West Virginia, which is closer in absolute distance to the large cities of the eastern seaboard than Florida is, had a warm, subtropical climate appropriate for growing citrus fruits, West Virginia would probably ship those products to the Northeast. In this case, West Virginia would represent an intervening opportunity for people in the Northeast to acquire fresh fruits and vegetables. This opportunity does not exist since, however, West Virginia's climate is too cold during the winter for the production of warm-climate crops like oranges.

Transferability refers to the costs involved in moving goods from one place to another. When the costs of moving people or goods from one place to another are high, transferability decreases. When costs are low, transferability increases. Transferability can be a function of the product itself. For example, shipping heavy and cheap fertilizer to distant markets makes less sense than shipping computer chips, which are small, lightweight, and expensive to purchase. Computer chips are cheap to ship and profitable to sell. This, in part, explains why so many computer chips can be made in fewer places, like California's Silicon Valley, and then shipped all over the world, while fertilizer is usually produced and sold locally. Transferability also depends on other factors such as transportation. Interstate highways, railroads, and shipping routes make some places highly accessible and decrease the transferability costs of transporting products to those places. People who live in Alaska often complain that food in their grocery stores is more expensive than food in the contiguous United States. Food in Alaska is expensive partly due to the high costs associated with transporting goods over great distances to "the last frontier." Thus, **accessibility** is an important factor in the interaction between places, and in the cost of goods and services traded among them.

TIP

In recent years, geographers have spent much time thinking about the ways that information technologies have altered space-time relationships. You should be prepared to answer a question about the changing nature of space, time, and communication in a world full of cellular phones, satellite links, and high-speed internet access.

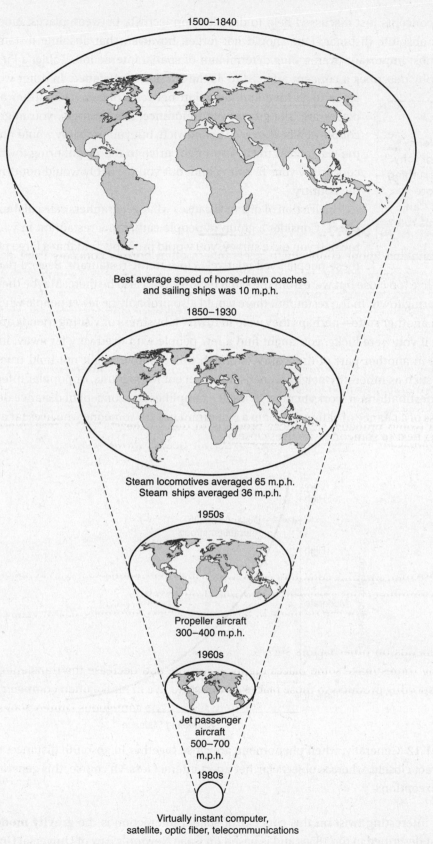

1500–1840

Best average speed of horse-drawn coaches and sailing ships was 10 m.p.h.

1850–1930

Steam locomotives averaged 65 m.p.h.
Steam ships averaged 36 m.p.h.

1950s

Propeller aircraft
300–400 m.p.h.

1960s

Jet passenger
aircraft
500–700
m.p.h.

1980s

Virtually instant computer,
satellite, optic fiber, telecommunications

Figure 1.11 Throughout history, improvements in technology have "shrunk" the size of Earth in terms of the amount of time it takes to travel to distant places and communicate with distant people.

The concepts just discussed help to describe connectivity between places, independent of their absolute distance. One should not forget, however, that absolute distance is the single most important overarching determinant of spatial interaction. Tobler's First Law of Geography describes a concept referred to as the **friction of distance**. In other words, distance itself hinders interaction between places. The farther apart two places are, the greater the hindrance. For example, you might walk a couple of blocks to grab a sandwich, but you probably would not walk all the way across town. You might drive to the neighboring town to see a concert by your favorite band, but you probably would not drive across the country.

The friction of distance causes what geographers call a **distance decay effect**. Consider a group of people eating in a restaurant in your hometown. If you did a survey, you would probably find that a large number of these people lived relatively close to the restaurant. Several people who did not live too close but were from the same region would also be there. Maybe they live in a neighboring town. In the restaurant there would also probably be fewer people who lived far away in another state—perhaps they were in town on business or visiting friends and family. Finally, if you were lucky, you might find a few people who lived very far away, in another country in another part of the world. Although this example might not hold true in some places, such as inner-city neighborhoods full of recent immigrants, or popular international tourist destinations, it works in most cases. It exemplifies the concept of distance decay. You have less of a chance of eating dinner in a restaurant next to someone who lives far away than of eating next to someone who lives close by.

NOTE

According to Tobler's "First Law of Geography," everything is related to everything else, but near things are more related than distant things. Can you think of examples of this rule? How about exceptions?

TIP

The College Board has defined five major areas of student proficiency to be tested on the AP exam. The FIFTH GOAL of the AP exam in Human Geography is to test students' ability to "characterize and analyze changing interconnections among places." As you work through the remainder of this book, you should always keep the five main goals of the AP exam in mind.

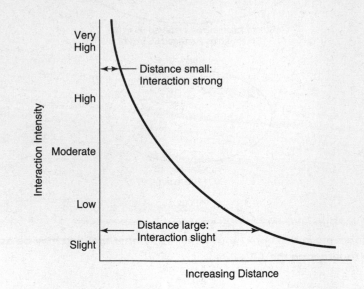

Figure 1.12 Generally, when phenomena are close together in absolute distance, they tend to interact closely, whereas objects farther apart interact less. Of course, this general rule has many exceptions.

One interesting twist on this concept of spatial interaction is the **gravity model**, which was first described in the 1850s and is based on Isaac Newton's Law of Universal Gravitation. According to Newton, the degree to which objects are attracted to each other by gravity is a result of the product of their respective masses divided by the square of their distance apart.

The gravity model in geography applies this exact formulation to the spatial interaction of various population centers. In the gravity model, population substitutes for gravity. Thus, the interaction between two places is equal to the product of the places' populations divided by the square of their distance apart.

THE GRAVITY MODEL

$$I_{ij} = \frac{P_i P_j}{D_{ij}{}^2}$$

where

I_{ij} = the interaction between places i and j

P_i = the population of place i

P_j = the population of place j

$D_{ij}{}^2$ = the distance between places i and j squared

One implication of this equation is that large cities may still have extensive and important interactions despite being separated by great distances. Los Angeles and New York are excellent examples of two highly connected large cities located on opposite sides of a continent. Their extremely large populations, when multiplied together in the numerator, are big enough to overcome the square of the distance between them, represented in the denominator. Put another way, the absolute distance between them is great, but the relative distance is much smaller because their populations are far more connected.

Another insight that arises from this general line of thinking is that large cities seem to have a greater "gravitational pull" for individual people than do small ones. This makes intuitive sense since large cities provide a diversity of opportunities for employment, education, products, and services that smaller towns usually do not. This phenomenon was formally described in 1931 in the **law of retail gravitation**. This law basically states that people will be drawn to large cities to conduct their business since large cities have a wide influence on the areas that surround them. The outer edge of a city's sphere of influence is called the **breaking point**. Beyond the breaking point, another city's sphere of influence begins.

STRATEGY

Free-response questions sometimes ask students to compare two geographic concepts and provide examples of each. Whenever you see two comparable or contrasting concepts in this book, you should remember that you may be asked to discuss them together on the AP exam.

Spatial diffusion is an extremely important concept in human geography because it describes the ways in which phenomena, such as technological innovations, cultural trends, or even outbreaks of disease, travel over distances. Geographers have identified two main types of spatial diffusion processes: expansion diffusion and relocation diffusion.

In **expansion diffusion**, the thing that is traveling both remains in its area of origin and spreads to surrounding areas. Expansion diffusion takes two main forms: contagious and hierarchical diffusion. In **contagious diffusion**, something is transmitted over a distance because people who carry it are close to each other. The common cold is transmitted from place to place simply because people are close enough to each other to pass it along in airborne germs. **Hierarchical diffusion** involves the transmission of a phenomenon from one place to another because the level of interaction between places overcomes the actual distance between them. Hierarchical diffusion processes strongly correlate with the interaction levels calculated in the gravity model described in the previous section. In the United States, places with higher levels of interaction, such as New York and Los Angeles, are often the first to adopt new trends in music, fashion, and art. Not far behind are those who live in other cosmopolitan urban centers, such as Chicago, San Francisco, Seattle, and Boston. These cities have much in common, including large populations of young people, diverse ethnic neighborhoods, and large numbers of important cultural venues, such as theaters, concert halls, recording studios, museums, and universities.

The second major diffusion type, **relocation diffusion**, occurs when people migrate from one place to another, bringing with them cultural traditions from their previous homelands. In the early 1900s, millions of people migrated to the United States from Europe, bringing with them the diverse languages, culinary delights, and social traditions that still characterize the American cultural landscape today.

MAP FUNDAMENTALS

TIP

The FIRST GOAL of the AP exam in Human Geography is to test students' ability to "use and think about maps and spatial data."

Since geography's early beginnings, mapmaking has been one of the main techniques geographers use to display spatial information. Even if you are not a geographer or are not at all interested in geography, you still probably encounter maps on a daily basis.

Maps come in all different shapes and sizes. Maps of local bus systems look quite different from maps depicting average income in a state. A map of national voting patterns will probably not help you find the nearest public library. Consequently, the ways that different maps look largely depend on their purpose. When making maps, cartographers base their scientific and artistic decisions on which information they are trying to communicate and on which data they have available. They must choose between types of projections, levels of simplification, levels of **aggregation**, map scale, and which symbols to use to depict information.

All maps are created by projecting Earth's true three-dimensional shape—a bumpy oblate spheroid or **geoid**—onto a two-dimensional surface. Transforming something spherical into something flat requires that a two-dimensional representation will never precisely represent three-dimensional reality. Maps were once produced by placing a light source, such as a candle or a bulb, inside of a translucent globe and then projecting the globe's features onto another shape surrounding it—such as a cylinder or cone—which could later be unrolled into a flat map. Today, geographers use various mathematical equations to produce **map projections**. No matter the method or the equation used, it is important to remember that three-dimensional shapes can *never* be transferred to two-dimensional surfaces without losing some detail or distorting some features. All flat maps have some distortion in the way they represent distance, shape, area, or direction.

The **Mercator projection**, on page 51, which preserves accurate compass direction, distorts the areas of landmasses relative to each other. In the Mercator projection, landmasses become amplified, or artificially large in size, at high latitudes near the North and South Poles. The Mercator projection was originally created by projecting Earth's features onto a cylinder. Unfortunately, when a geoid is projected onto a cylinder and then unrolled, lines of longitude, which on a globe intersect at the North and South Poles, all become parallel. This distortion causes space to get increasingly stretched out the closer you get to the poles.

This is why, on Mercator projections, Greenland, Alaska, Antarctica, and other high-latitude landmasses look so big.

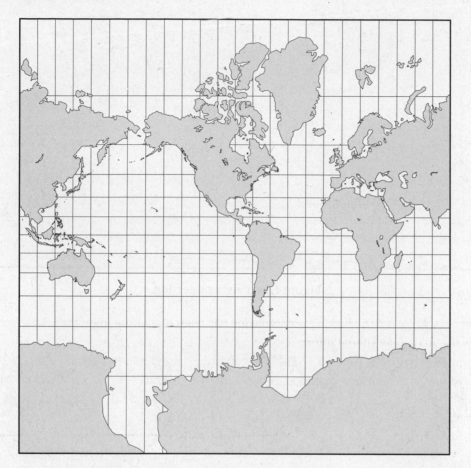

Figure 1.13 Mercator projection

The **Fuller projection** strikes a different compromise. It maintains the accurate size and shape of landmasses but completely rearranges direction. The cardinal directions—north, south, east, and west—no longer have any meaning. The **Robinson projection** is an example of an attempt to balance projection errors. It does not maintain accurate area, shape, distance, or direction, but it minimizes errors in each. The Robinson projection provides an aesthetically pleasing balance and, as a result, is frequently used by cartographers at organizations such as the National Geographic Society. The **Peters projection** is an equal-area projection purposefully centered on Africa in an attempt to treat all regions of Earth equally. **Azimuthal projections** provide a different perspective than most people are used to seeing. Azimuthal projections are planar, meaning they are formed when a flat piece of paper is placed on top of the globe and, as described earlier, a light source projects the surrounding areas onto the map. Thus, in an azimuthal projection, either the North Pole or the South Pole is oriented at the center of the map, giving the viewer an impression of looking up or down at Earth.

The important thing to remember here is there is no one *best* map projection, but certain projections are more suited than others to portraying certain aspects of Earth's surface or data types than others. It is up to the mapmaker to decide which projection to use for which purpose, with an understanding that every projection will involve some compromise.

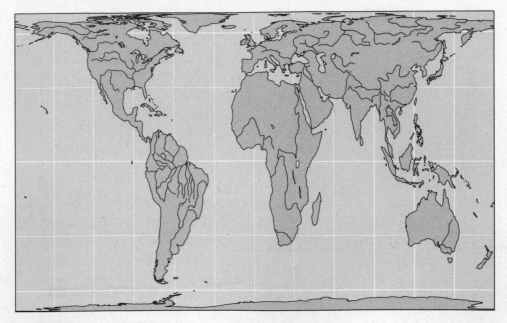

Figure 1.14 Peters projection

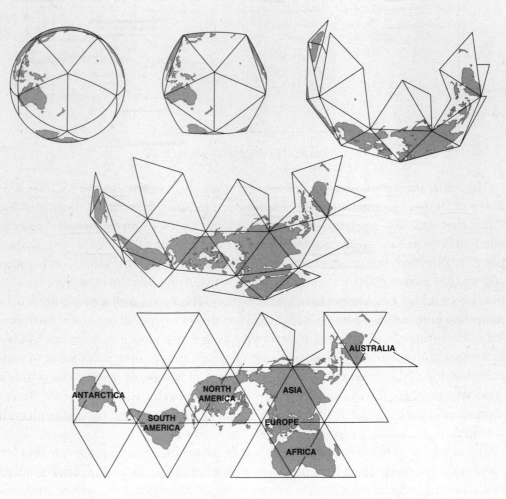

Figure 1.15 Fuller projection

In addition to projection, the cartographer must make decisions regarding which details are displayed on the map. All maps are simplifications of the things they represent. When designing a map of the entire United States, for example, minute details such as the locations of towns with populations less than 10,000 people might not be included. However, a map of a shopping mall will probably include such details as the location of the nearest restroom. Again, what cartographers choose to display on a map depends on the purpose of the map and the size of the area covered. The degree of aggregation cartographers choose to show on a map will also depend on the map's purpose. Level of aggregation just refers to the size of the unit under investigation, such as cities, counties, states, or countries. When showing patterns of population density across the United States, a cartographer may use the county as the basic unit of organization. If showing the population patterns for a single state, however, the same cartographer may zoom in to use the local zip code. In some cases, the level of aggregation may fundamentally affect the patterns displayed on the map.

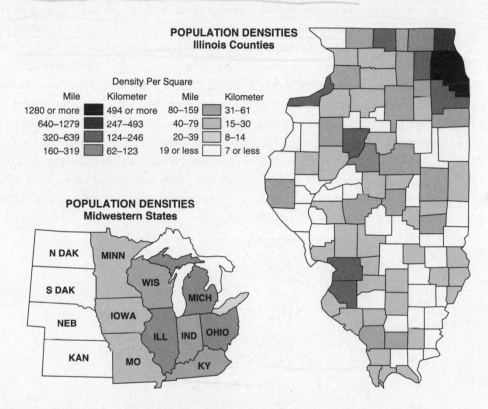

Figure 1.16 Population densities of midwestern states and Illinois by county. Notice that state-level categories obscure differences by county.

All maps have a scale and resolution. **Map scale** refers to the ratio between the distance on a map and the actual distance on Earth's surface. The United States Geological Survey (USGS), for example, produces standard quadrangle maps at the scale of 1:24,000, pronounced "one-to-twenty-four-thousand." On these maps, one unit—an inch, a foot, a finger, whatever—equals exactly 24,000 of those same units on the ground. In a **small-scale** map, the ratio between map units and ground units is small, such as 1:100,000. Since one map unit equals so many of those same units on Earth's surface, these maps tend to cover large regions. Maps of the whole world are always small-scale maps since they cover such an enormous area. **Large-scale** maps have large-scale ratios, such as 1:5,000, and cover much

TIP

The SECOND GOAL of the AP exam in Human Geography is to test students' ability to "understand and interpret the implications of associations among phenomena in places."

The THIRD GOAL
of the AP exam in
Human Geography
is to test students'
ability to "recognize
and interpret . . .
relationships
among patterns
and processes" at
different scales of
organization.

smaller regions. A large-scale map, at a one-to-five thousand scale, might depict a farm or a neighborhood.

Resolution is another extremely important concept. Resolution refers to a map's smallest discernable unit. For our purposes, you can also think of resolution as the smallest feature you can see on a map. If, for example, an object has to be one hundred meters long in order to show up on a map, then that map's resolution is one hundred meters. For many world maps, the smallest discernable detail may be the size of a large city, such as New York. If you have a globe or a world atlas, take a look at New York City—can you see Manhattan Island? How about Brooklyn? These places are pretty close to the resolution of a standard globe. At the other extreme, some highly accurate, large-scale satellite photos have a resolution of less than one meter. In these images, which can act as maps, you can often identify objects as small as automobiles, cattle troughs, and shrubs.

Finally, cartographers must choose the appropriate map type according to the information they are trying to communicate. On a very general level, maps fall into one of two categories: reference maps and thematic maps. **Reference maps** work well for locating and navigating between places. **Thematic maps** display one or more variables across a particular area. Whichever the map type, human geographers use many different types of symbols on their maps to depict spatial data. One important map symbol is the contour line, or **isoline**. Isolines are lines that represent quantities of equal value and are familiar to those who use **topographic maps** for navigation. On a topographic map, the path of each isoline indicates a constant elevation. If you took a topographic map out into the field and walked exactly along the path represented by an isoline on your map, you would always stay at the same elevation. If you turned and walked perpendicular to the isolines, then you would be walking either straight uphill or straight downhill. Isolines are also commonly used to represent values, such as population density, that vary continuously over space.

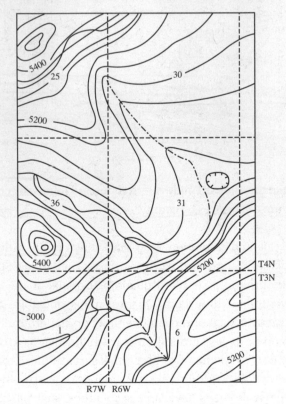

Figure 1.17 Topographic maps use isolines to represent points of common elevation.

Several other types of map symbols are common in thematic maps. In a **proportional symbols** map, the size of the chosen symbol—such as a circle or triangle—indicates the relative magnitude of some value for a given geographic region. Bigger circles, stars, dots, or icons represent more of some feature, such as churches, crimes, or baseball fans. **Location charts** convey a large amount of information by associating charts with specific mapped locations. In a map of Canada, each province may have a chart within it depicting the number of native French speakers versus the number of native English speakers. **Dot maps** use dots to show the precise locations of specific observations or occurrences, such as crimes, car accidents, or births. If two car accidents occur at the corner of Main and Horizon Streets, then two dots will show up there. **Choropleth maps** use colors or tonal shadings to represent categories of data for given geographic areas. A choropleth map of Africa may use five different colors to show levels of income by country. **Cartograms** transform space such that the political unit—a state or a country, for example—with the greatest value for some type of data is represented by the largest relative area. In a population cartogram of Asia, China would be the largest country on the map and India the second largest because of their populations, even though Russia has the most land area. These are just a few examples of the many different types of maps and the many ways spatial information can be portrayed.

Technological innovations in the area of mapmaking are radically changing the methods of modern cartography. One interesting class of maps that has become increasingly popular in recent years falls under the general category of **visualizations**. Visualizations use sophisticated software to create dynamic computer maps, some of which are three dimensional or interactive. Some visualizations allow geographers to investigate features that cannot be seen with the naked eye. Others use models to show how landscapes change over time. In others, you can even walk through or fly over the landscape. Visualizations are also extremely helpful learning tools for aspiring young geographers.

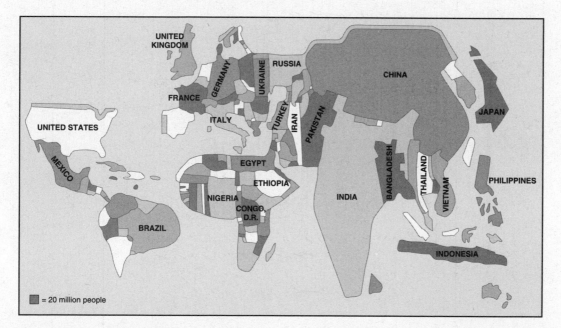

Figure 1.18 A cartogram of world population. Countries that have the greatest populations are also depicted as being the largest in size.

TIP

The FOURTH GOAL of the AP exam in Human Geography is to test students' ability to "define regions and evaluate the regionalization process."

Maps can be powerful tools. If a geographer makes even one error on his or her map, it can have dramatic implications. It thus is useful to think of maps as texts with the capacity to communicate important ideas and information. Maps can be powerful tools for examining spatial processes. For example, many spatial epidemiologists, or geographers who study the spread of disease, use maps to investigate the correlation between the incidence of disease and the proximity to harmful environmental factors. The most famous example is Dr. John Snow, who mapped cholera cases against water pumps in London in the 1800s to determine which pump was leading to the epidemic's outbreak. Such examples illustrate the power maps have as tools for research, teaching, and serving society.

An entire class of maps falls outside of the purview of traditional cartography as described above. **Cognitive maps**, which many behavioral geographers believe guide people's spatial behavior, are one example. A cognitive map is an individual's internal, geographic understanding of a place. Cognitive maps are formed when people perceive information about their surroundings and then process this information into a mental image that reflects both the physical environment and that individual's social, cultural, and psychological framework. In essence, people are internal cartographers of the places they encounter on a daily basis. They organize streets, landmarks, and districts in their mind to form dynamic, coherent mental maps of the places where they live and interact. When people are asked to draw maps of the places where they live, they include some details they personally consider important but leave out others. As a result, no two cognitive maps of the same place will look exactly alike. What people include on their maps can reveal much about their perceptions of and ideas about space and place. Cognitive maps are extremely informative. They give us important clues for understanding how people interpret and understand the places in which they live.

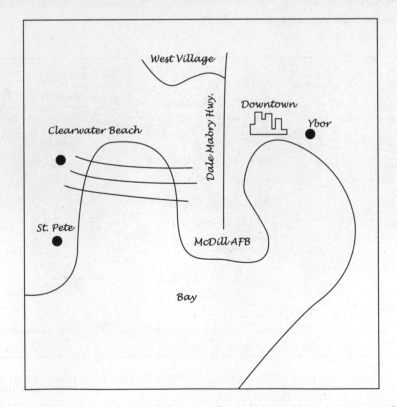

Figure 1.19 Cognitive map of Tampa, Florida, drawn by a former resident

Cognitive maps reflect much about the person who draws them. Different people's cognitive maps show different levels of engagement with the landscape. For example, Eskimos living in the North American Arctic have drawn amazingly detailed maps of the regions in which they live without ever seeing professionally produced maps of the area. One can only assume that this knowledge is essential for their daily lives, which often include long-distance travel in remote areas with few stationary landmarks. Cognitive maps also give clues to different people's levels of access to things like education, language, and transportation. In Los Angeles, affluent college students living in Westwood are often able to draw quite detailed maps of the city. In contrast, poor Latinos in East L.A., many of whom do not speak English, frequently possess spatial information about only the neighborhoods in their immediate vicinity. These urban disparities speak volumes about the relationship among geographic knowledge, economics, and access.

Another interesting type of cognitive mapping involves people's preferences for certain places over others. **Preference maps** show people's ideas about the environmental, social, or economic quality of life in various places. Although these maps are not necessarily directly related to spatial behavior, there is some correlation in the United States between growth rates and perceived quality of life among various states. When asked to rate certain states in terms of quality of life, most Americans give their home state a high ranking no matter where they live. However, California, Florida, and Colorado generally receive high scores regardless of where the individual being surveyed actually lives. Big cities tend to score higher than rural areas; places with beautiful scenery and sunny weather get high marks as well. It is interesting to note that high-ranking states are not only mentally associated with high levels of economic opportunity and quality of life but are also among the fastest growing in the country.

TIP

Geographers often think of hazards in terms of "risk." Risk is a valuable concept because it allows researchers to evaluate the potential exposure to or danger of certain environmental hazards. A coastal city, for example, may have a 1 in 100 chance of getting hit by a Category 3 hurricane in a given year. You should be prepared to answer an essay question on risk and environmental hazards.

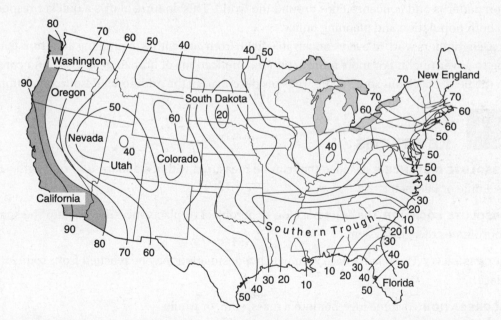

Figure 1.20 This preference map of the United States shows that most people would generally prefer to live on the West Coast or in New England.

As you are now aware, there are many different kinds of maps and many different ways to design them. This makes cartography and the field of geography, in general, much more complicated than it otherwise might seem. Furthermore, the act of mapping—whether it is a cartographer making decisions about how to produce a population map of a specific region or an individual developing a cognitive map of a new neighborhood—requires a thorough understanding not only of the technology of mapmaking, but also of conceptual complexities of space and place.

APPLICATIONS OF GEOGRAPHY

In recent years, geographers have moved to the forefront of debates regarding the world's social and environmental problems. This is, in part, because all such problems have spatial characteristics. For example, human populations have exploded during the last half century, leading to social strife and ecological degradation in many areas of the world. Because this growth has been limited primarily to less-developed countries in the tropics and Southern Hemisphere, the world's wealthiest countries have been largely isolated from its effects. Geographic analyses of population data can help policymakers pinpoint areas of rapid growth and design policies to deal with the consequences of an increasing population.

Geographers have also benefited from technological advances, such as improved remote-sensing imagery and increased computer capability. These advances have allowed geographers to analyze information in complex ways that were never before possible. One important example of this involves climatologists, who now use powerful computers to create models that predict potential changes in Earth's climate as a result of global warming. These same physical geographers have also worked with human geographers to predict some of the possible effects these changes could have on people. Furthermore, the ease of obtaining much more detailed information via the internet, such as United Nations demographic statistics, allows population geographers greater ability to understand and predict population patterns and concentrations around the world. This, in turn, allows a quicker response in both population and planning policy.

Geographers work in every conceivable field, from consulting to retailing and from teaching to government. For more information on applications of human geography and careers in the field, check out some of the websites described in the Additional Resources section.

KEY TERMS

ABSOLUTE DISTANCE A distance that can be measured with a standard unit of length, such as a mile or kilometer.

ABSOLUTE LOCATION The exact position of an object or place, measured within the spatial coordinates of a grid system.

ACCESSIBILITY The relative ease with which a destination may be reached from some other place.

AGGREGATION To come together into a mass, sum, or whole.

ANTHROPOGENIC Human-induced changes on the natural environment.

AZIMUTHAL PROJECTION A map projection in which the plane is the most developable surface.

BREAKING POINT The outer edge of a city's sphere of influence, used in the law of retail gravitation to describe the area of a city's hinterlands that depend on that city for its retail supplies.

CARTOGRAMS A type of thematic map that transforms space such that the political unit with the greatest value for some type of data is represented by the largest relative area.

CARTOGRAPHY The theory and practice of making visual representations of Earth's surface in the form of maps.

CHOROPLETH MAP A thematic map that uses tones or colors to represent spatial data as average values per unit area.

COGNITIVE MAP An image of a portion of Earth's surface that an individual creates in his or her mind. Cognitive maps can include knowledge of actual locations and relationships among locations as well as personal perceptions and preferences of particular places.

COMPLEMENTARITY The actual or potential relationship between two places, usually referring to economic interactions.

CONNECTIVITY The degree of economic, social, cultural, or political connection between two places.

CONTAGIOUS DIFFUSION The spread of a disease, an innovation, or cultural traits through direct contact with another person or another place.

COORDINATE SYSTEM A standard grid, composed of lines of latitude and longitude, used to determine the absolute location of any object, place, or feature on Earth's surface.

CULTURAL ECOLOGY Also called **NATURE-SOCIETY GEOGRAPHY**, the study of the interactions between societies and the natural environments in which they live.

CULTURAL LANDSCAPE The human-modified natural landscape specifically containing the imprint of a particular culture or society.

DISTANCE DECAY EFFECT The decrease in interaction between two phenomena, places, or people as the distance between them increases.

DOT MAPS Thematic maps that use points to show the precise locations of specific observations or occurrences, such as crimes, car accidents, or births.

EARTH SYSTEM SCIENCE A systematic approach to physical geography that looks at the interaction between Earth's physical systems and processes on a global scale.

ENVIRONMENTAL GEOGRAPHY The intersection between human and physical geography, which explores the spatial impacts humans have on the physical environment and vice versa.

EXPANSION DIFFUSION The spread of ideas, innovations, fashion, or other phenomena to surrounding areas through contact and exchange.

FORMAL REGION Definition of regions based on common themes such as similarities in language, climate, land use, etc.

FRICTION OF DISTANCE A measure of how much absolute distance affects the interaction between two places.

FULLER PROJECTION A type of map projection that maintains the accurate size and shape of landmasses but completely rearranges direction such that the four cardinal directions—north, south, east, and west—no longer have any meaning.

FUNCTIONAL REGION Definition of regions based on common interaction (or function), for example, a boundary line drawn around the circulation of a particular newspaper.

GEOGRAPHIC INFORMATION SYSTEMS (GIS) A set of computer tools used to capture, store, transform, analyze, and display geographic data.

GEOGRAPHIC SCALE The scale at which a geographer analyzes a particular phenomenon—for example, global, national, census tract, neighborhood, etc. Generally, the finer the scale of analysis, the richer the level of detail in the findings.

GEOID The actual shape of Earth, which is rough and oblate, or slightly squashed. Earth's diameter is longer around the equator than along the north-south meridians.

GLOBAL POSITIONING SYSTEM (GPS) A set of satellites used to help determine location anywhere on Earth's surface with a portable electronic device.

GRAVITY MODEL A mathematical formula that describes the level of interaction between two places, based on the size of their populations and their distance from each other.

HIERARCHICAL DIFFUSION A type of diffusion in which something is transmitted between places because of a physical or cultural community between those places.

HUMAN GEOGRAPHY The study of the spatial variation in the patterns and processes related to human activity.

INTERNATIONAL DATE LINE The line of longitude that marks where each new day begins, centered on the 180th meridian.

INTERVENING OPPORTUNITY If one place has a demand for some good or service and two places have a supply of equal price and quality, the supplier closer to the buyer will represent an intervening opportunity, thereby blocking the third from being able to share its supply of goods or services. Intervening opportunities are frequently used because transportation costs usually decrease with proximity.

ISOLINE A map line that connects points of equal or very similar values.

LARGE SCALE A relatively small ratio between map units and ground units. Large-scale maps usually have higher resolution and cover much smaller regions than small-scale maps.

LATITUDE The angular distance north or south of the equator, defined by lines of latitude or parallels.

LAW OF RETAIL GRAVITATION A law stating that people will be drawn to larger cities to conduct their business since larger cities have a wider influence on the surrounding hinterlands.

LOCATION CHARTS On a map, a chart or graph that gives specific statistical information about a particular political unit or jurisdiction.

LONGITUDE The angular distance east or west of the Prime Meridian, defined by lines of longitude, or meridians.

MAP PROJECTION A mathematical method that involves transferring Earth's sphere onto a flat surface. This term can also be used to describe the type of map that results from the process of projecting. All map projections have distortions in area, direction, distance, or shape.

MAP SCALE The ratio between the size of an area on a map and the actual size of that same area on Earth's surface.

MERCATOR PROJECTION A true conformal cylindrical map projection, the Mercator projection is particularly useful for navigation since it maintains accurate direction. Mercator projections are famous for their distortion in area that makes landmasses at the poles appear oversized.

MERIDIAN A line of longitude that runs north-south. All lines of longitude are equal in length and intersect at the poles.

NATURAL LANDSCAPE The physical landscape or environment that has not been affected by human activities.

PARALLEL An east-west line of latitude that runs parallel to the equator and that marks distance north or south of the equator.

W. D. PATTISON Geographer who claimed that geography drew from four distinct traditions: the earth-science tradition, the culture-environment tradition, the locational tradition, and the area-analysis tradition.

PERCEPTUAL REGION Highly individualized definition of regions based on perceived commonalities in culture and landscape.

PETERS PROJECTION An equal-area projection purposely centered on Africa in an attempt to treat all regions of Earth equally.

PHYSICAL GEOGRAPHY The realm of geography that studies the structures, processes, distributions, and changes through time of the natural phenomena of Earth's surface.

PREFERENCE MAP A map that displays individual preferences for certain places.

PRIME MERIDIAN An imaginary line passing through the Royal Observatory in Greenwich, England, that marks the 0° line of longitude.

PROJECTION The system used to transfer locations from Earth's surface to a flat map.

PROPORTIONAL SYMBOLS MAP A thematic map in which the size of a chosen symbol—such as a circle or triangle—indicates the relative magnitude of some statistical value for a given geographic region.

PTOLEMY Roman geographer-astronomer, author of *Guide to Geography*, which included maps containing a grid system of latitude and longitude.

QUALITATIVE DATA Data associated with a more humanistic approach to geography, often collected through interviews, empirical observations, or the interpretation of texts, artwork, old maps, and other archives.

QUANTITATIVE DATA Data associated with mathematical models and statistical techniques used to analyze spatial location and association.

REFERENCE MAP A map type that shows reference information for a particular place, making it useful for finding landmarks and for navigation.

REGION A territory that encompasses many places that share similar physical and/or cultural attributes.

REGIONAL GEOGRAPHY The study of geographic regions.

RELATIVE DISTANCE A measure of distance that includes the costs of overcoming the friction of absolute distance separating two places. Relative distance often describes the amount of social, cultural, or economic connectivity between two places.

RELATIVE LOCATION The position of a place relative to the places around it.

RELOCATION DIFFUSION The diffusion of ideas, innovations, behaviors, and so on from one place to another through migration.

REMOTE SENSING The observation and mathematical measurement of Earth's surface using aircraft and satellites. The sensors include photographic images, thermal images, multispectral scanners, and radar images.

RESOLUTION A map's smallest discernable unit. If, for example, an object has to be one kilometer long in order to show up on a map, that map's resolution is one kilometer.

ROBINSON PROJECTION A projection that attempts to balance several possible projection errors. It does not maintain area, shape, distance, or direction completely accurately, but it minimizes errors in each.

CARL SAUER Geographer from the University of California at Berkeley who defined the concept of cultural landscape as the fundamental unit of geographical analysis. This landscape results from the interaction between humans and the physical environment. Sauer argued that virtually no landscape has escaped alteration by human activities.

SENSE OF PLACE Feelings evoked by people as a result of certain experiences and memories associated with a particular place.

SITE The absolute location of a place, described by local relief, landforms, and other cultural or physical characteristics.

SITUATION The relative location of a place in relation to the physical and cultural characteristics of the surrounding area and the connections and interdependencies within that system; a place's spatial context.

SMALL SCALE A map scale ratio in which the ratio of units on the map to units on Earth is quite small. Small-scale maps usually depict large areas.

SPATIAL DIFFUSION The ways in which phenomena, such as technological innovations, cultural trends, or even outbreaks of disease, travel over space.

SPATIAL PERSPECTIVE An intellectual framework that looks at the particular locations of a specific phenomenon, how and why that phenomenon is where it is, and, finally, how it is spatially related to phenomena in other places.

SUSTAINABILITY The concept of using Earth's resources in such a way that they provide for people's needs in the present without diminishing Earth's ability to provide for future generations.

THEMATIC LAYERS Individual maps of specific features that are overlaid on one another in a Geographical Information System to understand and analyze a spatial relationship.

THEMATIC MAP A type of map that displays one or more variables—such as population or income level—within a specific area.

TIME-SPACE CONVERGENCE The idea that distance between some places is actually shrinking as technology enables more rapid communication and increased interaction among those places.

TOPOGRAPHIC MAPS Maps that use isolines to represent constant elevations. If you took a topographic map out into the field and walked exactly along the path of an isoline on your map, you would always stay at the same elevation.

TRANSFERABILITY The costs involved in moving goods from one place to another.

VISUALIZATION Use of sophisticated software to create dynamic computer maps, some of which are three dimensional or interactive.

CHAPTER SUMMARY

Human geography is the study of human activities on Earth's surface. Since the first scholars began studying geography some 3,000 years ago, the field has matured into an important and wide-ranging area of academic and applied research. One thing that binds all geographers together is the spatial perspective. Looking at Earth from a spatial perspective means looking at how objects, processes, and patterns change over the earth's surface. Geographers describe these variations by creating visual representations of spatial data in the form of maps. All maps are based on a projection. They have a characteristic scale and resolution. All maps use symbols to depict spatial information. Geographers use a diverse set of concepts, tools, technologies, and mathematical equations to study places, regions, and the processes that link them. In general, places that are closer to each other in absolute distance tend to interact more. However, the interaction among places is also determined by the size of each place, their level of connectivity, and the diffusion processes that carry information and cultural traditions from one place to another.

PRACTICE QUESTIONS AND ANSWERS

Geography as a Field of Inquiry

MULTIPLE-CHOICE QUESTIONS

1. Human-induced environmental change is often referred to as

    (A) anthropomorphic.
    (B) anthropocentric.
    (C) anthropogenic.
    (D) unsustainable.
    (E) environmental determinism.

2. Conserving resources to ensure enough for future generations is called

    (A) subsistence agriculture.
    (B) sustainability.
    (C) cultural ecology.
    (D) environmental determinism.
    (E) the organic movement.

3. _____ argued that cultural landscapes should form the basic unit of geographic inquiry.

    (A) Ptolemy
    (B) George Perkins Marsh
    (C) Eratosthenes
    (D) Carl Sauer
    (E) W. D. Pattison

4. A thematic layer is

    (A) a method used in cartography to produce mathematically accurate map projections.
    (B) a map portraying a particular feature that is used in a GIS.
    (C) used in GPS systems to provide more accurate navigational information.
    (D) a map used by early explorers to find particular resources in new regions of the earth.
    (E) used as a method to analyze thematic regions.

5. Which of the following is the oldest field of geography?

    (A) Cultural ecology
    (B) Conservation biology
    (C) Cartography
    (D) Environmental geography
    (E) Physical geography

FREE-RESPONSE QUESTION

1. Technological innovations have greatly influenced the methods by which geography can be done today.

   (A) Describe three technological advances that have dramatically changed the capabilities of the discipline of geography.

   (B) List an application for each type of technology.

Geography Basics

MULTIPLE-CHOICE QUESTIONS

1. A perceptual region's boundaries are

   (A) determined by a set of uniform physical or cultural characteristics across a particular area.
   (B) drawn around the functions that occur between a particular place and the surrounding area.
   (C) determined by the portion of a particular area that has been modified by human activities.
   (D) fuzzy because they allow for individual interpretation.
   (E) designated by the inclusion of a particular cultural characteristic.

2. If a geographer performs a study on people's perceptions of the Deep South using interviews as the primary data source, the geographer's method is

   (A) quantitative.
   (B) systematic.
   (C) anthropogenic.
   (D) qualitative.
   (E) idiographic.

3. Which of the following is true concerning regions?

   (A) They are strict functional units.
   (B) They are usually defined by a standard mathematical formula.
   (C) They are figments of the imagination.
   (D) They are conceptual units.
   (E) They all have well-defined boundaries.

4. Geographic scale refers to

   (A) the ratio between distance on a map and distance on Earth's surface.
   (B) a conceptual hierarchy of spaces.
   (C) a notion of place based on an individual's perception of space.
   (D) the many ways that people define regions.
   (E) the level of aggregation at which geographers investigate a particular process.

1. Geography is unique from other disciplines in that it applies a spatial perspective to different phenomena and processes that occur on Earth's surface.

    (A) Define the spatial perspective. Include in your definition what it means to think geographically. Include descriptions of the types of data that geographers analyze.
    (B) Provide an example of a problem that can be solved only from a spatial perspective.

2. The region is a highly contested yet critical concept in the study of human geography.

    (A) Why and how do geographers perform the regionalization process?
    (B) What is regional geography?
    (C) Discuss the different types of regions that human geographers study, and provide an example of each type.

Describing Location

MULTIPLE-CHOICE QUESTIONS

1. Seattle is located on Puget Sound in northwestern Washington. It has a large university, a famous downtown market, and a moist, marine climate. Seattle's primary economic activities include ship and aircraft construction and high-technology enterprises. This information gives us a description of Seattle's

    (A) situation.
    (B) cognitive image.
    (C) site.
    (D) landscape.
    (E) relative distance.

2. Lines of longitude

    (A) never meet.
    (B) begin at the equator.
    (C) are referred to as parallels.
    (D) intersect at the poles.
    (E) contain the two tropics.

3. Even though some cities are far apart in terms of absolute distance, they are actually quite connected economically and socially. This is representative of

    (A) topographic space.
    (B) cognitive space.
    (C) relative distance.
    (D) relative location.
    (E) situation.

4. Which of the following is a true statement regarding time-space convergence?

(A) Places seem to all look the same.
(B) Places seem to be getting closer together.
(C) Places are increasingly concentrated on maintaining their histories.
(D) Places are making more of an effort to converge activities to save time.
(E) Places are implementing more rapid forms of transportation.

5. Which of the following is NOT a measure of relative distance?

(A) 2,339 centimeters
(B) 35 seconds
(C) Two dollars and fifty cents
(D) 216 footsteps
(E) 15 minutes

FREE-RESPONSE QUESTION

1. The notion of time-space convergence has had dramatic impacts on how geographers think of distance.

(A) Define time-space convergence and give an example of this process at work in the world today.
(B) Describe the effects of this convergence on the level of connectivity between places. Does the process connect all areas of the globe?
(C) Discuss Tobler's First Law of Geography as it relates to the notion of time-space convergence. Does this law still apply and/or will it apply in the future?

Space and Spatial Processes

MULTIPLE-CHOICE QUESTIONS

1. Tobler's First Law of Geography states, "Everything is related to everything else, but

(A) distant things are generally unrelated."
(B) near things are more closely related than you might think."
(C) distance is always a factor."
(D) near things are more related than distant things."
(E) distance is relative."

2. Rap music first appeared in New York in the 1970s. Later it spread to large cities with vibrant African American populations—such as Los Angeles, Oakland, Chicago, Philadelphia, and Detroit—without being absorbed by the smaller cities and rural areas in between. This type of spatial diffusion is called

(A) relocation potential.
(B) hierarchical diffusion.
(C) contagious diffusion.
(D) cultural diffusion.
(E) cascade diffusion.

3. Stores and restaurants in Oregon that find it cheaper to buy fresh vegetables grown in California than to buy those grown in Florida are taking advantage of

    (A) expansion diffusion.
    (B) distance decay.
    (C) economies of scale.
    (D) intervening opportunities.
    (E) retail gravitation.

4. According to the gravity model, which two places are most likely to have a high level of interaction?

    (A) Two cities with very large populations but separated by the Atlantic Ocean like New York and London
    (B) Two cities with medium populations separated by a whole continent like Grand Rapids, Michigan, and Gulf Shores, Alabama
    (C) Two cities with small populations that are relatively close together like Richmond, Virginia, and Winchester, Kentucky
    (D) Two cities, one with a large population and the other with a medium population that are very close in distance like Seattle and Tacoma, Washington
    (E) Two cities with medium populations that are relatively close to each other like Akron, Ohio, and Springfield, Missouri

5. Which of the following is NOT a good example of a barrier to spatial diffusion?

    (A) A mountain range
    (B) A different language
    (C) A different dietary preference
    (D) A highway system
    (E) A strict religious system

FREE-RESPONSE QUESTION

1. Geographers define space, location, and distance according to both absolute and relative measures.

    (A) Describe the difference between absolute and relative measures of distance.
    (B) Give two examples of instances where the degree of interaction between places is more related to connectivity than to absolute distance.
    (C) Describe the difference between absolute and relative measures of location. Give examples.

Map Fundamentals

MULTIPLE-CHOICE QUESTIONS

1. The ratio between distance on a map and distance on Earth's surface is called the

    (A) projection.
    (B) resolution.
    (C) scale.
    (D) azimuth.
    (E) aggregation.

2. Cartography is the art and science of

    (A) demographics.
    (B) mapmaking.
    (C) spatial orientation.
    (D) cognitive imagery.
    (E) making visualizations.

3. Map projections attempt to correct for errors in

    (A) transferability.
    (B) area, distance, scale, and proportion.
    (C) area, distance, shape, and direction.
    (D) distance, proximity, and topology.
    (E) distance, shape, and lines of latitude and longitude.

4. The Mercator projection preserves

    (A) direction.
    (B) area.
    (C) shape.
    (D) scale.
    (E) distance.

5. Topographic maps use which of the following symbols to convey change over space?

    (A) Tonal shadings
    (B) Isolines
    (C) Proportional symbols
    (D) Location charts
    (E) Cartograms

6. Which of the following map projections preserves the correct shape of Earth's landmasses?

   (A) Fuller projection
   (B) Mercator
   (C) Robinson
   (D) Mollewide
   (E) Smithsonian

7. The size of a map's smallest discernable unit is its

   (A) scale.
   (B) density.
   (C) region.
   (D) resolution.
   (E) projection.

FREE-RESPONSE QUESTIONS

1. Scale is an extremely important concept in geography because spatial relationships appear to vary depending upon the scale at which they are measured.

   (A) Define scale. Discuss the relationship of scale to resolution.
   (B) Discuss the role of scale in interpreting geographical information.
   (C) In the 2000 US presidential election, George W. Bush won the electoral votes of every southern state. Explain how an analysis of these results at the county level could yield additional valuable information about voting patterns at finer geographical scales.

2. Preference maps are a unique type of isoline map used in human geography.

   (A) Use your knowledge of preference maps to describe why some places might be more attractive than others as places to live.
   (B) Use your knowledge of preference maps to describe why certain places may or may not be more attractive for certain cohorts.

Answers for Multiple-Choice Questions

GEOGRAPHY AS A FIELD OF INQUIRY

1. **(C)** Anthropogenic, by definition, means human-induced changes on the physical environment. The other options are tricky only because they look similar to this term.

2. **(B)** Sustainability is the idea of using Earth's resources in such a way that the needs of the current population are provided for without compromising the ability of future generations to use the same resources.

3. **(D)** Carl Sauer developed the notion of a cultural landscape to describe the parts of Earth's surface that have been modified by human activities. He also argued that virtually all of Earth's surface has in some way been affected by human activity and thus paved the way for environmental geography.

4. **(B)** The purpose of GIS is to provide a tool to understand and analyze spatial relationships between different phenomena better. This goal is accomplished by overlaying different thematic layers, or maps with different geographic information, on top of one another.

5. **(C)** The first geographers were primarily interested in exploration. Cartography allowed these first geographers to map the information gleaned from their expeditions.

GEOGRAPHY BASICS

1. **(D)** Perceptual regions are determined by commonly perceived characteristics of particular places on the earth's surface. Different individuals have different ideas about where these characteristics begin and end. Thus, the boundaries are necessarily fuzzy.

2. **(D)** Qualitative data are more humanistic, often collected through interviews, empirical observations, and interpretation of texts, artwork, old maps, and other archives. Quantitative data are typically numerically based, understood, and evaluated with statistical methods.

3. **(D)** Regions are an organizing tool that allow geographers to combine areas with similar features into one conceptual unit that provides them with a more manageable unit for analysis.

4. **(B)** Geographic scale refers to a scale of analysis. It looks at phenomena through a hierarchy of scale such as neighborhood, city, state, and nation.

DESCRIBING LOCATION

1. **(C)** Site is a description of the qualities of a place, independent of that place's relationship to other places around it. Situation refers to a place's relationship to the other places around it.

2. **(D)** All latitudes are parallel, but all lines of longitude converge at the North and South Poles. In many map projections—such as the Mercator projection—longitude lines do not converge at the poles as they do on the globe. This causes geographical features at high latitudes to be warped and to appear larger than they actually are.

3. **(C)** Relative distance the level of connectivity between places. Although some places may be quite far from each other in absolute space, in actuality they might be quite close in terms of the economic, social, and cultural relationships between them. New York and Los Angeles are much more connected in terms of economic and cultural relationships than are Los Angeles and Lincoln, Nebraska, even though Nebraska is much closer in absolute distance to Los Angeles.

4. **(B)** Time-space convergence is the notion that distance between places seems to be "shrinking" with increased improvements in transportation and communication technologies.

5. **(A)** Since the centimeter is a standard unit of measurement, it is a measure of absolute distance that can be universally understood across the globe.

SPACE AND SPATIAL PROCESSES

1. **(D)** Tobler's First Law of Geography expresses the concept of distance decay. Generally, things are less related the farther away they are from each other in absolute space.

2. **(B)** Hierarchical diffusion is the form of spatial diffusion that occurs when a phenomenon spreads from one place to another because the places have something in common. Large cities were the first to adopt rap music. Rap later spread to medium and small cities across the United States, but it has yet to be fully absorbed into many rural areas.

3. **(D)** An intervening opportunity exists when a closer source is available for the supply of some desired good or service. All else being equal, people tend to prefer closer sources of goods and services to those farther away.

4. **(D)** The gravity model predicts the level of interaction between two cities in terms of their populations and distances apart from one another. The correct option has a very large numerator (product of both populations) and a very small denominator (distance between Seattle and Tacoma). Thus, the resulting level of interaction will be quite high.

5. **(D)** A highway system facilitates diffusion because it connects places. In contrast, the other four options prevent certain innovations and cultural traits from spreading.

MAP FUNDAMENTALS

1. **(C)** In cartography, scale refers to the ratio of map distance to the distance on the earth's surface. Large-scale maps have a large ratio, such as 1:2,000. Small-scale maps have a small ratio, such as 1:200,000.

2. **(B)** Cartography is the art and science of mapmaking. Cartography is an art because of the challenge of effectively communicating spatial information in an aesthetically pleasing way. It is a science because of the difficulty of transforming the three-dimensional Earth onto a two-dimensional piece of paper.

3. **(C)** Map projections represent attempts by cartographers to correct for the simple geometrical fact that a spherical or geoidal surface, such as Earth, cannot be accurately

depicted on a two-dimensional surface. Map projections attempt to correct for errors in the area and shape of features on Earth's surface, errors in the distance between places, and errors in the compass direction from one place to another.

4. **(A)** The Mercator projection accurately preserves compass direction. However, because the lines of longitude do not meet at the poles in the Mercator projection as they do on the globe, area is distorted, with increasing inaccuracy at high latitudes.

5. **(B)** Isolines, or contours, are lines of equal value. Isolines are used on topographic maps to show the locations of places with equal elevation. Isolines are also commonly used for maps that represent spatial densities, such as population density or pollution concentration.

6. **(A)** The Fuller projection correctly preserves area and shape, although distance and direction are severely distorted.

7. **(D)** Resolution is important in geography. The smallest discernable unit is directly related to both the map's scale and to the amount of spatial data that can be displayed. Generally, large-scale maps have a greater resolution, although this is not always the case.

Answers for Free-Response Questions

GEOGRAPHY AS A FIELD OF INQUIRY

1. **Main points:**
   - Three technological innovations that have dramatically changed the methods of geography are remote sensing, the Global Positioning System (GPS), and Geographical Information Systems.
   - Remote sensing is mostly used to determine land-use change. Is it one of the most effective tools for quantifying the amount of tropical rain forest being deforested. GPS is a tool used by geographers (and the general public) for navigational purposes. Geographical Information Systems (GIS) are an extremely important tool for geographers today. GIS allow geographers to apply the spatial perspective to different phenomena more quickly through the overlay of thematic maps—each containing information about different spatial features. The map overlay allows geographers to determine quickly whether a spatial relationship exists among different spatial phenomena. A GIS of New York City with thematic layers of street systems, fire hydrants, and hospital locations would be a useful tool for determining evacuation routes and getting people to hospitals as quickly as possible.

GEOGRAPHY BASICS

1. **Main points:**
   - The spatial perspective is an intellectual framework in which the spatial characteristics of particular phenomena are analyzed in terms of location and relationships with other spatial phenomena. Thinking geographically involves recognizing the location of particular processes or features and also understanding how and why those processes occur at specific locations.

- Spatial data are any type of data that pertain to a particular location. Spatial data can be either quantitative or qualitative. Quantitative data are usually numerical and are manipulated through mathematical or statistical models. Qualitative data are more descriptive and are usually obtained from interviews, texts, or archives of particular places or phenomena.
- Any problem that is spatial in nature is best solved using a spatial perspective. An example that was given in this chapter was that of the Amish community and the measles epidemic in the United States in the mid-1980s. The Amish, because of their belief system, refused to get measles vaccinations. As a result, the disease ran rampant through this community and the surrounding Amish communities of the eastern United States. Understanding the spread of this disease could not be done without understanding the spatial relationship between the Amish communities and the areas of outbreak.

2. **Main points:**
- The regionalization process involves grouping similar characteristics of different places into a more manageable unit of study, which is the region. The region is a conceptual unit that is bounded based on whatever feature a particular geographer wants to include within a particular region. These features can be either physical or cultural.
- Regional geography is concerned with the study of regions and the characteristics of those particular regions that make them different from other places on Earth's surface. Generally, regional geography takes an idiographic approach because it looks at the unique characteristics of particular places without generalizing the processes occurring within those regions beyond their boundaries.
- Functional regions are defined by the connections and interactions that occur between a central place and its surrounding area. An example could be Chicago (or any other large city) and all the various transportation, economic, cultural, and recreational connections that exist within that area.
- The boundary of a formal region contains an area of similar cultural or physical characteristics. An example of a formal region would be Friesland in the northern part of the Netherlands. In this part of the country, people speak a different language and perceive themselves quite differently from the other citizens of the country.

DESCRIBING LOCATION

1. **Main points:**
- Time-space convergence is the idea that distance between places is, in effect, shrinking due to certain improvements in transportation and communications technologies. In today's world, instant communication with many parts of the globe is possible thanks to the telephone and the computer. If you ask individuals in developed parts of the world where they have traveled, it is not nearly as uncommon as it once was to hear that people are traveling all across their countries and even all across the globe. In the past, traveling to other countries used to take days and even months. Now it can happen at very rapid speeds.
- Connectivity between places increases rapidly as the result of the development of better mechanisms for establishing connections between places. Communication technologies such as the telephone, the internet, and fax machines allow cities all

across the globe to connect with one another instantaneously, thereby increasing and strengthening the level of interaction and connectivity between places. Distance between places used to prohibit interactions, but now rapid transportation removes this obstacle, allowing greater connectivity between distances. However, it is important to recognize that only a certain network of places is, in fact, converging. Large parts of the developing world do not have access to these technologies, making interaction difficult to occur.

- Tobler's First Law states that everything is related but near things are more closely related than far things. With the increasing levels of connectivity occurring between certain places as a result of time-reducing technologies, certain places that are far apart are becoming more related than some places that might be closer together. As the world becomes increasingly interconnected, we will probably see this law continue to lose potency as distance loses its ability to impede interactions between certain places.

SPACE AND SPATIAL PROCESSES

1. **Main points:**
   - Absolute measures of distance involve numerical computations in units such as meters or miles, unlike relative measures, which are dependent (typically on transportation type), often less precise, and typically not a standard of measure of space. Time is the most common relative measure of distance. If you ask someone how far their school is from their home, they are much more likely to say something like, "about 10 minutes" as opposed to "2.2 miles." The latter is an absolute measure of the separation between an origin and a destination. The former (10 minutes) is relative to what type of transportation you have access to, how fast that transportation moves, traffic conditions, etc.
   - Connectivity, often referred to as relative distance, describes the degree of interaction between places, independent of absolute distance between them.
   - Two places that are relatively far apart in absolute distance can have a high degree of connectivity. Connectivity can result from the easy transferability of goods and services, from economic complementarity, or from the absence of intervening opportunities for economic interaction. Other cultural, historical, or political factors also contribute to connectivity.
   - One example of the interaction between places being more related to connectivity than to absolute distance involves the production and distribution of agricultural products. Historically, the vast agricultural regions of the Midwest and Great Plains have had a high degree of economic complementarity with the city of Chicago. This is due to the fact that during the 19th century, Chicago became the primary center for the sale, processing, and distribution of agricultural products grown in the rural hinterlands—the region that was then called the Great West. In this way, Chicago had a high degree of connectivity with agricultural regions that were quite far away in absolute distance.
   - A second example of high connectivity independent of absolute distance is the establishment of ethnic enclaves in cities such as New York and Boston during the great period of immigration that occurred in America around the beginning of the 20th century. New York, for instance, had large neighborhoods dominated by immi-

grants of common ethnicity, heritage, and national origin, such as Chinatown and Little Italy. The people living in these neighborhoods often retained a high degree of social, cultural, and economic connectivity with the families and friends they had left behind in Europe and Asia.

- A place's absolute location, or site, is defined by its characteristics, independent of the place's relationship to other places around it. For example, Mexico City is located in southcentral Mexico at approximately 19° N latitude and at about 7,200 feet of elevation. Mexico City is a large, sprawling, and crowded metropolis. It is the center of Mexico's economic activities and the capital of the Mexican federal government.

- Situation refers to a place's location and function relative to other places around it. Tijuana, Mexico, is located in Baja California Norte on the Pacific Ocean, south of the US/Mexico border. As a major center of international trade and manufacturing, Tijuana serves as the industrial capital of northwestern Mexico, with close economic ties to large US cities such as Los Angeles, San Diego, and Phoenix.

MAP FUNDAMENTALS

1. **Main points:**
   - Map scale is the ratio between distance on a map and distance on Earth's surface. Small-scale maps have a small distance ratio and tend to represent larger geographical areas. Large-scale maps have a larger distance ratio and tend to depict smaller areas.
   - Resolution, which is the smallest discernable unit on a map, tends to decrease in smaller-scale maps and tends to increase in larger-scale maps. This is why large-scale maps, which tend to cover small regions, also have more detail. Small-scale maps that cover large regions tend to have low resolution and, as a result, less detail.
   - Scale is very important in interpreting geographical information for several reasons. One example of the importance of scale in geography involves the relationship between scale and resolution. If you analyze geographical information at smaller scales, you might lose valuable information apparent only at smaller scales and higher resolution. Conversely, by looking at geographical information from too large a scale, you can lose sight of the bigger context. Consider a map of the dominant regional religions drawn at a relatively small scale and depicting the entire United States. This map will show that in much of southern California, Catholicism is the dominant regional religion. However, a larger-scale, higher-resolution map of southern California will show that the region contains an extremely diverse population with a multitude of religious beliefs and practices displaying significant spatial variation. In this way, the larger-scale, higher-resolution map gives more detailed information within a more general regional context.
   - In the United States, electoral votes are apportioned by state. When a candidate wins a state, all of that state's electoral votes are given to the single winning candidate, no matter how close the total vote count was or how significant the spatial variation in voting patterns across the state. An examination of voting patterns in the 2000 presidential election in the southeastern states by county renders some extremely interesting geographical information. First, Al Gore won many urban counties in and around large metropolitan areas such as Miami, Atlanta, Memphis, and Raleigh-Durham. Second, Gore won many rural counties with predominantly

African American populations. Although George W. Bush won all the southern states, voting patterns within those states were often uneven, and Gore succeeded in many urban and minority-dominated districts. However, Bush won in enough counties to win the electoral votes for the entire state. A closer look at the southeastern states by county reveals that some geographic patterns in voting are not evident at the level of the state. This is an example of the importance of geographic scale in voting patterns.

2. **Main points:**
- Several factors—economic, recreational, and educational opportunities; cultural attractions; and climate—combine to make some places more attractive to live in than others.
- Most people tend to like the place in which they live. However, most people also usually rate the areas neighboring their own as low in terms of desirability. This is called the *boundary effect* and is part of a sense of place.
- In the United States, people tend to rate certain regions as highly desirable, no matter where they live. The West Coast states of California, Oregon, and Washington usually rank high, as do Colorado, Arizona, Texas, Florida, and New York.
- Young people tend to rate urban areas higher because they offer more jobs and cultural amenities.

 ADDITIONAL RESOURCES

Agnew, John, David Livingstone, Alisdair Rogers (eds.). 1996. *Human Geography: An Essential Anthology.* Cambridge, Massachusetts: Blackwell.

This text contains numerous essays written by academic human geographers, both in the past and the present, describing many different facets of human geography. The essays are organized into five main sections: Recounting Geography's History; The Enterprise; Nature, Culture and Landscape; Region, Place and Locality; and Space, Time and Space-Time.

Hanson, Susan (ed.). 1997. *Ten Geographic Ideas That Changed the World.* New Brunswick, New Jersey: Rutgers University Press.

An excellent book that covers various types of maps and various types of mapmaking. Consisting of ten chapters, by ten different authors, a wide variety of perspectives discuss a range of cartographic issues from reading weather maps to understanding GIS.

Johnston, R. J. 1997. *Geography and Geographers: Anglo-American Human Geography since 1945.* New York: Arnold.

Johnston discusses the history of the discipline of human geography beginning in 1945. He begins by discussing the nature of an academic discipline and then discusses geography's many different foci and some of the difficulty the discipline has encountered in struggling to both define itself and maintain academic rigor.

Monmonier, Mark. 1995. *Drawing the Line: Tales of Maps and Cartocontroversy.* New York: Henry Holt.

Monmonier loves to discuss the deceptive powers of maps. In this book, he looks at some historical decisions based on maps that had false details. He discusses how the cartographer's bias manifests itself in many different historically popular maps, and how many political decisions have been made using maps that portrayed information in an inaccurate manner.

Monmonier, Mark. 1991. *How to Lie With Maps.* Chicago: University of Chicago Press.

Similar to his other book, *Drawing the Line: Tales of Maps and Cartocontroversy,* Monmonier discusses how people accept maps as objective models of reality, when in actuality they present many forms of deception that most map-readers never recognize. He encourages a healthy skepticism of maps and discusses how to be a critical evaluator of cartographical information.

Population and Migration Patterns and Processes

2

IN THIS CHAPTER

→ **HUMAN POPULATION: A GLOBAL PERSPECTIVE**

→ **POPULATION PARAMETERS AND PROCESSES**

→ **HUMAN MIGRATION**

→ **POPULATION STRUCTURE AND COMPOSITION**

→ **POPULATION AND SUSTAINABILITY**

 Key Terms

Age-sex distribution	Dependency ratio	Migration
Agricultural density	Doubling time	Natural increase rate
Arithmetic density	Emigration	Neo-Malthusian
Baby boom	Epidemiological	Overpopulation
Baby bust	transition	Physiologic density
Carrying capacity	Exponential growth	Population density
Chain migration	Forced migration	Population geography
Child mortality rate	Generation X	Population pyramid
Cohort	Geodemography	Pull factor
Cotton Belt	Immigration	Push factor
Crude birth rate	Infant mortality rate	Refugees
Crude death rate	Internal migration	Rust Belt
Demographic accounting	Intervening obstacles	Sun Belt
equation	Involuntary migration	Total fertility rate
Demographic transition	Life expectancy	Voluntary migration
model	Thomas Malthus	Zero population growth
Demography	Maternal mortality rate	

HUMAN POPULATION: A GLOBAL PERSPECTIVE

The world's population is now currently over 7.5 billion people. It is hard to comprehend a number like 7.5 billion, but even dizzying numbers do little to illustrate the immensity and diversity of the human endeavor here on Earth. Perhaps even more astounding than the sheer number of people alive today is the rate at which human population has increased during the past 200 years. According to the United Nations Population Division, it wasn't until 1804 that world population reached 1 billion. It took just 123 years for that population to double, reaching 2 billion by 1927. Forty-seven years later, in 1974, the population had once again doubled, and by 2000, 6 billion people inhabited the earth. The world population reached 7 billion in 2011.

Staggering human population growth is one of the defining characteristics of our present era in world history; it is also one of the most important issues in all of human geography. To explain this amazing and rapid growth, and to begin to understand its implications, we must ask difficult questions about politics, culture, economics, and history. Why is the population growing so quickly? Which areas are growing fastest? What effects has this growth had on social and ecological systems? And what can we expect in terms of future growth? These questions all lie at the heart of **population geography**, which is sometimes referred to as **geodemography**.

On a global scale, human population shows several distinct geographic characteristics. First, approximately 80% of the world's population lives in the less-developed countries, which includes all of Africa, Asia (excluding Japan), Latin America, and the island nations of the Caribbean and Pacific. Two countries, China and India, each have over a billion people and together hold one-third of the world's current population! Less-developed countries also contain the fastest-growing populations. Of the approximately 80 million people that were being added to the world's population each year during the 1990s, 95% lived in the less-developed world. People are also living longer. During the past 50 years, the global average life expectancy has increased by 20 years, from 45 to 65. In the less-developed countries, where most people reside, the basic equation is relatively simple—more babies are being born, and people are living longer.

When people read news reports about growth rates in the less-developed countries, they often see figures like 3% and think, "Hey, that's not too much." However, if you look a little more closely at the math, you will quickly realize that even growth rates that seem low can cause rapid population increases over time. Like interest in a bank, population growth is *compounded*. In other words, if a population grows by 3% both this year and next year, then next year's 3% will actually include more people than this year's. This is because next year's growth will be 3% of a population that is bigger than it was just a year ago. For example, if you live in a country with a population of 1 million people, after one

THE WORLD'S 20 FASTEST-GROWING COUNTRIES BY POPULATION

Rank	Country	Population growth rate (%)
1	Syria	7.9
2	South Sudan	3.8
3	Angola	3.5
4	Burundi	3.5
5	Malawi	3.3
6	Uganda	3.2
7	Niger	3.2
8	Burkina Faso	3.0
9	Mali	3.0
10	Zambia	2.9
11	Ethiopia	2.9
12	Tanzania	2.8
13	Benin	2.7
14	Western Sahara	2.7
15	Iraq	2.6
16	Togo	2.6
17	Guinea	2.6
18	Cameroon	2.6
19	Egypt	2.5
20	Madagascar	2.5

year, the population will grow by 3% for a total of 1,030,000. You've added 30,000 more people to your country in one year. The next year, when you apply a 3% growth rate to your current population of 1,030,000, you get a new total of 1,060,900. In year two, 3% ended by including 900 more people than in year one! You can now see how, at a 3% rate of increase, it would not take long for your country to grow dramatically. In fact, one of the most surprising implications of this concept is that, at a 3% growth rate, the time it will take for a population to double, known simply as **doubling time**, is less than 25 years! Growth rates currently exceed 3% in parts of sub-Saharan and tropical Africa, the Middle East, and Central America.

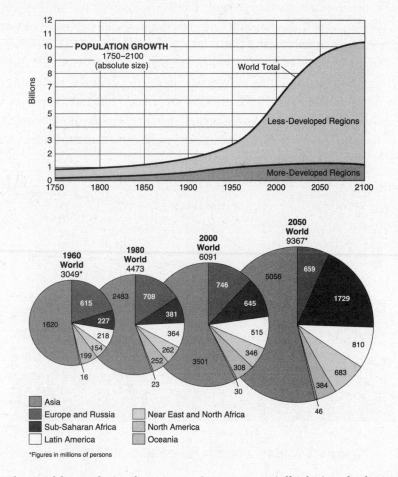

Figure 2.1 The world population began growing exponentially during the late 20th century, with particularly high rates of growth in the world's less-developed countries. Global population will probably plateau sometime in the 21st century.

The study of human populations, or **demography**, is not static, because the world's population will not keep growing forever. Most **demographers**, or people who study population, agree that growth is already showing signs of slowing down. Many current models based on demographic data predict that the world's population will plateau at around 12 billion people sometime in the 21st century. Some of the most recent predictions have population leveling off even lower, at about 9 billion people by the end of this century. In the future, as now, most of the people on Earth will be living in Africa or Asia.

While the overall pattern of population growth throughout human history shows steady and even rapid increase, certain events and environmental limitations have also served to check population growth at different periods in history. With increasing technology, the human species has enabled itself to adapt to many of these constraints, explaining why the number of people on Earth is continually growing. The innovations that occurred as a result of the domestication of plants and animals and the Industrial Revolution had dramatic impacts on the number of people the earth could sustain. However, in the 1300s, the Black Plague wiped out between 30% and 40% of the entire European continent. The Irish potato famine in 1845 eliminated half of the country's population in just 50 years as millions of people died of starvation or left the country. Natural disasters such as earthquakes, floods, and hurricanes regularly reduce at-risk populations. Of all the forces acting to check world population, none has been more effective than epidemic disease. When European explorers and settlers arrived in the Americas and the Pacific Islands, they introduced devastating diseases, such as smallpox, that formerly had been known only in the Old World. Indeed, disease—not war—was the main factor that led to the toppling of native cultures throughout the New World. Today, in many African countries, AIDS is contributing to an escalating death rate and constraining population growth within those nations. In 1999 in sub-Saharan Africa, 23.20 million people were living with the AIDS virus, and over 4 million were becoming infected each year. While this part of the world is one of the areas mentioned as experiencing rapid growth, in the future we may see its growth severely limited by the devastating effects of this virus.

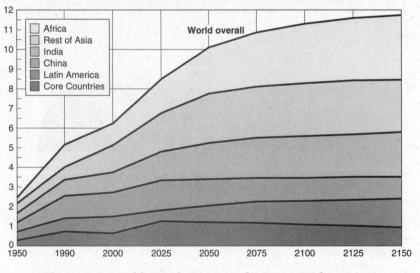

Figure 2.2 World population growth since 1950 by region

POPULATION PARAMETERS AND PROCESSES

All the population processes that have been discussed so far can be broken down into a set of parameters, each of which holds vital information about the history and future of a given population. One such statistic is the **total fertility rate** (TFR), which is a measure of the average number of children born to a woman over her entire life. Fertility rates vary both over time and between places in response to a number of factors that affect women's lives. In some countries with poor health-care systems, the total fertility rate may be very high, but it might also be offset by equally high **infant**, **child**, and **maternal mortality rates**. Infant mortality rate, which refers to the percentage of children who die before their first birthday, may be a significant factor limiting population growth. Sadly, the infant mortality rate exceeds 10% in some less-developed countries. Infant, child, and maternal mortality rates can have a significant effect on **life expectancy**—the average length of someone's life. Life expectancy varies dramatically from place to place and even within populations. In 1996, African American males in the United States had a life expectancy of about 66 years, while Anglo-American males, on average, lived to be almost 74. Violence, infant mortality, poor health care, epidemic disease, and risk factors such as smoking all contribute to life expectancy.

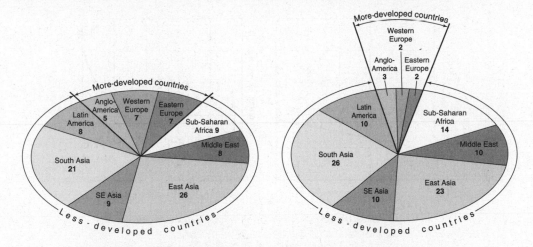

Figure 2.3 Population distribution percentage by world region

Annual Percentage Increase	Doubling Time (years)
0.5	140
1.0	70
2.0	35
3.0	24
4.0	17
5.0	14
10.0	7

Figure 2.4 Doubling time as a function of a population's annual percentage increase

Each of the parameters described here can be encapsulated within two aggregate variables, the **crude birth rate** (CBR) and **crude death rate** (CDR), which are statistical terms that refer to the number of live births and deaths, respectively, per thousand people. The difference between the CBR and CDR is called the **natural increase rate** (NIR). This term is a bit misleading because there is nothing really "natural" about natural increase and because natural "increase" can be either positive or negative. A negative increase rate indicates that the number of babies being born is not high enough to make up for deaths and, as a result, the population is declining. It is also important to recognize that the NIR is an internal measure that does not account for migration into or out of a country.

A few countries are currently showing negative rates of natural increase. In the more-developed regions of the world, which include North America, Japan, Europe, and Australia/New Zealand, the natural increase rate has decreased markedly, in some cases actually leading to a stable or even declining population. This is particularly evident in some European countries with aging populations, where death rates now outpace birth rates. As a result, much of the growth occurring in the more-developed world is associated with immigration from less-developed countries. In the United States, natural rates of increase are low; however, immigration and high fertility rates among some groups of newer immigrants are causing the overall population to continue growing. It is also interesting to note that, in the many developing countries, such as Mexico, Brazil, and Indonesia, rates of natural increase are actually beginning to decline. Although the populations of these countries are still growing, they are no longer growing as fast as they once were.

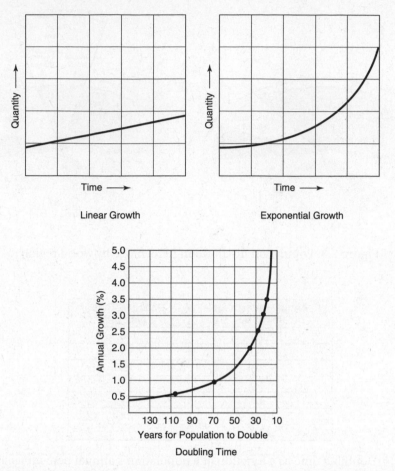

Figure 2.5 When a population is growing exponentially, the rate of growth increases over time, whereas linear growth connotes a steady rate.

Year	Estimated Population	Doubling Time (years)
1	250 million	
1650	500 million	1650
1804	1 billion	154
1927	2 billion	123
1974	4 billion	47
World population may reach		
2021	8 billion	47[a]

[a]The leveling of doubling time reflects assumptions of decreasing and stabilizing fertility rates. No current projections contemplate a further doubling to 16 billion people.

Source: United Nations.

Figure 2.6 The earth's human population is currently growing exponentially. However, there are already signs that the global growth rate is slowing down, which confirms the predictions of many demographers.

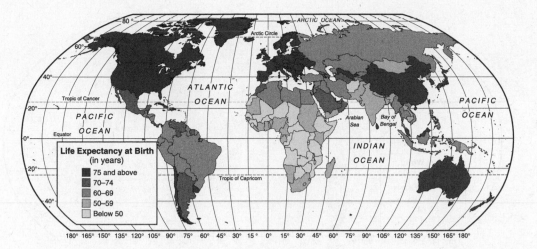

Figure 2.7 Life expectancy by country

What determines a population's natural increase rate? Several factors have been identified as affecting this rate, most of which are related to economic development, culture, or public policy. The following factors are considered important in determining a population's rate of natural increase.

- **ECONOMIC DEVELOPMENT** has profound implications on the quality of available health care, employment opportunities, nutrition, and many other factors that affect population growth. Generally, increases in economic development lead to decreases in fertility and growth rate.
- **EDUCATION** affects every aspect of population growth, from fertility rates to prenatal care to the use of contraception. Populations with better education tend to have lower fertility rates and lower rates of natural increase.
- **GENDER EMPOWERMENT** refers to the relative status and opportunities available to women in a given population. When women have more economic and political access, power, and education, fertility rates inevitably drop.
- **HEALTH CARE** can have contradictory effects on the rate of natural increase. Improved health care in the less-developed countries has decreased the infant mortality rate and increased the life expectancy, thus contributing to population growth. Conversely, the same health care services are often effective at providing desperately needed contraception and family-planning education.
- **CULTURAL TRADITIONS** in many parts of the world encourage high fertility rates by limiting women's employment opportunities outside of the home, by elevating motherhood to a high post and deterring women from doing anything else, or by discouraging the use of contraception.
- **PUBLIC POLICY** can have important implications for population growth in places like China, where the "one couple, one child" program, in place from 1979 to 2018, provided economic incentives favoring families who have fewer children and legal penalties for those who have too many.

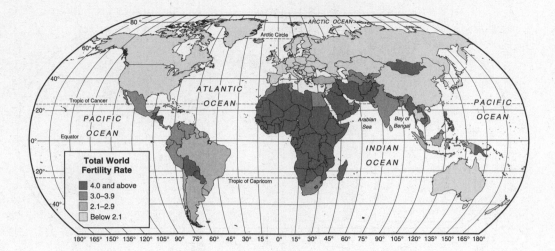

Figure 2.8 World fertility rate by country

Countries that have low levels of economic development, education, and gender empowerment, as well as newly reduced infant mortality rates because of improving health care, cultural traditions favoring fertility, and little or no public policy limiting population growth, tend to have the highest growth rates. These countries, again, are found mainly in sub-Saharan Africa, parts of the Middle East, and Latin America.

Will a given population continue to grow indefinitely? If not, why, when, and at what level will it stabilize? An enormous body of research surrounds these questions, and many careers have been dedicated to projecting and planning for population growth. To predict how much a population will grow, population geographers start with a single, basic formula called the **demographic accounting equation**.

$$P(t+1) = P(t) \ldots$$

$$+ B\,(t,t+1) - D\,(t,t+1) \qquad \text{(natural change)}$$
$$+ I\,(t,t+1) - E\,(t,t+1) \qquad \text{(net migration)}$$

Where P = population

            B = births

            D = deaths

            I = immigration

            E = emigration

            t = time now

     $t + 1$ = some time in the future

The demographic accounting equation states that if you want to predict the population at some time in the future, you need to start with the population now, add the number of births you expect between now and then, subtract the number of deaths, add immigration, and subtract emigration. The part of the equation regarding births and deaths is the natural increase rate and is, in large part, the subject of this chapter. The part of the equation that computes net migration is covered in the next section. For now, just remember that immigration refers to people moving into some place and emigration refers to people moving out of some place. In the demographic accounting equation, you add people coming in and subtract people leaving.

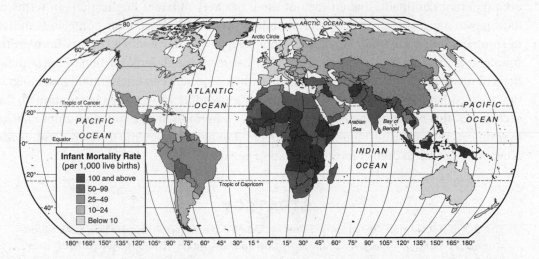

Figure 2.9 Infant mortality rate by country

The demographic accounting equation offers what at first appears to be a very simple way of predicting future population. Unfortunately, as we have already seen, many factors affect the parameters that go into the equation, and small differences in these basic inputs can radically change the final answer. You now know that parameters, such as the CBR and CDR, vary geographically from place to place, but to make things even more complicated, these variables also change over time. Furthermore, as we will now discuss, another critically important component of the equation is the movement of human populations between and within countries, or migration, which also varies dramatically over both space and time.

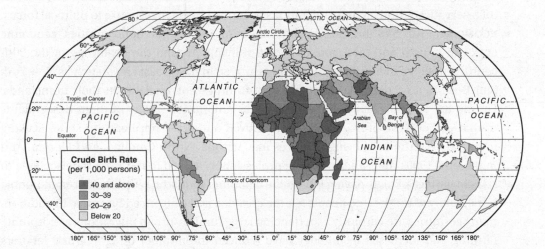

Figure 2.10 Crude birth rate by country

HUMAN MIGRATION

Migration is defined as a long-term move of a person from one political jurisdiction to another. Migration can include a move to a neighboring city or a move to another country on a different continent, though each of these has very different implications in terms of local governance, social systems, and planning. People who leave their homelands to live in another country are said to **emigrate**, while people who move into a country **immigrate**. Immigrants from the less-developed world form an increasingly large portion of the populations of many more-developed countries. The difference between immigration and emigration is considered in the demographic accounting equation. As was already mentioned, the accuracy of this equation tends to be somewhat compromised as population rates change over time. With increasing immigration from developing countries into developed countries, it becomes harder to predict future population growth.

STRATEGY

The 2005 AP Human Geography exam included a free-response question that asked students to explain patterns of immigration to the United States between 1900 and 2000, based on a graph depicting immigration over time. How would you answer such a question?

Several factors cause people to migrate. The following list outlines some of the main reasons why people might leave homelands for new places.

- **POLITICAL ISSUES**, such as armed conflict and the policies of oppressive regimes, have been important historical forces leading to migration. The Pilgrims, who sailed on the *Mayflower* to America, fled oppressive governments that had imposed limitations on their religious freedom in Europe. Later, as the United States expanded westward toward the Pacific Ocean, Americans forcibly removed thousands of Native Americans from the lands that their ancestors had inhabited for millennia, relocating them to far-off reservations. Both of these migrations occurred largely in response to political forces.

- **ECONOMIC FACTORS** that may lead to migration include job opportunities, economic cycles of growth and recession, and cost of living. Around the beginning of the 20th century, millions of European immigrants arrived in East Coast cities, such as New York and Boston, searching for economic opportunities not available in their homelands. During the last 40 years, thousands of older Americans—many of them descendants of that earlier wave of migration—have moved out of the northeastern states seeking inexpensive retirement living in places like Arizona, North Carolina, and Florida. And in the mid-1990s thousands of young professionals moved to the San Francisco area to take advantage of high-paying jobs in the computer industry. When the dot-com bubble burst, many of those same people left the crowded and overpriced Bay Area. In addition, many countries are currently experiencing large rural-to-urban migrations as corporate farming and increased technology have reduced the number of agricultural laborers needed in rural areas. Many of these former farm workers have migrated to cities in hope of finding new economic opportunities.

- **ENVIRONMENTAL ISSUES** can be an important cause of migration in both the less-developed and the more-developed world. For example, in African countries such as Ethiopia, Sudan, and Kenya, many nomadic herders have been forced to breach the boundaries of their former rangelands, searching for more-fertile areas that have not been adversely impacted by drought or overgrazing. In the United States, regions such as the Sierra Nevada range in California have experienced dramatic population growth as people living in the state's crowded coastal cities have sought cleaner air, cheaper houses, less traffic, and a perceived higher quality of life in the mountains. The irony is that, as urbanites leave places like Los Angeles and San Francisco in favor of smaller towns and rural areas, they frequently bring big-city problems—such as crime, pollution, traffic, and high costs of living—with them.

- **CULTURAL ISSUES** can also cause people to move to places where they feel more at home or where they are able to take advantage of certain institutions. For example, after World War II, many Jews from Europe, the Americas, and elsewhere relocated to the newly formed state of Israel. Israel was the ancestral hearth of Jewish culture and religion, and many Jews feel a strong sense of attachment to it. Israel also served as a place where the Jewish people could regroup in safety, reestablish social ties, and create a sense of political unity after the tragedy of the Holocaust.

- **TRANSPORTATION ROUTES** can enable and entice people to migrate to new areas. Throughout history, improved transportation technology and improved routes between places have allowed many people to move within countries and across borders. During the 17th and 18th centuries, better ships and more reliable navigation systems made safe travel across the Atlantic Ocean to the Americas a real possibility for many aspiring European immigrants. Similarly, during the 19th century, new stagecoach routes enabled many white settlers to move westward across the American Great Plains and Rocky Mountains to California and Oregon. Finally, new roads into the Amazon constructed by the Brazilian government during the second half of the 20th century encouraged thousands of people to leave Brazil's densely populated southeastern coast for a life of farming in the country's largely unsettled interior.

STRATEGY

The APHG exam has included a free-response question that asked students to consider why, over the past 150 years, Europe has gone from a source to a destination for migration. Can you account for this transition?

As you can see, many factors cause people to migrate, despite the inherent costs involved with picking up and moving one's entire life to a different place.

Another way that geographers have sought to understand the nature of these motivating factors behind migration is by dividing them up into push factors and pull factors. **Push factors** include anything that would cause someone to want to move *from* somewhere, such as an economic recession or a lack of religious freedom. **Pull factors** induce people to move *to* someplace because that place has something enticing to offer them, such as a pleasant climate or an abundance of jobs in their chosen field. One important pull factor that is probably on your mind right now is educational opportunity. If you go away to college, then the prospect of getting an education will have "pulled" you from your current home to a new place.

Geographers also think of migration as being either voluntary or forced. **Voluntary migration** occurs when someone chooses to leave a place, as a result of either push factors or pull factors. For example, most of the Mexican immigrants who have come to the United States over the decades have done so voluntarily, in order to take advantage of the economic opportunities available north of the border. One type of migratory pattern that is usually voluntary is called **chain migration.** In chain migration, people follow others in succession from one place to another. Distinct ethnic neighborhoods in American cities are often the result of this process because new immigrants often move to places where family members and friends from their home country have already established themselves. In **forced** or **involuntary migration**, someone is removed from his or her home and must leave the area without any choice. Yet, even those who want to migrate to another area may encounter **intervening obstacles**, such as national immigration policies, that limit their movement.

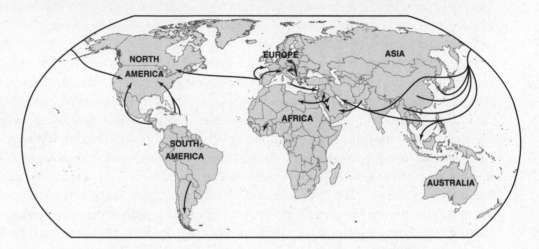

Figure 2.11 Recent global flows of voluntary migration

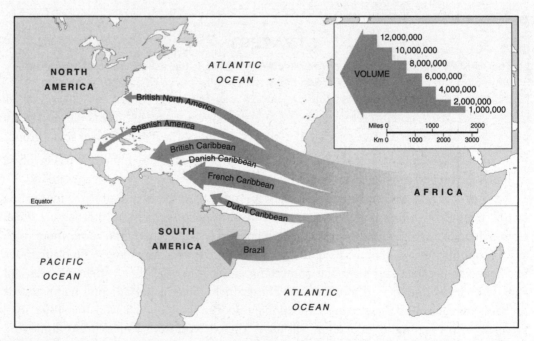

Figure 2.12 The slave trade of the Colonial Period resulted in a mass forced migration of Africans to the New World.

The removal of Native Americans from their traditional homelands in the eastern United States and the African slave trade of the Colonial Period are two particularly tragic examples of forced migration. In 1830, the US Congress, under the direction of President Andrew Jackson, passed the Indian Removal Act, which forced about 100,000 Cherokees, Chickasaws, Choctaws, Creeks, and Seminoles to move west of the Mississippi. Their route of forced migration is now known as the Trail of Tears. Another dreadful example of forced migration was the African slave trade, which, between the 15th and 18th centuries, removed hundreds of thousands of Africans from their homelands and transported them against their will to the Americas. Those who survived the voyage were condemned to a life of bondage.

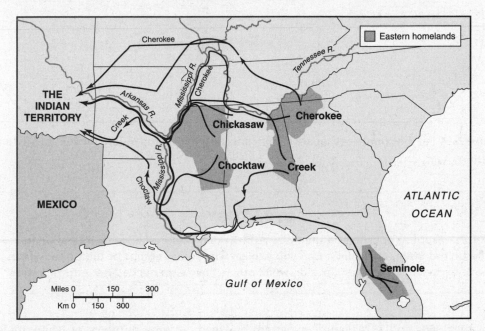

Figure 2.13 In the 1830s, thousands of Native Americans were forced to migrate from their ancestral homelands in the southeastern United States to the High Plains. Their route is remembered as the Trail of Tears.

People who leave their homes because they are forced out, but not because they are being officially relocated or enslaved, are said to be **refugees**. The 1951 Convention Relating to the Status of Refugees defined a refugee as someone who, "owing to a well-founded fear of being persecuted for reasons of race, religion, nationality, membership in a particular social group, or political opinion, is outside the country of his nationality, and is unable to or, owing to such fear, is unwilling to avail himself of the protection of that country." According to the United Nations High Commissioner for Refugees, the global refugee population, as of 2002, was over 21 million people. In recent years, Asia and Africa together have accounted for more than two-thirds of those people, with countries like Pakistan, Afghanistan, Sri Lanka, Angola, and the Democratic Republic of the Congo heading up the list.

Within the United States, **internal migration** patterns have had a tremendous impact on the ethnic composition of large urban areas and on the relative economic dominance of various cities and regions. For example, beginning in the early 20th century, large numbers of African Americans moved from the rural South to large cities in the Northeast and Midwest to join the growing industrial workforces located in places like Chicago, Detroit, and New York. For African Americans, the South had offered racial oppression and little economic opportunity, whereas the North promised a new start and a better way of life. It is also important to remember that in today's increasingly mobile society, people often make many short-term moves for professional or personal reasons.

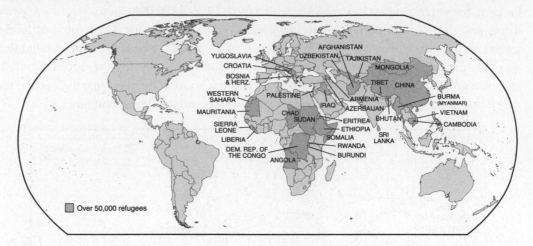

Figure 2.14 All the countries that are significant sources of refugees are located in Africa and southern Asia.

STRATEGY

A free-response question on the 2008 APHG exam explored internal migration in the United States by county. Can you explain why some regions of the United States have experienced net in-migration while others have experienced net out-migration?

In the 1960s and 1970s, another pattern emerged, as large numbers of white, middle-class Americans moved from older northeastern and midwestern cities to the South and to the West Coast. At this time, the northern industrial states, such as Ohio, Michigan, and Pennsylvania, were becoming known as the **Rust Belt**. These states, which had previously been industrial powerhouses with vibrant economies, were now losing much of their economic base to other parts of the country and the world. Factories were closing down, and people were losing their jobs. For would-be migrants, the South offered job opportunities in new high-tech industries, such as software development and aerospace engineering, a pleasant climate, and a relatively affordable cost of living. As a result, during the mid-20th century, the South ceased to be known as the **Cotton Belt**, with its connotations of agrarian poverty and backwardness, and instead became the new land of opportunity—the **Sun Belt**.

Today's Sun Belt includes the "New South" states of Florida, Georgia, Tennessee, and North Carolina, and areas of the Southwest, including portions of Texas, Arizona, Nevada, and Southern California. Cities such as Houston, Los Angeles, and San Diego were among the first to experience the rapid population growth associated with the development of the Sun Belt, but beginning in the 1960s, growth also spread rapidly to places like Phoenix, Las Vegas, Dallas, Miami, Tampa, Austin, and Nashville. Some parts of the South and West, such as Louisiana, Mississippi, Alabama, and New Mexico, have yet to benefit significantly from the Sun Belt phenomenon. However, the economy, culture, and landscape of much of the southern and western United States have been dramatically transformed.

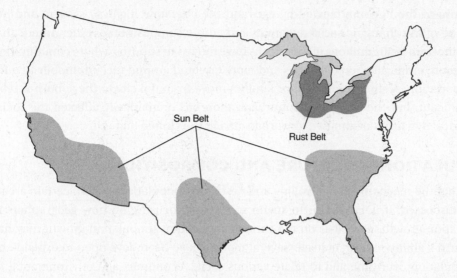

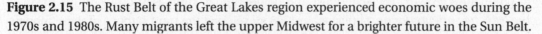

Figure 2.15 The Rust Belt of the Great Lakes region experienced economic woes during the 1970s and 1980s. Many migrants left the upper Midwest for a brighter future in the Sun Belt.

Internal migration patterns have also radically altered the balance of political and economic power. California, Texas, and Florida are now three of the four most populous states in the country (New York is the other). These states, all three of which are at least partly located in the Sun Belt, carry a disproportionate number of electoral votes, have large congressional delegations, and are dominant in many economic sectors, such as technology, energy production, and agriculture. One interesting side note is that the centroid, or geographic center, of the US population is now much farther west and south than it was during the early part of the 20th century, indicating an overall change in the geographic distribution of the US population.

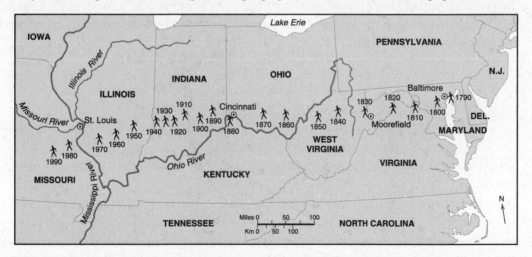

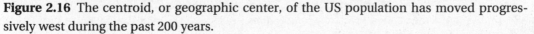

Figure 2.16 The centroid, or geographic center, of the US population has moved progressively west during the past 200 years.

On a smaller geographic scale, suburbanization is one of the most important geographic phenomena affecting the cultural landscape of the United States in the last century. Suburbanization involves migration from the inner city to outlying neighborhoods near the perimeters of urban areas. Suburbanization, which will be covered more extensively in Chapter 6, is partly a response to increasing wealth among the middle classes, partly a response to the freedom created by cars and the interstate highway system, and partly a function of the changing social dynamics and ethnic composition of older American cities. Since the mid-1940s millions of Americans have moved to suburbs, where communities tend to be more ethnically homogeneous and more oriented around the automobile as a form of transportation. A single individual or family's move from the city to the suburb is relatively insignificant, but the process of suburbanization has dramatically affected the social and ecological dynamics of almost every urban area in the United States.

POPULATION STRUCTURE AND COMPOSITION

Now that the mechanisms controlling and describing populations within certain areas have been discussed, it is important to spend some time focusing on how geographers model population growth, as well as on the various implications population growth may have on the earth's ability to sustain itself. Several models have been developed to explain changes in population over time and to relate various social, economic, and environmental factors to population growth. The economist and demographer **Thomas Malthus**, in his *Essay on the Principle of Population*, published in 1798, developed the most famous of these models. Malthus based his argument on two claims: (1) people need food to survive, and (2) people have a natural desire to reproduce. He also noted that food production increases arithmetically, but population increases geometrically. What he meant by this was that food production grows by the *addition* of more acreage into cultivation, whereas population grows by the *multiplication* of human beings. Malthus's geometric growth is now commonly referred to as **exponential growth**. Based in this premise, Malthus argued that human population growth would eventually outpace people's ability to produce food, leading to widespread starvation and disease, what he called "negative checks" on the population. Malthus's theory has been revisited over the decades by many prominent scholars, most recently in 1968 by the **neo-Malthusian** Stanford ecologist Paul Ehrlich. In *The Population Bomb*, Ehrlich made a similar argument about the ability of the earth to sustainably provide resources for an exponentially growing population.

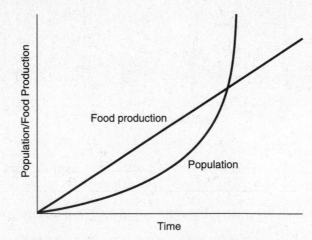

Figure 2.17 Thomas Malthus predicted the population, which grows geometrically (exponentially), would eventually outpace food production, which grows arithmetically (linearly).

There are several problems with the Malthusian perspective. First, although Malthus foresaw the development of new agricultural technologies, he did not fully account for the ability of people to increase food production dramatically with these technologies. Second, Malthus assumed that humans have no control over their reproductive behavior. He did not foresee that population growth would slow down over time because of effective contraception, the changing roles of women in society, and individual people's reproductive decisions. Finally, he did not recognize that famine is usually related not to a lack of food, but to the unequal distribution of food. For example, famine struck various parts of Africa repeatedly throughout the 20th century, despite the fact that an abundance of food existed in other parts of the world. These problems have caused many people to discount Malthusian theory altogether. However, Malthus's theories have helped to bring attention to issues of sustainability and have informed the work of numerous other influential scholars, such as the great ecologist and evolutionary theorist Charles Darwin.

The **demographic transition model**, which explains changes in the natural increase rate as a function of economic development, provides another interpretation of population growth. According to this model, at low levels of economic development, birth and death rates will both be high, but births will significantly outpace deaths. As a country progresses through several stages of economic development, birth rates and death rates will both decrease, ultimately flattening out at some low level. The total population, which increased markedly during the early and middle stages, eventually plateaus as birth rates and death rates converge. This model seems to describe accurately the paths taken by some countries that are already highly developed. It also effectively characterizes various states of population change and development currently existing in many countries across the world.

The subject of Malthusian population growth provides many possibilities for essay questions. You should familiarize yourself with arguments for and against Malthus's theory, and be ready to write an effective essay.

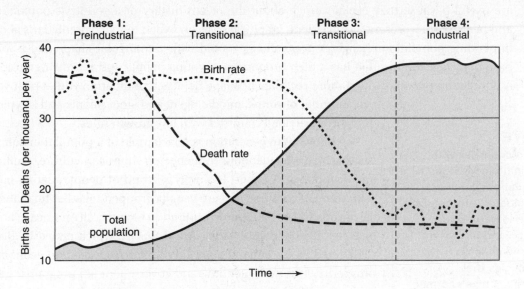

Figure 2.18 The demographic transition model

Whether this model is universally applicable is another matter. Many geographers have argued that the demographic transition model is too simplistic in its portrayal of the relationship between economic development and population and that factors such as culture, religion, geopolitics, migration, and the structure of the global economic system itself may prevent many of today's less-developed countries from ever taking the path described. Moreover, geographers have also pointed out regional variations in demographic transitions, such as those demonstrated in different countries throughout Western Europe. Thus, the relationship between national economics and population is complicated by other factors, such as the structure of the world economic system and cultural constraints.

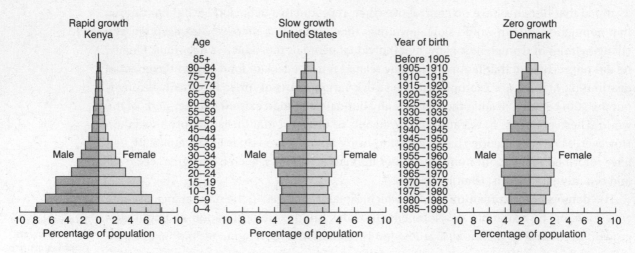

Figure 2.19 Population pyramids for three different countries: Kenya, the United States, and Denmark. Compare Kenya's rapidly growing, youthful population to Denmark's older, stable population.

Another way of looking at population growth is by analyzing a country's **age-sex distribution** through the use of **population pyramids**. A population pyramid shows how a country's populace is distributed between males and females of various ages. Population pyramids are useful because they explain much about the recent history of a country's population and because they present a convenient, graphical basis on which to make predictions about impending population change. A population pyramid with a triangular shape and a wide base depicts a country that has a high proportion of young people and is growing rapidly. A population pyramid with a more rectangular shape depicts a population with a relatively even number of young, middle-aged, and older people and is typical of highly developed countries with low growth rates.

> **NOTE**
>
> Some countries with declining indigenous populations, such as France, have initiated pro-natalist policies that encourage births. Others with high and increasing populations have enacted anti-natalist policies that discourage births. China's former "one child" policy provides an example of the latter.

The **baby boom** generation is an example of a population **cohort** that has had a tremendous influence on American culture, politics, and economics. A cohort is simply a group of people that all have something in common and are usually grouped together for statistical purposes. The baby boom generation includes all Americans born between 1946 and 1964. After World War II, which ended in 1945, the United States entered a period of relative peace and economic prosperity. Jobs were plentiful, and government aid programs, such as the G.I. Bill, helped many veterans reestablish themselves in the postwar economy. War veterans were able to begin a new life at home in the States with an education, a home, and a secure job. These conditions, combined with the relatively conservative social environment of the day, encouraged high rates of marriage and fertility. Although the total fertility rate had already peaked before 1950, the baby boom still produced the largest, best-educated, and most financially secure generation in all of American history.

The generation that followed the baby boom is another story. People born between the years of 1965 and 1980 are sometimes referred to as **Generation X**. The term "Generation X" was coined by the off-beat author and artist Douglas Coupland to describe a generation without the overwhelming numbers and unifying identity enjoyed by the baby boomers. By the mid-1970s the American psyche had been severely damaged by the tragedy of the Vietnam War and the debacle of the Watergate affair. In addition, the women of the baby boom generation, many of whom reached mothering age during this period, were seeking more education, pursuing more demanding careers, waiting longer to marry, and having fewer children than the generation that had come before them. As a result, the natural increase rate declined significantly during the 1960s and 1970s, in what some demographers have called a **baby bust**.

The baby boomers are currently leaving Generation X with another interesting problem. As they get older, they will place unprecedented pressure on health care, social security, and other services that will cost billions of dollars. As a result, the **dependency ratio**, which refers to the percentage of people in a population who are either too old or too young to work and, thus, must be supported by others, will increase dramatically over the next two decades. It remains to be seen what Gen-Xers will be able to do with the complex world they have inherited, as they come of age and emerge out of the shadow of the baby boom.

POPULATION AND SUSTAINABILITY

To truly appreciate the impact of population on people and the landscape, geographers must also consider **population density**. Population density, or **arithmetic density**, defined as the number of people living in a given unit area, varies dramatically from place to place. It tends to be the greatest in large urban areas and least in regions with harsh environments, such as deserts and polar regions. **Physiologic density** compares human population to the area of cropland in less-developed countries, where production is geared mainly toward subsistence agriculture. On a global scale, the greatest population densities currently occur in eastern China, Japan, Southeast Asia, the Indian subcontinent, Western Europe, and the northeastern United States. The countries with the highest population densities on Earth are generally considered to be Bangladesh and the Netherlands. Within the United States, the highest population densities occur in crowded cities such as New York and San Francisco.

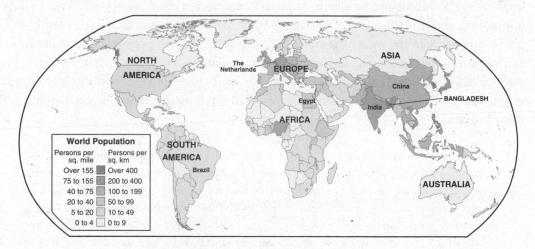

Figure 2.20 Population density by country

In many of the world's most populous regions, population density has become so great that some people have raised the thorny issue of **carrying capacity**. In human geography, as in ecology, carrying capacity refers to the number of individuals a given area is capable of maintaining. Carrying capacity is a tricky subject because most people do not live exclusively off products produced locally. Average consumption of resources varies dramatically from place to place, and social and technological changes constantly alter people's resource demands. These problems make carrying capacity a moving target that is impossible to ever really compute. One way of thinking about carrying capacity is to use the concept of the limiting factor. This concept, which was originally used to describe the way in which plants use resources in the environment, states that the only truly limiting resource is the one that is in the scarcest supply. Las Vegas, for example, may have an ample amount of electrical power and inexpensive land, but in a place that receives just a few inches of rain a year, water should guide planning and limit growth. By this rule, the population of Las Vegas, like so many other cities in arid and semiarid regions, has probably already surpassed its sustainable carrying capacity. The result of artificially surpassing such natural limits is almost always ecological degradation, loss of arable land, and harm to native ecosystems, not to mention the profound social and psychological impacts of living in an overcrowded, overused landscape. Yet, for the past 20 years, Las Vegas has been the fastest-growing city in the United States.

With so many potential problems resulting from population growth, many countries and some international organizations have begun to think about and address **overpopulation**. Overpopulation is not easy to define, because many of today's most pressing international problems related to overpopulation also involve a host of other issues, from overconsumption of resources to the inefficient allocation of goods and services, to unsustainable land-use practices. Generally, overpopulation in a particular area implies a breach of that area's carrying capacity. A given geographical area has the ability to sustainably support a specific population. As that population increases, the region may employ new agricultural techniques or cultivate more land to support more people. At some point, the area is no longer able to return enough goods to sustain a certain population; at this point, the area has extended beyond its carrying capacity. In addition, the mechanisms used to try to increase carrying capacity often have environmental and social repercussions. Many activists in the more-developed countries, who believe that overpopulation is the root cause of the world's social and environmental problems, support policies that they believe will lead to **zero population growth**. However, most leaders from the less-developed countries tend to lay blame on people in the wealthiest countries, who consume a disproportionate share of the earth's natural resources. The following statement—submitted by the Indian delegation to the UN International Conference on Population and Development, held in Cairo, Egypt, in 1994—regarding the development of international population policy is representative of that sentiment.

> *The Indian delegation agrees . . . that the draft document focuses disproportionately on the linkages of population and environment. There is little mention of the pattern of consumption in the industrialized nations which cannot long be supported without serious damage to the biosphere. The industrialized nations with approximately 20% of the world's people are currently responsible for three-quarters of the world's energy use, two-thirds of all greenhouse gases and 90 percent of the chlorofluorocarbons. Population growth alone is not responsible for environmental degradation. The inter-linkage between population, poverty, environment and sustainable development is complex.*

As the world population swells and as urban areas and developing countries experience increasing population density, it is increasingly apparent that certain policy initiatives will be necessary to combat future problems stemming from overpopulation. Some population policies seek to limit population growth in particular countries either through fertility control or by imposing immigration restrictions. In both cases, the type of policy implemented is largely determined by the politics of public perception and by prevailing ideologies. Many seemingly unsolvable questions surround the design and implementation of population policy. Is it possible to humanely limit an individual's right to reproduce? How should a government determine who will be permitted to enter? Furthermore, when designing an appropriate policy, countries must determine the number of people they think they can sustain. Of course, this involves even trickier questions regarding an appropriate standard of living and people's level of consumption.

The world's two most populated countries, India and China, both have implemented policies to curb population growth within their borders. The policies implemented in each country reflect differing cultures and political situations. In communist-controlled China, up until 2018 families were encouraged to have only one child by rewards given to families who follow this dictate and punishments given to those who do not. In the more democratic India, individuals are also encouraged to limit the number of children they bear, but instead the means of encouragement come through education and improved access to family planning.

The numerous questions surrounding population policy illustrate that it is an issue that is continually fraught with controversy. Some leaders see international population policy as infringing upon the fundamental human right to reproduce. For others, fertility has special cultural or religious value, and any attempts to limit it are seen as cultural persecution. The Catholic Church officially believes that practices of birth control and abortion are sinful, thus implementing any form of fertility control can be difficult in predominantly Catholic countries. However, even in predominantly Catholic countries, where there is a high standard of living and plentiful opportunities for women, growth is always low. Italy, which has an extremely low growth rate, is a case in point.

Immigration policy is also frequently contentious. Many people are concerned that policies specifically limiting the number of immigrants into a particular area are inhumane and may be the product of racist desires to limit the population of particular types of people. Within a given state, it is hard to predict how citizens may react to seeing new immigrants move into their neighborhood. Whether or not you believe that the Indian delegation is correct in its assessment of the population debate, it is obvious that there are no easy solutions, and that the delegation does make an excellent point when it states that the "inter-linkage between population, poverty, environment and sustainable development is complex." India has firsthand experience as a country with over a billion people.

As you have probably surmised, population geography is a wide-ranging and multifaceted field, and one that keeps geographers extremely busy! Geographers know that it is critical to understand population dynamics and growth patterns throughout the world, in order both to provide for current populations and to plan for the future. At the same time, issues of mobility, health care, education, policy, and economics render making steadfast predictions extremely difficult. With these challenges in mind, geographers work to improve their models, to test their theories, and to better understand human populations throughout the world.

TIP

The relative importance of population versus consumption in environmental problems provides another possible essay question. You should be able to describe the various arguments, give appropriate examples, and offer a critical appraisal.

KEY TERMS

AGE-SEX DISTRIBUTION A model used in population geography that describes the ages and numbers of males and females within a given population; also called a POPULATION PYRAMID.

AGRICULTURAL DENSITY The number of farmers per unit area of farmland.

ARITHMETIC DENSITY The number of people living in a given unit area.

BABY BOOM A cohort of individuals born in the United States between 1946 and 1964, which was just after World War II in a time of relative peace and prosperity. These conditions allowed for better education and job opportunities, encouraging high rates of both marriage and fertility.

BABY BUST Period during the 1960s and 1970s when fertility rates in the United States dropped as large numbers of women from the baby boom generation sought higher levels of education and more competitive jobs, causing them to marry later in life. As such, the fertility rate dropped considerably, in contrast to the baby boom, in which fertility rates were quite high.

CARRYING CAPACITY The largest number of people that the environment of a particular area can sustainably support.

CHAIN MIGRATION The migration event in which individuals follow the migratory path of preceding friends or family members to an existing community.

CHILD MORTALITY RATE Number of deaths per thousand children within the first five years of life.

COHORT A population group unified by a specific common characteristic, such as age, and subsequently treated as a statistical unit.

COTTON BELT The term by which the American South used to be known, as cotton historically dominated the agricultural economy of the region. The same area is now known as the New South or Sun Belt because people have migrated here from older cities in the industrial north for a better climate and new job opportunities.

CRUDE BIRTH RATE The number of live births per year per thousand people.

CRUDE DEATH RATE The number of deaths per year per thousand people.

DEMOGRAPHIC ACCOUNTING EQUATION An equation that summarizes the amount of growth or decline in a population within a country during a particular time period, taking into account both natural increase and net migration.

DEMOGRAPHIC TRANSITION MODEL A sequence of demographic changes in which a country moves from high birth and death rates to low birth and death rates through time.

DEMOGRAPHY The study of human populations, including their temporal and spatial dynamics.

DEPENDENCY RATIO The ratio of the number of people who are either too old or too young to provide for themselves to the number of people who must support them through their own labor. This is usually expressed in the form $n:100$, where n equals the number of dependents.

DOUBLING TIME Time period required for a population experiencing exponential growth to double in size completely.

EMIGRATION The process of moving out of a particular country, usually the individual person's country of origin.

EPIDEMIOLOGICAL TRANSITION Sudden population growth as a result of improved food security and health care, followed by a plateau in growth because of subsequent declines in fertility rates.

EXPONENTIAL GROWTH Growth that occurs when a fixed percentage of new people is added to a population each year. Exponential growth is compound because the fixed growth rate applies to an ever-increasing population.

FORCED MIGRATION The migration event in which individuals are forced to leave a country against their will.

GENERATION X A term coined by artist and author Douglas Coupland to describe people born in the United States between the years 1965 and 1980. This post-baby-boom generation will have to support the baby boom cohort as they head into their retirement years.

GEODEMOGRAPHY See POPULATION GEOGRAPHY.

IMMIGRATION The process of individuals moving into a new country with the intention of remaining there.

INFANT MORTALITY RATE The percentage of children who die before their first birthday within a particular area or country.

INTERNAL MIGRATION The permanent or semipermanent movement of individuals within a particular country.

INTERVENING OBSTACLES Any forces or factors that may limit human migration.

INVOLUNTARY MIGRATION See FORCED MIGRATION.

LIFE EXPECTANCY The average age individuals are expected to live, which varies across space, between genders, and even between races.

THOMAS MALTHUS Author of *Essay on the Principle of Population* (1798) who claimed that population grows at an exponential rate while food production increases arithmetically, and thereby that, eventually, population growth would outpace food production.

MATERNAL MORTALITY RATE Number of deaths per thousand of women giving birth.

MIGRATION A long-term move of a person from one political jurisdiction to another.

NATURAL INCREASE RATE The difference between the number of births and number of deaths within a particular country.

NEO-MALTHUSIAN Advocacy of population-control programs to ensure enough resources for current and future populations.

OVERPOPULATION A value judgment based on the notion that the resources of a particular area are not great enough to support that area's current population.

PHYSIOLOGIC DENSITY A ratio of human population to the area of cropland, used in less-developed countries dominated by subsistence agriculture.

POPULATION DENSITY A measurement of the number of persons per unit of land area.

POPULATION GEOGRAPHY A division of human geography concerned with spatial variations in distribution, composition, growth, and movements of population.

POPULATION PYRAMID A model used in population geography to show the age and sex distribution of a particular population.

PULL FACTORS Attractions that draw migrants to a certain place, such as a pleasant climate and employment or educational opportunities.

PUSH FACTORS Incentives for potential migrants to leave a place, such as a harsh climate, economic recession, or political turmoil.

REFUGEES People who leave their home because they are forced out, but not because they are being officially relocated or enslaved.

RUST BELT The northern industrial states of the United States, including Ohio, Michigan, and Pennsylvania, in which heavy industry was once the dominant economic activity. In the 1960s, 1970s, and 1980s, these states lost much of their economic base to economically attractive regions of the United States and to countries where labor was cheaper, leaving old machinery to rust in the moist northern climate.

SUN BELT US region, mostly comprising of southeastern and southwestern states, which has grown most dramatically since World War II.

TOTAL FERTILITY RATE The average number of children born to a woman during her childbearing years.

VOLUNTARY MIGRATION Movement of an individual who consciously and voluntarily decides to locate to a new area—the opposite of forced migration.

ZERO POPULATION GROWTH Proposal to end population growth through a variety of official and nongovernmental family-planning programs.

CHAPTER SUMMARY

During the past two centuries, the world's human-population growth has exploded. The causes, mechanisms, and consequences of this tremendous increase are multifaceted. Population geographers collect demographic data and use mathematical equations to understand trends in population growth and to make predictions about the future. Population growth and population density both vary dramatically over space—some countries are not growing at all, while others are doubling their populations every couple of decades. The fastest-growing countries are also among the poorest, and in many cases rapid population growth has led to severe environmental degradation and human suffering. However, environmental problems are not simply a function of population growth; they also result from the overconsumption of resources. Consequently, issues regarding population, consumption, social justice, and environmental sustainability provide geographers with many fascinating and complex questions.

PRACTICE QUESTIONS AND ANSWERS

Human Population: A Global Perspective

MULTIPLE-CHOICE QUESTIONS

1. Which of the following regions is currently experiencing the fastest population growth?

   (A) Northern Asia
   (B) Tropical Africa
   (C) Eastern Europe
   (D) Sun Belt
   (E) Northeast United States

2. Most of the world's people live in

   (A) the world's poorest countries.
   (B) the southern hemisphere.
   (C) the developed world.
   (D) China.
   (E) urban areas in the developed world.

3. Throughout human history, world population has

   (A) grown at a steady rate.
   (B) experienced numerous periods of dramatic decline.
   (C) been confined to countries in the southern hemisphere.
   (D) grown most rapidly over the last 200 years.
   (E) grown most rapidly in the developed world.

4. _____ occurs when a population is adding a fixed percentage of people to a growing population each year.

   (A) Doubling
   (B) Arithmetic growth
   (C) Overpopulation
   (D) Exponential growth
   (E) Demographic accounting

5. Life expectancy has increased

   (A) only in the most-developed countries.
   (B) only in the least-developed countries.
   (C) due to increased food production.
   (D) worldwide.
   (E) due to the Green Revolution.

1. Exponential world population growth is one of the defining characteristics of contemporary human geography. Explain this growth in terms of

   (A) historical trends.
   (B) global geographic patterns.
   (C) economic development.

Population Parameters and Processes

MULTIPLE-CHOICE QUESTIONS

1. The number of live births per thousand people per year is called the

   (A) total fertility rate.
   (B) natural increase rate.
   (C) crude birth rate.
   (D) exponential growth rate.
   (E) infant growth rate.

2. Which of the following countries is most likely to be showing the lowest natural increase rate?

   (A) Afghanistan
   (B) Liechtenstein
   (C) United States
   (D) Japan
   (E) Chile

3. Total fertility rate is NOT closely correlated with which of the following?

   (A) Industrial output
   (B) Gender empowerment
   (C) Education
   (D) Economic development
   (E) Literacy

4. The demographic accounting equation does NOT take into account _____ when calculating a country's population.

   (A) the death rates
   (B) emigration
   (C) natural increase over time
   (D) instances when natural increase is negative
   (E) immigration

5. Within the United States, overall life expectancy

   (A) is limited by an unusually high infant mortality rate.
   (B) varies between various cohorts within the larger population.
   (C) varies between regions, with people in the Southwest living longer on average.
   (D) All of the above
   (E) Both (B) and (C)

Human Migration

MULTIPLE-CHOICE QUESTIONS

1. Millions of _____ came to the United States during the early years of the 20th century.

   (A) suburbanites
   (B) emigrants
   (C) immigrants
   (D) refugees
   (E) colonists

2. In the 1930s, thousands of "Okies" fled the Dust Bowl of the southern Great Plains and moved to the fertile agricultural regions of California to start a new life. This is an example of

   (A) external migration.
   (B) eco-migration.
   (C) political migration.
   (D) economic migration.
   (E) forced migration.

3. Which of the following is the result of chain migration?

   (A) The African slave trade
   (B) French colonial rule
   (C) The formation of Israel
   (D) San Francisco's Chinatown
   (E) Colonization of the American frontier

4. Refugees are produced through

   (A) cultural migration.
   (B) forced migration.
   (C) internal migration.
   (D) economic migration.
   (E) chain migration.

5. Many recent college graduates and young professionals move to large, vibrant cities—such as New York, Chicago, and Los Angeles—with nightlife, cultural amenities, and job opportunities. These attractions are examples of

(A) economic factors.

(B) mobility opportunities.

(C) suburban amenities.

(D) pull factors.

(E) push factors.

6. Suburbanization is most evident in

(A) older American cities like Boston.

(B) large European cities like Madrid.

(C) regionally planned Canadian cities like Toronto.

(D) newer American cities like Las Vegas.

(E) large South American cities like Sao Paulo.

7. The Sun Belt includes

(A) the Rocky Mountain States.

(B) Alabama and Louisiana.

(C) Texas and New Mexico.

(D) Southern Nevada, southern California, and South Florida.

(E) Florida, Georgia, Alabama, and South Carolina.

Population Structure and Composition

MULTIPLE-CHOICE QUESTIONS

1. Thomas Malthus predicted that

(A) technology will offset population growth.

(B) the distribution of resources would be a continuing problem.

(C) population would outpace food production.

(D) the environment would allow less food to be grown in the future.

(E) the Green Revolution would provide agricultural technology to support increasing populations.

2. Which of the following countries is at stage two of the demographic transition model?

(A) San Marino

(B) Nigeria

(C) Denmark

(D) Russia

(E) Finland

3. A rectangle-shaped population pyramid indicates a country that is

(A) growing slowly or not at all.

(B) growing rapidly.

(C) experiencing high immigration rates.

(D) composed mainly of the older age classes.

(E) highly dependent on the economically productive generations.

4. The baby boom

(A) occurred in the years following World War I.

(B) was a result of free love during the late 1960s.

(C) was fostered by economic prosperity and relative peace.

(D) was limited to California and the West.

(E) was described by the off-beat author Douglas Coupland.

5. When baby boomers have reached retirement age, what will the population pyramid for the United States look like?

(A) An hourglass, wide at both top and bottom but narrow in the middle

(B) Relatively rectangular, with a slight bulge near the top

(C) Carrot-shaped, a narrow bottom and wide top

(D) Pear-shaped, wide at the bottom, but narrow at the top

(E) None of these

Population and Sustainability

MULTIPLE-CHOICE QUESTIONS

1. Which of the following countries would you expect to have the densest population?

(A) China

(B) Peru

(C) Mexico

(D) Belgium

(E) Colombia

2. Carrying capacity is a function of

(A) technology.

(B) natural resources.

(C) resource allocation.

(D) limiting factors.

(E) (A), (B), and (D)

3. Population policy usually involves limitations on

(A) fertility levels.

(B) immigration levels.

(C) education levels.

(D) All of the above

(E) Both (A) and (B)

4. India and China are the world's two most populous countries. While China has instituted a strict population policy, India

(A) for cultural reasons, encourages women to continue to reproduce.

(B) does not endorse birth control because of the Catholic majority.

(C) encourages lower fertility through education and access to family planning.

(D) has a similar policy to China.

(E) because of its agricultural system, encourages reproduction.

FREE-RESPONSE QUESTIONS

1. According to the demographic transition model, population growth should slow down as a country becomes more developed.

(A) Where is the United States according to the demographic transition model?

(B) In the 1990s, the United States experienced increased population growth; explain this recent growth and compare it to slow growth patterns in other highly developed countries.

2. At the United Nations International Conference on Population and Development held in Cairo, Egypt, in 1994, the Indian delegation claimed that population policy involved much more than limiting growth.

(A) Describe India's claim that consumption of natural resources must enter global conversations on population policy.

(B) Discuss the ways that population, technology, and affluence have affected the environments of three countries: Costa Rica, China, and Canada.

Answers for Multiple-Choice Questions

HUMAN POPULATION: A GLOBAL PERSPECTIVE

1. **(B)** Tropical Africa is one of the fastest-growing areas in the world. Increases in crop production and better access to medical care, combined with high fertility rates, have caused tremendous population growth throughout the region. Cities, like Lagos, Nigeria, are also among the fastest-growing urban areas anywhere. Although the Sun Belt region of the United States has grown rapidly since World War II, this growth is nowhere near as sudden or as dramatic as that of tropical Africa.

2. **(A)** Approximately 80% of the world's population lives in the less-developed countries, which include all of Africa, Asia (excluding Japan), Latin America, and the island nations of the Caribbean and Pacific.

3. **(D)** Human population has demonstrated overall steady growth throughout human history; however, in the last 200 years, population has been growing at exponential rates.

4. **(D)** When you add a fixed number of people to a growing population each year, it is called arithmetic growth, but when you add a fixed percentage of people each year, it is called exponential growth. Exponential growth is compound, since the same percentage is being added to an increasing population each year; if the population is growing, then the same percentage will include more people next year than it does this year.

5. **(D)** Life expectancy varies both between countries and within countries and is related to many factors, including race, sex, and wealth.

POPULATION PARAMETERS AND PROCESSES

1. **(C)** Many people get the total fertility rate (TFR) and the crude birth rate (CBR) mixed up. The TFR refers to the average number of children born to a woman over the course of her life.

2. **(B)** Liechtenstein is a small, wealthy, and highly developed country located in the European Alps between Switzerland and Austria. Like other wealthy European countries with aging populations, Liechtenstein's growth rate is currently less than 1%.

3. **(A)** Although the total fertility rate is correlated with overall development, its relationship to industrial output is less direct. Small, highly developed countries, like Liechtenstein, may have little heavy industry but low fertility rates, while countries like Brazil may have much more industrial activity but much higher fertility rates. Gender empowerment, education, and general economic development are all closely correlated with fertility.

4. **(C)** The demographic accounting equation predicts a country's future population on the basis of current birth rates, death rates, immigration rates, and emigration rates. It is not always a very accurate prediction, because these rates can change dramatically over time.

5. **(B)** On average, white American males live longer than males of other races within the United States. Generally, life expectancy is calculated for particular countries; thus, it would be hard to determine regional variations in this statistic.

HUMAN MIGRATION

1. **(C)** During the late 19th and early 20th centuries, millions of immigrants came to the United States from all over the world. Many came from southern and Western Europe in search of new economic opportunities.

2. **(B)** The Dust Bowl refugees, often called Okies, left the southern Great Plains because of the environmental disaster of the Dust Bowl. When people leave an area because of environmental factors, it is called eco-migration.

3. **(D)** Ethnic urban enclaves, such as San Francisco's Chinatown, result when people follow those who went before them in migrating from one region to another. Chinatown has been an attractive place for many Chinese immigrants to settle after arriving in the United States because of the neighborhood's familiar language and customs.

4. **(B)** Refugees, by definition, are people who are forced to leave their homes and move to a new place. Refugees can be produced through any type of forced migration, such as religious persecution, environmental degradation, or even natural disasters.

5. **(D)** Pull factors include anything that draws someone from one place to another. Cities provide strong pull factors for young people looking for economic opportunities and recreational diversions.

6. **(D)** Most newer American cities, and newer areas of older American cities, have been designed to accommodate sprawling suburban housing communities and wide highways geared for automobile transportation. Although some European cities have been affected by sprawl, it is much less of a problem in countries like France and Germany, where people tend to live in older neighborhoods and the car is less important for transportation. Some Canadian cities, such as Toronto, have limited sprawl through well-coordinated urban planning.

7. **(D)** Parts of the Rocky Mountains, the Old South, and the Southwest are all considered to be within the Sun Belt. However, each of these regions also contains areas that have not benefited from the economic growth associated from the Sun Belt phenomenon. Southern Nevada, southern California, and South Florida are all classic Sun Belt regions.

POPULATION STRUCTURE AND COMPOSITION

1. **(C)** Thomas Malthus predicted that food production would grow arithmetically while population would grow geometrically (exponentially). Malthus's theory has been criticized because it does not account for other factors affecting food production and population, such as technology, the distribution of food resources, personal choice, and environmental change.

2. **(B)** In stage two of the demographic transition model, a country's population growth is high because death rates have decreased but birth rates have not. Nigeria's explosive population growth is an example of this situation. Denmark, Russia, and San Marino have all advanced to stage three, in which fertility decreases and population growth slows down.

3. **(A)** A triangle-shaped population pyramid indicates that there is a high percentage of young people in the population and that it is growing rapidly. A rectangle-shaped age-sex distribution means that the population is composed of a more even range of older and younger people and that the population is growing slowly or not at all.

4. **(C)** The baby boom, which occurred during the years following World War II, was a national phenomenon in which economic prosperity and relative peace were accompanied by high fertility rates. Although fertility began to decline as early as the late 1940s, the baby boom generation became the most numerous, wealthiest, and most prosperous generation in American history.

5. **(B)** The baby boom is a large cohort of the American population, larger than the generation behind them. When they reach retirement age, their bracket, near the top of the pyramid, will most likely be larger than other segments of the pyramid. This demonstrates the magnitude of the dependency ratio, or number of individuals relying on younger, economically productive generations for support.

POPULATION AND SUSTAINABILITY

1. **(D)** Small northern European countries, such as the Netherlands and Belgium, are some of the most densely populated in the world. China has the largest population of any country on Earth, yet its immense size allows for large, sparsely populated rural areas to remain.

2. **(E)** The quality and quantity of natural resources available in an area and technological innovations that help people to use those resources both affect carrying capacity. The limiting factor describes the resource in scarcest supply within a region and thus provides a method for understanding how big a population an area can adequately sustain. However, the limiting factor within a region can be extended by use of technological innovations. Because of this, carrying capacity changes over time and is notoriously hard to pin down.

3. **(E)** These two forces, births and immigration, are the largest contributors to population increase within a country. Humane population policy usually seeks to limit or decrease somehow the fertility level within a country, or it seeks to limit the number of individuals allowed into a particular country.

4. **(C)** Population policies implemented within countries usually reflect the country's prevailing ideologies and, thus, usually its political system. China is a communist-controlled country, which was demonstrated in their strict enforcement of their one-child policy. However, India is a democratic nation that encourages, rather than demands, lower fertility through increased education and access to family planning.

Answers for Free-Response Questions

HUMAN POPULATION: A GLOBAL PERSPECTIVE

1. **Main points:**
   - For the great majority of human history, population grew very slowly. It was not until the 19th century that world population reached 1 billion.
   - During the 19th and 20th centuries the world's human population grew exponentially because of increases in crop yields, advances in health care, and a variety of other factors.
   - During the 20th century, developing countries experienced the most dramatic population growth. Many of the least-developed countries are still among the fastest growing.
   - The current world population is over 7 billion. Most of the world's people live in poverty in the less-developed countries. Many of these people have inadequate access to health care, social services, and healthy environments with clean air and water.
   - Population growth is a gendered issue. In countries where women are given access to education, health care, family-planning services, and employment opportunities, population growth always drops, independent of other cultural factors such as religion.
   - China and India alone account for about a third of the world's population. With less than 350 million people, the United States has about 5% of the world's population.

POPULATION AND SUSTAINABILITY

1. **Main points:**
   - The demographic transition model (DTM) is based on a simplified, deterministic approach to population growth; it does not account for particular historical events, such as wars, international agreements, economic cycles, immigration, or a country's relations with its neighbors.
   - In the United States, population growth has fluctuated over time. In the 1990s, several factors that are not accounted for in the DTM caused growth to increase.
   - The most important of these factors was immigration. During the 1990s, economic prosperity—in part associated with a boom in the high-tech economy—initiated a need for more workers. American companies recruited workers from overseas, and thousands of construction and service-industry jobs opened for unskilled and semi-skilled laborers from Mexico and other developing countries.
   - Another factor causing population growth in the United States during the 1990s is that many of the new immigrants to the United States had higher overall fertility rates than citizens whose families had been established in the United States for several generations.
   - Although the DTM was correct in predicting that fertility rates would drop within established American families, economics and immigration caused the rate of growth to increase in the United States during the 1990s.

2. **Main points:**
   - Environmental degradation is a nearly universal feature of today's world; however, its causes are difficult to quantify. Large-scale economic and social forces drive environmental change, yet most conservation efforts are aimed at the specific, proximate causes of environmental problems.
   - In addition, the relationships between population, consumption, and technology are extremely complex.
   - Malthusians think that overpopulation is the main cause of environmental problems. Indeed, countries such as China, India, and Indonesia, and other developing countries with their enormous human populations, place huge stresses on their environments and natural resources.
   - However, as the Indian delegation has stated, overconsumption by the most-developed countries is also responsible for environmental problems. The United States, for example, consumes a disproportionately large share of the world's natural resources. Even though the United States has only about 330 million people, their impact is felt worldwide through their consumption of resources such as oil and forest products.

ADDITIONAL RESOURCES

Livi-Bacci, Massimo. 2017. *A Concise History of World Population, 6th Edition.* Malden, Massachusetts: Blackwell.

This book provides a concise history of the world's population, primarily by looking at the intersection between nature, culture, and population. By examining historic checks on growth, and projections for future growth, Livi-Bacci proposes, using historical patterns, methods for preventing future environmental and human catastrophes that may result from overpopulation.

Weinstein, Jay, and Vijayan Pillai. 2015. *Demography: The Science of Population, 2nd Edition.* Needham Heights, Massachusetts: Allyn and Bacon.

The authors begin by talking about population in general, but then further explore different demographic measures across the globe, including birth and fertility, mortality, and migration. They also look at several different models designed to understand population growth and change, and finally investigate population policy and some of the environmental repercussions of overpopulation.

Meadows, Dana. H., Dennis L. Meadows, and Jorgen Randers. 1992. *Beyond the Limits.* Post Mills, Vermont: Chelsea Green.

This book discusses some of the population limits the earth has already passed, forecasting global collapse if the trends do not change and people do not start living sustainably. The authors introduce a new model, called the World3, which is computer-based and uses different policy initiatives to predict different scenarios for the future of the world's resources.

Internet Resources

- Population Resources Bureau: *http://www.prb.org/*
- United States Census Bureau: *https://www.census.gov/*
- Population Pyramid Interactive: *https://www.populationpyramid.net/*
- Joel Cohen, "Malthus Miffed: An Introduction to Demography":
  https://www.youtube.com/watch?v=2vr44C_G0-o

Cultural Patterns and Processes

3

IN THIS CHAPTER

→ CULTURAL BASICS
→ LANGUAGE
→ RELIGION
→ GENDER
→ ETHNICITY
→ POPULAR CULTURE

 Key Terms

Acculturation	Ethnic cleansing	Local religions
Animism	Ethnic neighborhood	Minority
Artifact	Ethnic religion	Missionary
Buddhism	Ethnicity	Monotheism
Caste system	Ethnocentrism	Multiculturalism
Christianity	Evangelical religion	Official language
Creole	Folk culture	Pidgin
Cultural complex	Fundamentalism	Pilgrimage
Cultural extinction	Gender	Polyglot
Cultural geography	Gender Inequality Index	Polytheism
Cultural hearths	Genocide	Pop culture
Cultural imperialism	Ghetto	Possibilism
Cultural relativism	Global religion	Race
Cultural traits	Hinduism	Romance languages
Culture	Indo-European family	Shaman
Customs	Islam	Sino-Tibetan family
Denomination	Isoglosses	Stimulus diffusion
Dialects	Judaism	Syncretic
Diaspora	Language extinction	Toponyms
Dynamic	Language family	Tradition
Ecumene	Language group	Transculturation
Environmental determinism	Lingua franca	Universalizing religion
Esperanto	Literacy	
	Local culture	

CULTURAL BASICS

Culture means many things to many people. Linguistically, the English word *culture* derives from the Latin word *cultus,* which means "to care about." The concept of culture dates back at least to the age of Enlightenment (1630s–1780s), when culture referred to a variety of endeavors that were essentially human, such as agri*culture.* Later, the term came to connote differences between people's lifestyles in different areas of the world. The modern notion of culture, which includes all the ideas, practices, and material objects associated with a particular group of people, evolved from this notion of difference. The study of how cultures vary over space is called **cultural geography**. Cultural geographers use techniques from sociology, anthropology, psychology, history, and numerous other disciplines to better understand the spatial dimensions of human cultures throughout the inhabited world, known as the **ecumene**.

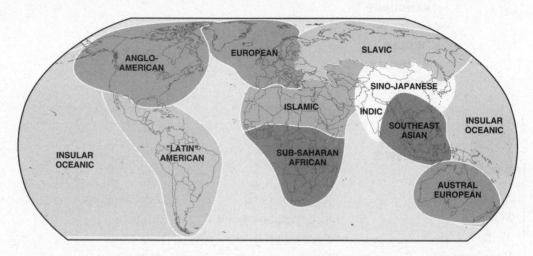

Figure 3.1 This map depicts the world's **macrocultural regions**—the geographic areas encompassed by different cultural realms around the globe.

When geographers think about culture, they include both the material things that a group of people cares for and the suite of beliefs, values, and characteristics that define their collective identity and set them apart from others. The material aspects of culture are called **artifacts** and include such things as clothing, tools, and artwork, whereas the practices followed by the people of a particular group are called **customs**. A cohesive collection of customs is called a **tradition**. Traditions are often **syncretic**, in that they blend cultural traits from different contemporary sources, and **dynamic**, since they change over time. On a broader scale, **cultural traits** are specific customs that are part of everyday life, such as language, religion, ethnicity, social institutions, and aspects of popular culture. The group of traits that define a particular culture is called a **cultural complex**. All cultural traits have **hearths**, or places of origin. But they may also expand broadly through processes of diffusion, adoption, and assimilation, referred to collectively as **transculturation**.

Cultural geographers study the spatial distribution of cultural traits and the intricate relationships between cultures and the natural environment. The doctrine of **environmental determinism**, which had enjoyed considerable popularity at several points in history, was a particularly important impediment in modern geographers' early efforts to study culture-environment relations. Environmental determinists claim that cultural traits are formed and controlled by environmental conditions. Certain types of people, who come from cultures that arose in certain physical environments, may be smarter, more attractive, or more able to govern themselves as a result. This doctrine, the racist implications of which are obvious, was often used by European states to justify their colonization of native peoples in Africa, South America, Australia, and elsewhere because the people who lived in those hot, muggy, and seemingly oppressive environments were deemed less intelligent and industrious than individuals coming from more temperate climates characteristic of the European continent.

A different conception of environment-culture relations has found favor among some geographers. In this version, called **possibilism**, different environmental conditions offer both restraints and opportunities to people living in various regions. However, people control their own destinies and deal with these various environmental factors in ways that are dynamic and contingent and that unfold unpredictably over time. Many possibilists argue that the degree to which a particular culture is influenced by environmental forces depends on the level of technology prevalent within that society. Cultures in the more-developed parts of the globe have designed various technologies to counteract environmental limitations. For instance, automobiles with four-wheel drive make traveling in virtually any weather condition possible. These and other similar types of technologies are not readily available in developing societies; thus, these cultures may encounter greater limitations from certain environmental constraints. Possibilism offers excellent new opportunities to explore the fascinating ways in which cultures have interacted with their environments over time, without the intellectually limiting and clearly racist overtones of environmental determinism.

Finally, beyond studying how cultures vary over space and how different cultures interact with the environment, cultural geographers also explore the various ways cultural qualities diffuse to other parts of the world. Recall from Chapter 1 the discussion of the different ways certain phenomena diffuse across space. Cultural geographers study how language, religion, and other cultural artifacts such as fashion, music, and culinary traditions move from their areas of origin to different parts of the world. How is it that Americans can choose from pad thai, Indian curry, pasta primavera, or Japanese sushi for dinner on almost any given evening of the week? Or, why are Europeans wearing Levi's jeans as they walk into a Burger King in Amsterdam? In today's world, cities have become havens to numerous cultural diversities; cultural geographers study how and why specific cultural traits are so easily transmitted and accepted in other parts of the world.

LANGUAGE

Language is one of the oldest, most geographically diverse, and most complex cultural traits on Earth. Although language is basic to the human experience, it is clear that people from different parts of the world have found very different ways to express themselves. In the prehistoric past, there were probably at least 10,000 languages spoken throughout the world. Currently, about 5,000 to 7,000 languages remain, with Africa and Asia being the most linguistically rich continents. The world's greatest concentration of linguistic diversity is on the island of New Guinea. In New Guinea, rugged terrain and social mores limit interaction between different tribal groups, enabling some 900 languages to persist into the present day. The linguistic diversity of New Guinea provides a sharp contrast to the modern global trend, in which a few languages are becoming increasingly dominant across the world. A knowledge of the geography of language is essential for understanding larger spatial patterns in human societies and for piecing together the common histories of people who have spoken to, written to, and learned from each other over time.

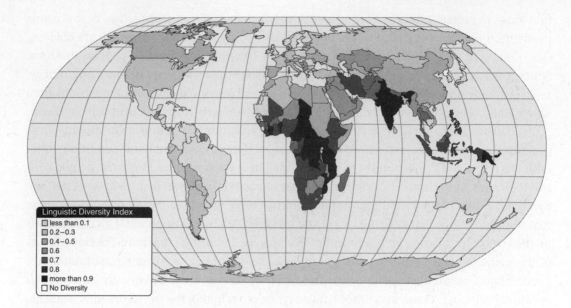

Figure 3.2 Linguistic diversity by country (high to low)

On the broadest scale, all languages belong to a **language family**. A language family is a collection of many languages, all of which came from the same original tongue long ago, but have since evolved different characteristics. Although all languages in a language family have a common origin, two members of the same family may sound very different depending on how long ago the two languages branched off, as well as the historical events that have altered them since.

About 50% of the world's people speak languages belonging to the **Indo-European family**. Languages from this family are spoken on all continents but are dominant in Europe, Russia, North and South America, Australia, and parts of southwestern Asia and India. This language family includes the Germanic and Romance languages, as well as Slavic, Indic, Celtic, and Iranic. Of the world's people, 20% speak languages from the **Sino-Tibetan family**. This language area spreads through most of Southeast Asia and China and comprises Chinese (which has the world's most speakers), Burmese, Tibetan, Japanese, and Korean. The final 30% of the world's population speak languages from the Afro-Asiatic, Niger-Congo, Altaic, or Austronesian language families.

Language families can also be divided into smaller **language groups**. A language group is a set of languages with a relatively recent common origin and many similar characteristics. Spanish and Italian, for example, are both part of the **Romance languages**—they are both derived from Latin, they have many related words, and they contain similar grammatical structures. Diversity also exists within individual languages. **Dialects** are geographically distinct versions of a single language that vary somewhat from the parent form. Italian and English are both languages that contain numerous dialects, reflecting the historical, social, and geographic differences between many diverse peoples. Anyone who has traveled to cities such as London, Toronto, New York, Houston, and Sydney is well aware of distinct variations in the English language. Different dialects may have different terms for the same thing, for example, an English speaker from the American South might call to his friends "y'all," whereas an English speaker from Australia might call them "mates." Dialectical differences are, however, often more easily recognized through differences in accent. Geographical boundary lines where different linguistic features meet are called **isoglosses**.

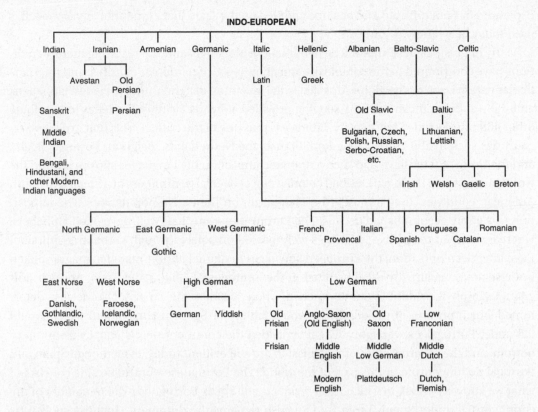

Figure 3.3 The Indo-European language family represents just one of the world's major language families, but its members include many of the languages spoken across the globe today.

Languages are carried over space by the same set of diffusion processes described in Chapter 2. Language diffusion occurs when migration, trade, war, or some other event exposes one group of people to the language of another. When two groups of people with different languages meet, a new language with some characteristics of each may result. This hodgepodge form is called a **pidgin**. If, over time, a pidgin evolves to the point at which it becomes the primary language of the people who speak it, then it is called a **Creole**. Interesting Creole languages have frequently developed in colonial settings where the linguistic traditions of indigenous peoples and colonizers have blended. Multiple other tongues have actually influenced some modern languages. Modern English, for example, contains aspects of half a dozen different languages because the British Isles have seen so many foreign conquerors and visitors over the centuries. The colonial history of West Africa offers a very different example of the ways in which linguistic interactions can affect other aspects of culture and economics. In Ghana, which was colonized by Britain during the late 1800s, many people speak English. In neighboring Togo, where the French exercised colonial control during the same period, most native people now speak French. In this case, patterns of colonial history have had important implications for trade, interpersonal interactions, power relationships, and international politics.

Esperanto is an example of a constructed international auxiliary language. Created by L. L. Zamenhof, Esperanto combines features of a number of different linguistic groups into a single synthetic language. Zamenhof's goal was to create a universal second language that would enable people around the world to communicate easily and effectively. Although

Esperanto has not diffused and become adopted on the scale that Zamenhof envisioned, it is used today in a variety of contexts.

Many linguists believe that the development of alphabets and the resulting literary traditions have contributed to the complexity and dominance of particular cultures, and thus particular nations, across the globe. Most likely the invention of agricultural societies, alphabets, and the resulting efficient record-keeping provided a means for these societies to dominate other illiterate societies more easily. Literacy is thus one of the critical tools that explains why countries such as the United Kingdom, France, the Netherlands, Belgium, Portugal, Spain, and the United States have had such a dramatic impact on the languages spoken around the world today. When these nations had colonial power over large numbers of African and South American countries, they imposed their languages on the native populations. The imposition was easily accomplished because the European nations had well-developed alphabets, whereas many of the native languages were passed on solely through verbal transmission. Even after decolonization, the European languages remained as dominant languages spoken in these areas, as already demonstrated in the examples of Ghana and Togo. Another poignant example is evident in South America, where, in 1494, the Treaty of Tordesillas determined which portion of the continent would fall under Spanish control and which would fall under Portugal's sovereignty. The line divided the continent nearly equally from top to bottom, and the repercussions of this division are still evident today, even though Spain and Portugal no longer rule in this part of the world. The Portuguese-controlled side consists of what we know as Brazil, in which most residents still speak Portuguese; the remainder of the continent fell under Spanish rule, and Spanish remains the dominant language spoken by most of the population.

When people who speak different languages need to communicate quickly and efficiently, a **lingua franca** frequently results. A lingua franca is an extremely simple language that combines aspects of two or more complex languages. For example, in Southeast Asia, where the residents of hundreds of little islands and mountain valleys each speak their own unique language or dialect, simple trade languages are used at ports and central markets. Lingua francas are usually very simple, often lacking fundamental features common to most full-fledged languages, such as verb tense. However, they provide an efficient and easy-to-learn means for diverse people to engage in trade despite the great distances and significant cultural barriers that normally separate them.

Although many countries have established one or more **official languages**, in which all government business occurs, most countries also contain significant linguistic diversity. In the United States, which has no official language, dozens of native tongues are in common usage. In many urban areas, such as Los Angeles, New York, Chicago, and San Francisco, a significant portion of the resident population speaks a first language other than English. A few countries, such as Switzerland and Belgium, have formally recognized their cultural diversity by establishing multiple official languages. In Canada, language and the cultural heritage associated with it has been a source of conflict between secession activists from French-speaking Quebec and the majority English-speaking Canadian population. Canada is an example of a **polyglot**, or multilingual, state.

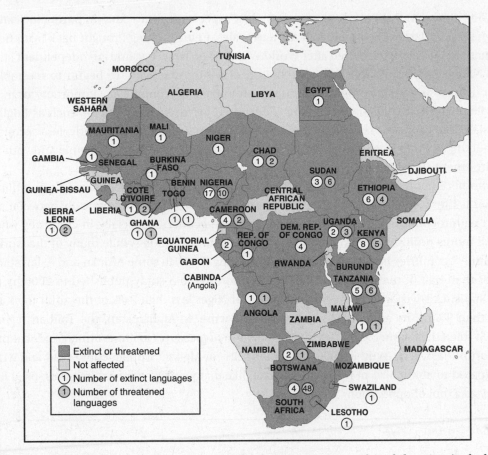

Figure 3.4 Indigenous languages are rapidly dying out across the globe, particularly in formerly colonized countries, as in Africa, where colonial languages wiped out many native tongues.

An important topic in the current scholarship surrounding linguistic diversity is **language extinction**. Language extinction occurs when a language is no longer in use by any living people. Thousands of languages have become extinct over the eons since language first developed, but the process of language extinction has accelerated greatly during the past 300 years. Colonialism in the 18th and 19th centuries and economic globalization in the 20th century have driven many languages to premature extinction. As in the examples of Ghana and Togo, languages such as English and French have replaced dozens of native tongues all over the world. Although languages can be lost through the extinction of an entire people or through linguistic evolution over time, the pressures of economic and social **acculturation** are responsible for most of today's losses. Acculturation refers to the adoption of cultural traits, such as language, by one group under the influence of another. Many of the languages that have been lost over the past few hundred years were spoken in now defunct Native American societies. It is important to note that some languages have also been lost as part of a greater **cultural extinction** in which an entire culture was obliterated by war, disease, acculturation, or a combination of the three. When a culture and its linguistic tradition disappear, it takes with it a tremendous amount of history and knowledge that might never be regained. For instance, language can provide clues to various historical human migration patterns through the study of the assimilation of certain words across differing historical cultural groups.

TIP

The 2007 APHG exam included a free-response question that offered the following observation: Even as English has become the world's dominant lingua franca, minority languages such as Welsh and Basque have experienced a resurgence. Can you explain these two apparently contradictory trends?

Today, movements have begun to revive lost aspects of culture and, in particular, native languages. In parts of Scotland, Ireland, and Wales, Celtic is being brought back from near-extinction; Hebrew was revived after World War II when Israel became an independent state; and Native Americans from Alaska to the tip of South America have begun to reestablish their distinct and unique linguistic heritage. Although it is economically important for many people to speak the languages that are widely used for international trade, such as English, Russian, and Chinese, the world's thousands of other languages all hold priceless secrets to human history and important insights into our relationships with other people and with the environments in which we live.

Another important issue having to do with the geography of language that was mentioned earlier is **literacy**. Literacy, or the ability to read and write, varies dramatically between and even within countries. Literacy also varies between genders, especially in countries where social mores prohibit women from receiving a formal education. While many of the world's wealthiest countries have literacy rates approaching 100%, in some African and Asian states, fewer than half of the population can read. According to a study published in 2000 by the UN Statistics Division, in the African country of Niger less than 24% of the adult men and less than 9% of the adult women can read and write. In Afghanistan, the Taliban regime, which ruled the country from 1996 to 2001, instituted measures to limit women's educational opportunities. As a result, in 2000, only about 22% of Afghani women could read and write, compared to about 52% of men. The Taliban, like many other tyrannical regimes, used illiteracy as a tool of oppression.

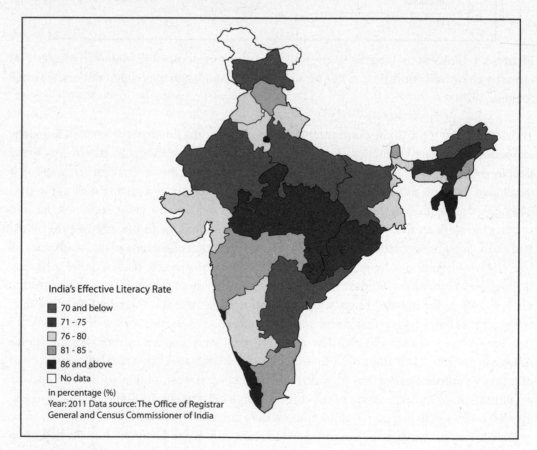

Figure 3.5 Literacy in India. Literacy rates vary among countries, but they also vary within countries. India is an example of a country with widely varying literacy rates by region.

One final aspect of language of interest to cultural geographers is that of how language manifests itself in the landscape. The names different cultures give to various features of the earth, such as settlements, terrain features, streams, and other land features, are called **toponyms** and can reveal interesting aspects of the spatial patterns of different languages and dialects. In the United States, many of the names given to American cities reveal the dominant cultures of their first inhabitants: "New York," "Baton Rouge," and "San Diego" reveal the English, French, and Spanish influence of some of the first settlers in these parts of the country. Many of the states in the United States were named after royalty in the settlers' countries of origin: Georgia for an English King, Louisiana for a French King, and Virginia for the Virgin Queen Elizabeth. Most place names in any culture contain two parts: the generic and the specific. The generic classifies whatever is being described, such as a lake, river, mountain, or street, while the specific term modifies the classification: Lake *Erie*, *Mississippi* River, Mount *Whitney*, *Wall* Street. The various ways different cultures have named the land throughout history can provide insights into historical cultural migration patterns and diffusion processes across the globe.

RELIGION

For many people, religion, more than any other cultural trait, defines who they are and how they understand the world around them. Because religion is tied to all aspects of human culture and social systems, studying the geography of religion can help us understand everything from population growth, to international politics, to the design and structure of cities. For these reasons, religion occupies a central place in the field of cultural geography.

There are many commonalities between the world's many religious traditions. First, all religions share some set of teachings that imply a value system. Second, all religions include some notion of the sacred, whether the sacred be a single divine being, a set of texts, or some powerful symbol. And third, all religions include some ideas about the place of human beings in the universe. Many religions also have a creation story to explain the origins of humans and the physical universe. Some religions also include teachings on law, politics, social mores, sexual relations, physical fitness, cleanliness, eating habits, and even interior decorating!

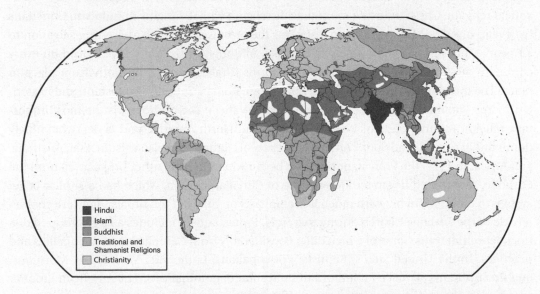

Figure 3.6 The geographic distribution of the world's major religions

TIP

The 2002 APHG exam included a free-response question that asked students to consider how religion shapes cultural landscapes, in terms of sacred sites, place names, burial practices, and architecture.

In thinking about the geography of religion, it is useful to first consider why some religions may have spread far from their hearths, while others have remained primarily local or regional in distribution. One explanation for this is that some religions, through their teachings, seek to unite people from diverse backgrounds, while others seek to ground people in local traditions or landscapes. While Buddhism seeks to explain ultimate realities for all people—such as the nature of suffering and the path toward self-realization—many Native American religions center on local environmental phenomena and use locally occurring plants and animals as religious figures or in religious ceremonies. Religions that seek to unite are called **universalizing religions**, and those that are more spiritually bound to particular regions are called **local religions**. Another useful distinction that may help to explain the global distribution of religions is that some religions are explicitly evangelical, while others are not. **Evangelical religions**, such as Christianity, expand their membership by using **missionaries** to recruit new followers actively. However, some nonevangelical religions, like Buddhism, are also widespread.

Most scholars think of religions as being divided into a few main categories. Some religions are **monotheistic**, meaning that they teach the primacy of a single god, whereas other religions are **polytheistic**, teaching that there are numerous gods or spiritual powers. Christianity, Islam, and Buddhism are **global religions** in the sense that their members are numerous and widespread and that their doctrines might appeal to different people from any region of the globe. **Ethnic religions** tend to appeal to smaller groups of people with a common heritage or to large groups of people living in a single region. Local religions, which were mentioned earlier, are also associated with particular places; tend to attract small, localized followings; and are often invested in the powers of particular living people or local natural phenomena. **Shamanism** is the term for a local religion in which a single person takes on the roles of priest, counselor, and physician and claims a conduit to the supernatural world. **Animism** is another class of local religious traditions, mostly from Africa and the Americas, in which the world is seen as being infused with spiritual and even supernatural powers. Although there are many other ways to classify the world's diverse religions, these distinctions can help us to understand the spatial distribution of religious belief and participation.

Now let's look at the three global religions already mentioned, because they affect the daily lives of so many people. With about 2 billion believers, **Christianity** is the world's most widespread religion. Christianity is a monotheistic religion with its origins in **Judaism**. Christians believe in one God and that his son Jesus was the promised Messiah, delivering salvation to all people, not just the chosen people of Israel. Although Christianity is practiced on every continent and in almost every country, the forms it takes vary significantly from place to place. The three major categories of Christianity are Roman Catholic, Protestant, and Eastern Orthodox. The Roman Catholic Church, based at Vatican City in Rome, is the most important religion in large parts of Western Europe and North America and is overwhelmingly dominant in Central and South America. Pockets of Catholicism also exist in Asia, Australia, and Africa. In 1517, the Protestant tradition began when Martin Luther broke away from the Catholic Church and began a different type of Christian church, which had a similar belief system to Catholicism but with much less emphasis on many of the rituals that were characteristic of the Catholic Church during that time. Protestantism includes a large group of distinct **denominations**, some of which differ considerably from Catholicism in their beliefs and practices. In the United States, Baptists, Episcopalians, Lutherans, Methodists, Mormons, and Presbyterians all make up important Protestant denominations. The American Frontier provided individualistic new settlers an atmosphere in which to freely express whatever type of faith they desired. While certain regionalizations of different denominations exist in America today, such as the Mormons in Utah or the Baptists in the South, in many American cities, you can find a diversity of different worship centers for different types of Protestant faiths. Finally, Eastern Orthodox is dominant only in Eastern Europe and Russia, although its adherents also live in smaller populations throughout the world.

Islam claims about 1 billion members worldwide. Although its distribution is centered in North Africa and the Middle East, Muslims (practitioners of Islam) are found throughout the world, including Europe, Southeast Asia, and the United States. Islam is a monotheistic religion, also stemming from Judaism, which is based on the belief that there is one God, Allah, and that Muhammad was Allah's prophet. Mecca, Saudi Arabia, is the birthplace of Muhammad and serves as the base for the nation of Islam. Observance of the Koran, or word of Allah revealed to Muhammad, along with the observation of the five pillars of the faith unite Muslims across the globe. The five pillars consist of repeated recital of the basic creed; prayers five times daily, facing Mecca; the observance of Ramadan, which is a month of daytime fasting; almsgiving; and, if possible, a **pilgrimage**, or journey, to the holy city of Mecca.

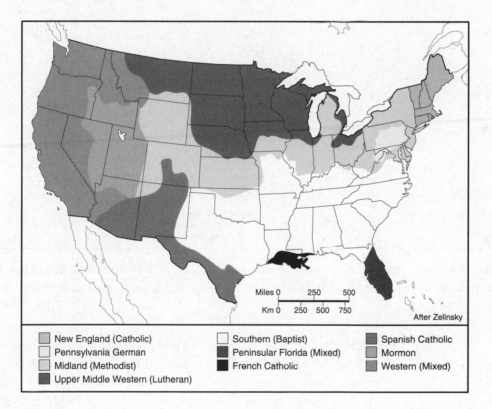

Figure 3.7 The geographic spread of Protestant denominations across the 48 contiguous states

In recent years, an increase in **fundamentalism** has resulted in religious division and conflict in many parts of the world. Islamic fundamentalism is associated with the September 11, 2001, terrorist attacks on the United States. Yet, it is also important to remember that Christian, Jewish, Hindu, and even Buddhist fundamentalism are also prevalent in today's world. Fundamentalism is sometimes associated with conflict between different religions or among different sects, and sometimes results in the emergence of new subgroups within a religious tradition. This has occurred in Mormonism (The Church of Jesus Christ of Latter Day Saints) and in many Christian Protestant denominations.

With more than 300 million adherents worldwide, **Buddhism** is the third great world religion. Buddhism, which originated in the 6th century B.C. in northern India, traces its origins and many of its traditions from Hinduism. Founded by Siddhartha Gautama (or, simply, the Buddha), Buddhism teaches that suffering originates from our attachment to life and to our worldly possessions. According to Buddhism, a state of Nirvana, or ultimate purification and happiness, can be achieved through the process known as the Eightfold Path. Although Buddhism is still centered in its ancient East Asian hearth, it has gained an increasingly large following in Europe and North America since the 1950s. This pattern can be attributed in part to emigration by Asian people to Western nations and in part to Buddhism's teachings, which seem to resonate with many westerners. Nearly half of the Buddhists in the United States live in southern California.

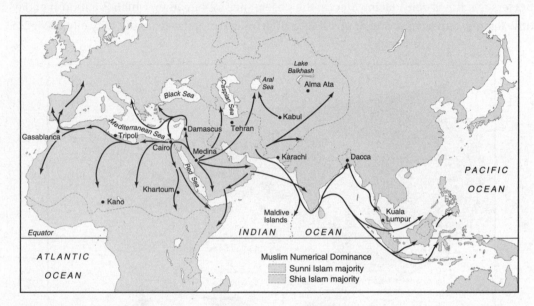

Figure 3.8 Islam originated in Saudi Arabia near Mecca and Medina and diffused originally through expansion diffusion to surrounding areas and then by relocation diffusion to Malaysia, Indonesia.

Although they are not truly global religions, Hinduism and Judaism are important ethnic religions that deserve to be mentioned briefly here. **Hinduism** is a religion closely tied to Indian culture. For over 4,000 years, people living on the Indian subcontinent have developed a cohesive and unique society that integrates their spiritual beliefs with their daily practices and official institutions. One important aspect of the Hindu culture is the **caste system**, which gives every Indian a particular place in the social hierarchy from birth. Each caste defines individuals' occupations along with their social connections, where they can live, the clothes they wear, and the food they eat. Individuals may improve the position they inherit in the caste system in their next life through their actions, or karma. After many lives of good karma, they may be relieved from the cycle of life and achieve salvation and eternal peace through union with the universal soul known as the Brahman. Hindus worship in temples or shrines that can be found in every Hindu village. Additionally, Hindus follow the doctrine of *ahimsa*, as do Buddhists, which instructs them to refrain from harming any living being. Thus, animals are an enduring presence in most Hindu societies.

Judaism was the first major monotheistic religion. It is based on a sense of ethnic identity, and its adherents tend to form tight-knit communities wherever they live. In 1948, after the catastrophe of the Holocaust and almost 2,000 years of existing as ethnic minorities in Christian- and Muslim-dominated countries, the Jewish people finally established their own state in Israel. Today, most Jews live in either Israel or the United States.

Beyond understanding the location and diffusion patterns of the world's major religions, geographers are also interested in how religious traditions manifest themselves in the landscape. The fact that Muslims face the direction of Mecca each time they pray provides an example of a sacred space, which is of special significance to geographers for it shows the geographical implications of a particular belief system. Another example from the Muslim community can be found in Jerusalem, where three of the world's major religions, Judaism, Christianity, and Islam, share, often with hostility, sacred space. Here stands the Wailing Wall, a remnant of the temple of the Jews destroyed by the Romans in A.D. 70, with the Dome of the Rock, where Muslims believe Muhammad rose into heaven, just in front. The significance each religion places on this shared sacred space largely contributes to much of the conflict currently prevalent in this part of the world. Other examples of sacred spaces in the landscape include places of worship, such as temples, synagogues, churches, mosques, or cathedrals. The spatial distribution of these buildings provides geographers with an understanding of the prevalence of particular religions in particular places; they also provide clues about the belief system simply through architectural style. If you have ever driven by a Mormon temple, you probably noted the near-regality of these establishments. It is obvious from the lighting, the gates, the stone, and the detail that these places are sacred houses of worship. In fact, you may not completely enter one of these temples or tabernacles unless you are a Mormon.

One final important topic regarding the geography of religion is how it contributes to various territorial tensions and conflicts across the globe. An extreme level of violence that relates to religion and territory currently plagues the Middle East. The Palestinians, who are predominantly Muslim, are working hard to establish their own state including the territories of the West Bank, Gaza Strip, and East Jerusalem in Israel. However, their requests conflict with the predominantly Jewish Israelites who currently have control over the land. This tension has caused major bloodshed between the two culturally and religiously different groups, particularly in the past few years, although the conflict has existed for almost a century. Another example of current territorial conflict is that of Northern Ireland. This country is under British rule, but a large portion of the Irish-Catholic population wishes to be under the sovereignty of culturally and religiously similar Ireland. This conflict manifests itself most poignantly in the city of Belfast, where a wall separates the two culturally and religiously different groups. In addition, flags from each country are displayed predominantly through the city, murals of Irish nationalists are painted on walls in the Irish-Catholic part of town, and even cemeteries choose the nationalities of those they will hold. As mentioned earlier, religion can be one of the most defining characteristics of a culture, and, as such, it becomes a unifying and sometimes violent force in securing a place where a particular culture can freely celebrate and maintain its belief system.

GENDER

Sex is generally considered a universal biological trait. However, **gender** also includes social constructs and cultural practices that vary across space and time. Since the 1980s, the study of gender has assumed an important place in the discipline of geography. Geographers who study gender may focus on such topics as cultural differences in the expression of gender roles, the relationships between gender and labor, the links between gender and social welfare, and the connections between gender and economic development. Gender studies in geography have usually focused on women's issues. However, other topics—such as homosexuality, masculinity, transgender sexuality, and other gender identity themes—have become more prominent in recent years. Scholarship in all areas of gender geography has been shaped by thinking in the diverse group of social movements known as feminism.

An underlying message in this research is that throughout most of the world, gender remains a crucial factor in determining access to health care, civil rights, welfare, security, and other forms of social and political equality. Another key message is that although gender is important almost everywhere, its role in society varies both historically and geographically. The history of gender roles, for any country or society, is far too complex to recount here. However, it is important to recognize that ideas of gender are flexible, are dynamic, and have changed considerably over time. Tremendous cultural variation still occurs despite inherent biological characteristics, such as the ability to bear children, that would seem to provide some "natural" constraints on gender roles.

One of the most active areas of research in gender geography involves gender and economic development. The United Nations tracks gender-related development indicators. In 2010, the United Nations introduced the **Gender Inequality Index** (UN index), which measures a country's loss of achievement due to gender inequality, based on reproductive health, employment, and general empowerment. Throughout the world, men score higher on gender empowerment than women, particularly in areas such as income and literacy. Women, however, score higher than men in life expectancy. Gender differences tend to be least in wealthy, highly developed areas, such as the Scandinavian countries, while differences tend to be greatest in developing countries. There are, however, exceptions to these patterns. Some primarily Muslim countries score lower in gender empowerment than would be expected given their level of wealth. Some relatively poor countries, such as Botswana, score surprisingly high. It should also be said that during times of acute stress, such as during food shortages and armed conflicts, women and children tend to suffer malnutrition, displacement, and other traumas at disproportionate rates compared with their male counterparts.

Another important aspect of this story is the relationship between women's empowerment and population growth. Chapter 2 discussed the relationship of population growth to economic development. Yet numerous scholars have pointed out that the gender component of economic development is far more important in determining a population's growth rate than other aspects. There is a simple reason for this important insight. Women who experience greater levels of political and economic empowerment—and particularly those who have greater access to health care, education, and employment opportunities—tend to delay child rearing until later in life and have lower birth rates. Some people believe that religion, not women's empowerment, is the key factor in determining fertility. Indeed, some religious groups do tend to have higher growth rates than other groups in the same society. In the United States, Orthodox Jews are one example of a religious group with a relatively high birth rate. Yet globally, economic development and women's empowerment play much more important roles in fertility rates than do religious beliefs. One example comes from Italy, an overwhelmingly Catholic but developed country with relatively high women's empowerment, which has an extremely low natural growth rate. Another example comes from the United States, where fertility rates among Latino immigrant populations tend to decline with succeeding generations as women in those families find more employment opportunities.

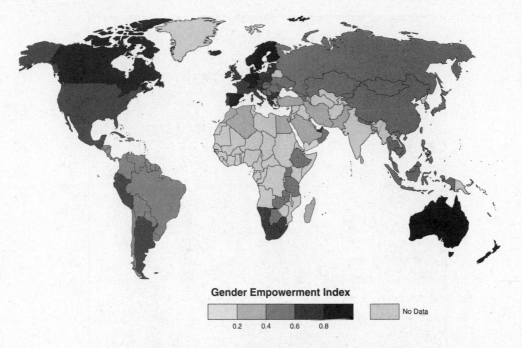

Gender Empowerment Index

0.2 0.4 0.6 0.8

No Data

Figure 3.9 Gender empowerment by country

Employment is yet another major area of gender studies in geography. During the Great Depression of the 1930s, 26 out of 48 states prohibited married women from working outside the home. These laws were, in part, legacies of earlier times when women's employment prospects were even more limited and, in part, an effort to provide greater opportunities for unemployed men. This changed dramatically during World War II when women entered the labor force in record numbers to participate in the war mobilization effort. The 1950s saw another effort to constrain women's opportunities, followed by a new feminist movement in the 1970s. Today, women are receiving college degrees at a rate nearly one and a half times that of men. As a result, they are entering the workforce in greater numbers. Yet, compensation for women still lags behind that of men, and women are wildly underrepresented in top corporate, political, and academic positions. Patterns of gendered labor vary in equally complex ways in other countries throughout the world. Geographers' ability to conduct historical and cross-cultural analyses of these trends enables them to understand large-scale trends, point to examples of injustice, and identify cases where progress is being made.

ETHNICITY

The word **ethnicity** originates from the ancient Greek root *ethnos*, which refers to a unique and cohesive group of people. Currently, the term ethnicity refers to a group of people who share a common *identity*. The term first came into popular usage during the 1940s as an alternative to the term **race**, which had become negatively associated with Hitler's Nazi regime, but the two terms do not mean exactly the same thing. Ethnicity involves more than simply the physical characteristics commonly associated with race—it also involves a person's perceived social and cultural identity. The meaning of the concept continues to be debated, however, since individual people express their common ethnicities in different ways because the lines that divide ethnic groups are almost always blurry, and the cultural traditions that characterize particular ethnic groups are notoriously flexible.

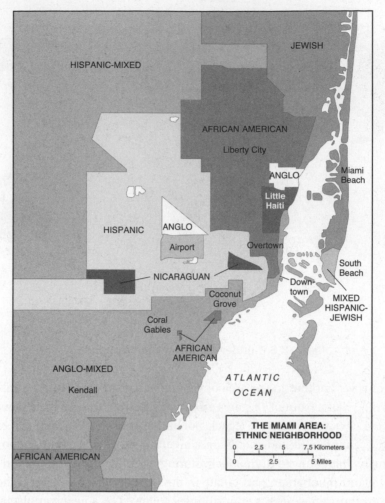

Figure 3.10 Miami, like many large coastal cities in the United States, demonstrates a strong ethnic diversity in the various neighborhoods that make up the urban landscape.

Today, many people place great value on ethnic diversity, realizing that communities are made richer by a variety of perspectives. However, disagreements between people of differing ethnic identities are also at the heart of many social and political conflicts throughout the world. Over 90% of the world's countries contain more than one ethnicity, most countries share borders with people of foreign ethnicities, and many otherwise cohesive ethnic groups have been artificially divided by political boundaries. Places where particularly sharp ethnic boundaries characterize the cultural landscape or where people of various ethnic identities lay claim to the same lands or resources are often marred by political unrest and violence. In recent years, the Middle East, the Balkans of Eastern Europe, eastern Africa, and Kashmir (on the border between India and Pakistan) have all experienced devastating ethnic violence. In the worst cases, this violence has taken the form of **ethnic cleansing**, which is the effort to rid a country or region of everyone of a particular ethnicity either through forced migration, or through **genocide**, which is a premeditated effort to kill everyone from a particular ethnic group.

In the United States, the 2000 census showed in dramatic fashion the extent to which the American cultural landscape has become truly **multicultural**. Hispanic Americans, Asian Americans, and other once **minority** ethnic groups are now part of a culturally diverse, polyglot nation in which a clear ethnic majority simply no longer exists. Geographers are now hard at work making sense of the tremendous amount of cultural data collected through the 2000 census. Perhaps the most striking ethnic pattern to emerge from early analyses of the 2000 census is that the population of Hispanic Americans, particularly Latinos of Mexican origin, has increased dramatically during the past ten years. Of particular interest was the discovery that Latino populations have increased not only in areas traditionally associated with Hispanic American culture—such as California, Texas, Florida, New York, and Illinois— but also in large cities and rural regions across the country. The relative openness of the US borders during the 1990s, combined with economic globalization, has literally changed the face of America. The American populace looks very different now.

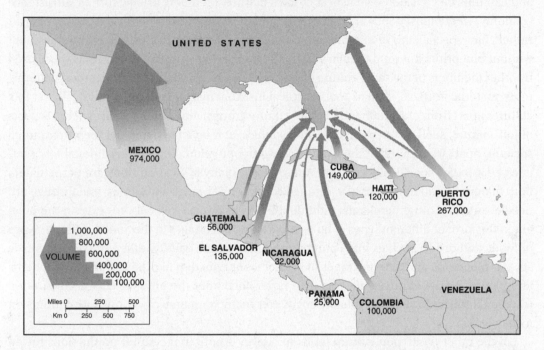

Figure 3.11 This map depicts legal immigration into the United States from 1981 to 1990. As shown, the majority of immigrants from Central and South America come through Mexico.

American cities are the best representatives of this diversity of ethnicities. The process of migration, particularly chain migration (discussed in Chapter 2), makes America's ethnic mosaic possible. Many American cities display their ethnic diversity in **ethnic neighborhoods**, or concentrations of people from the same ethnicity in certain pockets of the city. Common examples are the various "Chinatowns" that exist in cities such as New York, Chicago, and San Francisco. These clusterings result from friends and relatives who have immigrated to the United States, encouraging friends and relatives back home to join them where opportunities or freedom may be more abundant. Unfortunately, sometimes ethnic groups are essentially forced to live in certain segregated parts of the city. These ethnic neighborhoods are called **ghettos**, and their locations tend to be some of the least desirable within the city.

The experiences of people who come from a common ethnic background but who live in different regions or ethnic neighborhoods is called **diaspora**. This term is often used to refer to Jews or to blacks of African descent, who maintain aspects of their common heritage despite living in diverse communities throughout the world. Although blacks in the United States may be integrated into American culture, many also identify with a common

African heritage, and this might be illustrated through music, food, or religious traditions that allow these individuals to celebrate and maintain their common heritage outside of their native culture region.

POPULAR CULTURE

Folk and popular culture are also active areas of geographic research. **Folk culture** refers to a constellation of cultural practices that form the sights, smells, sounds, and rituals of everyday existence in the traditional societies in which they develop. A folk culture is usually rural, with strong family ties and strong interpersonal relationships leading to a cohesive group identity, and can thus be considered an aspect of **local culture**. Members usually form a subsistence economy, where most goods are handmade, and most individuals perform a variety of tasks rather than specializing in any one area. Buildings representative of folk culture are built without blueprints but tend to follow a similar plan and use similar materials as those used by other members of the same culture group. Along with their distinct architectural patterns, other material artifacts, such as tools, musical instruments, and clothing, physically set folk cultures apart from one another and other culture groups. Additionally, nonmaterial aspects of folk culture, such as songs, stories, philosophies, and belief systems, set these traditional societies apart from much of the world's current population. Very few traditional folk societies exist today, specifically in North America, but many of the traditions are perpetuated, both materially and nonmaterially, through collections of songs and stories, and through art, needlework, and other handcrafts. Additionally, relics of past folk cultures exist in the present in the form of different types of houses like shotgun cottages in the South, different types of foods and drinks such as hush puppies and moonshine whisky, different types of music such as bluegrass, and different kinds of medicines or remedies like the use of different herbs and plants. Consequently, while you may never encounter the physical establishment of a traditional folk culture, you probably encounter many remnants of various folk cultures on a daily basis.

On the other hand, **pop culture** tends to convey a notion of cultural productions fueled by mass media and consumerism. Included in this are the visual and performing arts (e.g., painting, sculpture, and dance), the culinary arts, architecture and city planning, music, fashion, sports, leisure activities, and other forms of entertainment. Unlike folk culture, pop culture does not reflect the local environment; it looks virtually the same anywhere it appears. Generally, elements of folk culture vary dramatically from place to place but do not change much over time. Conversely, pop culture is relatively uniform across space but rapidly changes over time as conveyed by terms such as "fad" or "trend" commonly used in pop culture lingo.

Artifacts of popular culture are those things that can be produced, transmitted, and accepted virtually anywhere on the earth's surface. They include music, food, entertainment, fashion, recreation, and various forms of art. Popular culture is easily diffused across national boundaries, primarily through advertising and now through the internet, enabling individuals all across the globe access to Big Macs, Levi's jeans, Taylor Swift, and other various exports of American pop culture. The increasingly globalized world system allows for the rapid diffusion and acceptance of elements of pop culture. In some cases, this process has led to increased access and improved economic and educational opportunities. However, in many instances, the invasion of Americanized pop culture has been seen by many as just another example of **cultural imperialism**, causing people to lose their traditional ways of life in favor of cheap entertainment and disposable goods. In some less-developed countries, cultural globalization associated with the introduction of fast food restaurants, automobiles, and other emblems of American economic power has resulted in social and environmental problems.

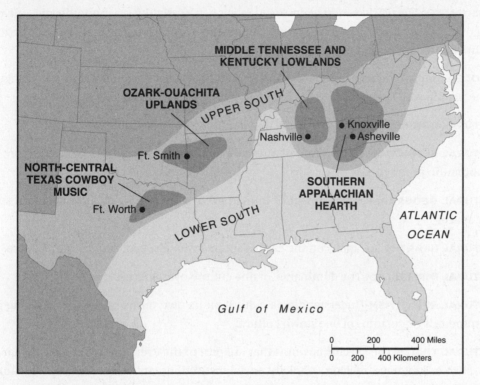

Figure 3.12 Country music has its origins in folk music that began in the southeastern United States, with different regions producing different types of music.

Language, religion, ethnicity, and popular culture are just some of the subjects that researchers study within the field of cultural geography. By mapping the distribution and diffusion of cultural traits, geographers can gain both a broad understanding of the ways in which aspects of culture are expressed over space and a deeper understanding of the relationship between various cultural groups. In addition, geographers can analyze the ways cultures have interacted with the environment, resulting in the creation of unique cultural landscapes.

 KEY TERMS

ACCULTURATION The adoption of cultural traits, such as language, by one group under the influence of another.

ANIMISM Most prevalent in Africa and the Americas, doctrine in which the world is seen as being infused with spiritual and even supernatural powers.

ARTIFACT Any item that represents a material aspect of culture.

BUDDHISM System of belief that seeks to explain ultimate realities for all people—such as the nature of suffering and the path toward self-realization.

CASTE SYSTEM System in India that gives every Indian a particular place in the social hierarchy from birth. Individuals may improve the position they inherit in the caste system in their next life through their actions, or karma. After many lives of good karma, they may be relieved from the cycle of life and win their place in heaven.

CHRISTIANITY The world's most widespread religion. Christianity is a monotheistic, universal religion that uses missionaries to expand its members worldwide. The three major categories of Christianity are Roman Catholic, Protestant, and Eastern Orthodox.

CREOLE A pidgin language that evolves to the point at which it becomes the primary language of the people who speak it.

CULTURAL COMPLEX The group of traits that define a particular culture.

CULTURAL EXTINCTION Obliteration of an entire culture by war, disease, acculturation, or a combination of the three.

CULTURAL GEOGRAPHY The subfield of human geography that looks at how cultures vary over space.

CULTURAL HEARTHS Locations on the earth's surface where specific cultures first arose.

CULTURAL IMPERIALISM The dominance of one culture over another.

CULTURAL RELATIVISM Understanding a culture on its own terms rather than judging it by the standards or customs of one's own culture.

CULTURAL TRAITS The specific customs that are part of the everyday life of a particular culture, such as language, religion, ethnicity, social institutions, and aspects of popular culture.

CULTURE A total way of life held in common by a group of people, including learned features such as language, ideology, behavior, technology, and government.

CUSTOMS Practices followed by the people of a particular cultural group.

DENOMINATION A particular religious group, usually associated with differing Protestant belief systems.

DIALECTS Geographically distinct versions of a single language that vary somewhat from the parent form.

DIASPORA People who come from a common ethnic background but who live in different regions outside of the home of their ethnicity.

ECUMENE The proportion of the earth inhabited by humans.

ENVIRONMENTAL DETERMINISM A doctrine that claims that cultural traits are formed and controlled by environmental conditions.

ESPERANTO A constructed international auxiliary language incorporating aspects of numerous linguistic traditions to create a universal means of communication.

ETHNIC CLEANSING The systematic attempt to remove all people of a particular ethnicity from a country or region either by forced migration or genocide.

ETHNIC NEIGHBORHOOD An area within a city containing members of the same ethnic background.

ETHNIC RELIGION Religion that is identified with a particular ethnic or tribal group and that does not seek new converts.

ETHNICITY Refers to a group of people who share a common identity.

ETHNOCENTRISM An evaluation of other cultures according to preconceptions of one's own cultural standards and traditions.

EVANGELICAL RELIGION Religion in which an effort is made to spread a particular belief system.

FOLK CULTURE Refers to a constellation of cultural practices that form the sights, smells, sounds, and rituals of everyday existence in the traditional societies in which they developed.

FUNDAMENTALISM The strict adherence to a particular doctrine.

GENDER INEQUALITY INDEX A United Nations index, introduced in 2010, which measures a country's loss of achievement due to gender inequality, based on reproductive health, employment, and general empowerment.

GENOCIDE A premeditated effort to kill everyone from a particular ethnic group.

GHETTO A segregated ethnic area within a city.

GLOBAL RELIGION Religion in which members are numerous and widespread and whose doctrines might appeal to different people from any region of the globe.

HINDUISM A cohesive and unique society, most prevalent in India, that integrates spiritual beliefs with daily practices and official institutions such as the caste system.

INDO-EUROPEAN Language family containing the Germanic and Romance languages that includes languages spoken by about 50% of the world's people.

ISLAM A monotheistic religion based on the belief that there is one God, Allah, and that Muhammad was Allah's prophet. Islam is based in the ancient city of Mecca, Saudi Arabia, the birthplace of Muhammad.

ISOGLOSSES Geographical boundary lines where different linguistic features meet.

JUDAISM The first major monotheistic religion. It is based on a sense of ethnic identity, and its adherents tend to form tight-knit communities wherever they live.

LANGUAGE EXTINCTION This occurs when a language is no longer in use by any living people. Thousands of languages have become extinct over the eons since language first developed, but the process of language extinction has accelerated greatly during the past 300 years.

LANGUAGE FAMILY A collection of many languages, all of which came from the same original tongue long ago, that have since evolved different characteristics.

LANGUAGE GROUP A set of languages with a relatively recent common origin and many similar characteristics.

LINGUA FRANCA An extremely simple language that combines aspects of two or more other, more-complex languages usually used for quick and efficient communication.

LITERACY The ability to read and write.

LOCAL CULTURE A set of common experiences or customs that shapes the identity of a place and the people who live there. Local cultures are often the subjects of preservation or economic development efforts.

LOCAL RELIGIONS Religions that are spiritually bound to particular regions.

MINORITY A racial or ethnic group smaller than and differing from the majority race or ethnicity in a particular area or region.

MISSIONARY A person of a particular faith who travels in order to recruit new members into the faith represented.

MONOTHEISM The worship of only one god.

MULTICULTURAL Having to do with many cultures.

OFFICIAL LANGUAGE Language in which all government business occurs in a country.

PIDGIN Language that may develop when two groups of people with different languages meet. The pidgin has some characteristics of each language.

PILGRIMAGE A journey to a place of religious importance.

POLYGLOT A multilingual state.

POLYTHEISM The worship of more than one god.

POP CULTURE (OR POPULAR CULTURE) Dynamic culture based in large, heterogeneous societies permitting considerable individualism, innovation, and change; having a money-based economy, division of labor into professions, secular institutions of control, and weak interpersonal ties; and producing and consuming machine-made goods.

RACE A group of human beings distinguished by physical traits, blood types, genetic code patterns, or genetically inherited characteristics.

ROMANCE LANGUAGES Any of the languages derived from Latin, including Italian, Spanish, French, and Romanian.

SHAMAN The single person who takes on the roles of priest, counselor, and physician and acts as a conduit to the supernatural world in a shamanist culture.

SINO-TIBETAN Language area that spreads through most of Southeast Asia and China and comprises Chinese, Burmese, Tibetan, Japanese, and Korean.

STIMULUS DIFFUSION When a specific cultural element is diffused to another culture that gives it a new and unique form.

SYNCRETIC Traditions that borrow from both the past and present.

TOPONYMS Place names given to certain features on the land, such as settlements, terrain features, and streams.

TRADITION A cohesive collection of customs within a cultural group.

TRANSCULTURATION The expansion of cultural traits through diffusion, adoption, and other related processes.

UNIVERSALIZING RELIGION Religion that seeks to unite people from all over the globe.

CHAPTER SUMMARY

Cultural geography is the study of how cultures vary over space. Cultural geographers also study the ways in which cultures interact with their environments. Possibilism, the notion that humans are the primary architects of culture and yet are limited somewhat by their environmental surroundings, is now a dominant paradigm in the field. Geographers study a wide diversity of cultural traits, including language, religion, and ethnicity. Geographers also study the everyday aspects of people's lives, such as folk traditions and popular culture, in order to better understand the many ways that diverse people make sense of a rapidly changing world.

PRACTICE QUESTIONS AND ANSWERS

Cultural Basics

MULTIPLE-CHOICE QUESTIONS

1. Cultural geography is the study of

    (A) global customs and artifacts.
    (B) cultural complexes.
    (C) the spatial distribution of cultural traits.
    (D) human-environment relationships.
    (E) how cultures change through time.

2. Throughout history, numerous colonial powers have argued that certain types of people, living in certain areas of the world, are less able to govern themselves because of the qualities they have developed due to their interactions with natural factors, such as climate. This is an example of

    (A) environmental determinism.
    (B) cultural ecology.
    (C) possibilism.
    (D) ecumenism.
    (E) positivism.

3. Cultural traditions, such as Christmas, are _____ since they borrow from the past and are continually reinvented in the present.

    (A) erratic
    (B) inauthentic
    (C) complex
    (D) syncretic
    (E) ecumenical

4. The cultural hearth of Christianity is in

    (A) New York.
    (B) Rome.
    (C) Israel.
    (D) South Carolina.
    (E) Turkey.

5. Wooden shoes characteristic of the Dutch culture are an example of a(n)

    (A) mentifact.
    (B) artifact.
    (C) custom.
    (D) syncretism.
    (E) complex.

FREE-RESPONSE QUESTION

1. Consider the impacts of colonialism on the world's cultural geography. Explain how colonialism affected global patterns of language and religion, using specific examples to support your argument.

Language

MULTIPLE-CHOICE QUESTIONS

1. The most widespread language family on Earth is the

    (A) Sino-Tibetan.
    (B) Romance.
    (C) Germanic.
    (D) Indo-European.
    (E) Mandarin Chinese.

2. People in London, Melbourne, Vancouver, and Mumbai all speak

    (A) a pidgin language.
    (B) lingua francas.
    (C) different dialects.
    (D) official languages.
    (E) different Creoles.

3. Acculturation is a common cause of

    (A) illiteracy.
    (B) language extinction.
    (C) assimilation.
    (D) creolization.
    (E) cultural diffusion.

4. A simple trade language is called a

    (A) lingua franca.

    (B) pidgin.

    (C) dialect.

    (D) Creole.

    (E) syncretic.

5. Literacy rates vary by

    (A) sex.

    (B) location.

    (C) education.

    (D) economic development.

    (E) All of the above

FREE-RESPONSE QUESTION

1. Language extinction, both currently and throughout history, has been a major concern for cultural geographers, linguists, anthropologists, and other academics.

    (A) What are some of the causes of language extinction?

    (B) What kind of repercussions exist as a result of the loss of linguistic diversity?

    (C) Discuss some current trends to revive endangered or extinct languages around the world.

Religion

MULTIPLE-CHOICE QUESTIONS

1. All evangelical religions are also

    (A) local religions.

    (B) universal religions.

    (C) animist religions.

    (D) ethnic religions.

    (E) polytheistic religions.

2. Local Native American and African religions that teach a belief in a natural world full of spiritual beings and supernatural powers are often referred to as

    (A) animist.

    (B) shamanistic.

    (C) missionary.

    (D) denominational.

    (E) local religions.

3. The world's most widespread religion is

   (A) Islam.
   (B) Animism.
   (C) Christianity.
   (D) Hinduism.
   (E) Buddhism.

4. The hearth and spiritual center of Islam is at

   (A) Baghdad.
   (B) Cairo.
   (C) Jakarta.
   (D) Mecca.
   (E) Jerusalem.

5. _____ is an excellent example of a nonevangelical, universalizing religion.

   (A) Christianity
   (B) Buddhism
   (C) Protestantism
   (D) Polytheism
   (E) Hinduism

6. In _____ religions, community, common history, and social relations are inextricably intertwined with spiritual beliefs.

   (A) monotheistic
   (B) local
   (C) evangelical
   (D) ethnic
   (E) universal

FREE-RESPONSE QUESTION

1. Religions exhibit several different patterns for diffusing across the globe.

   (A) Describe the major diffusion mechanisms for two dominant world religions: Buddhism and Christianity.
   (B) How has the difference between how these two religions are spread affected the current distribution of these two religious traditions?

Ethnicity and Popular Culture

MULTIPLE-CHOICE QUESTIONS

1. An ethnicity is defined as

   (A) a group of people with a common history.
   (B) a group of people with similar physical characteristics.
   (C) a group of people who share a common identity.
   (D) a group of people united against a common enemy.
   (E) a group of people with a similar religion.

2. In the 1990s, the United States

   (A) became less ethnically diverse.
   (B) decreased in overall population.
   (C) saw few changes in its ethnic composition.
   (D) saw dramatic changes in its ethnic composition.
   (E) remained relatively homogenous in its ethnic makeup.

3. A group of people, all of the same ethnicity, live in the same area of a city near a nuclear waste facility. This is an example of a(n)

   (A) diaspora.
   (B) ghetto.
   (C) cultural landscape.
   (D) ethnic neighborhood.
   (E) gentrified neighborhood.

4. Which is the most characteristic statement of a folk culture?

   (A) They look virtually the same anywhere on the globe.
   (B) Individuals within the culture specialize in producing specific goods for the community.
   (C) They quickly adopt new techniques useful for their community.
   (D) They have a subsistence economy.
   (E) They have weak ties to friends and family.

FREE-RESPONSE QUESTION

1. Geographers use the term "cultural imperialism" to describe a trend especially dominant in our current global society.

   (A) What is cultural imperialism?
   (B) Describe three global effects of the spread of Western popular culture to the rest of the world.

Answers for Multiple-Choice Questions

CULTURAL BASICS

1. **(C)** Cultural geographers do study customs, artifacts, cultural complexes, and human-environment relationships; however, what makes cultural geography different from other disciplines, like anthropology, is its focus on the spatial distribution and diffusion of human cultures.

2. **(A)** Environmental determinism is the notion that human traits or historical events are directly attributable to environmental factors, that people's actions are determined by environmental factors. This idea has been discredited as simplistic and racist. Possibilism, the notion that humans have agency and yet are limited somewhat by the environmental surroundings, provides an attractive alternative to environmental determinism.

3. **(D)** The term "syncretic" refers to something, such as a cultural tradition, that borrows from multiple sources.

4. **(C)** Rome is the center and headquarters of Catholicism, which is the largest wing of Christianity. However, the hearth, or birthplace, of Christianity is in Israel, where Jesus was born and lived his life.

5. **(B)** Artifacts are the material aspects of a particular culture and would include such things as wooden shoes or other fashion apparel, along with artwork or tools.

LANGUAGE

1. **(D)** The Indo-European family includes the Romance and Germanic groups. About 50% of the world's people speak Indo-European languages.

2. **(C)** Although many people in all of these cities speak English, the versions of English that they speak all vary somewhat in pronunciation, spelling, and other characteristics.

3. **(B)** Literally thousands of languages are currently in danger of becoming extinct. Reasons for language extinctions include genocide, cultural collapse, and acculturation.

4. **(A)** Simple trade languages are called lingua francas and use terms developed and understood by both cultures to complete economic transactions.

5. **(E)** Sex, geographic location, education, and economic development are all factors affecting literacy rates. In many countries where women are prevented from attaining education, women's literacy rates are considerably lower then men's.

RELIGION

1. **(B)** A universal religion is one that seeks to unite people from different backgrounds under one all-encompassing faith. Christianity is the best example of an evangelical, universalizing religion.

2. **(A)** Many animist religions also include a shamanistic aspect. However, shamanism itself refers specifically to beliefs in which a single person takes on supernatural and healing powers.

3. **(C)** With about 2 billion believers, Christianity is the most widespread world religion. Islam is the second largest, and Buddhism is the third.

4. **(D)** Mecca is the hearth and holy city of Islam. Many Muslims face Mecca and pray several times each day.

5. **(B)** Buddhism teaches beliefs about the nature of life and human suffering that are universally applicable, yet its adherents generally do not attempt to recruit followers.

6. **(D)** In ethnic religions, culture, history, public life, and spiritual beliefs are interwoven. Examples of ethnic religions include Judaism and Hinduism.

ETHNICITY AND POPULAR CULTURE

1. **(C)** Whereas race connotes common physical characteristics, ethnicity connotes a common identity. Because people's outward traits do not necessarily say anything about their personal identities, and because the notion of race is associated with prejudice and superficiality, ethnicity has largely replaced it as a way of grouping people.

2. **(D)** During the 1990s, large-scale immigration from Asia and Latin America dramatically changed the ethnic composition of the United States, making it a truly polyglot nation. In some areas, whites are no longer a majority.

3. **(B)** A ghetto is a form of an ethnic neighborhood where individuals of a particular ethnicity are essentially forced to live. They usually exist in areas of a city where most individuals would rather not live, such as near a nuclear waste facility.

4. **(D)** Folk cultures vary significantly over space as opposed to pop culture, which looks similar everywhere you encounter it. Also, pop culture is characterized by a consumer economy, whereas folk cultures practice a subsistence economy; individuals usually do not specialize in any one activity but instead provide multiple goods and skills for the community.

Answers for Free-Response Questions

CULTURAL BASICS

1. **Main points:**
   - Colonialism has had dramatic impacts on the cultural geography of language and religion.
   - Acculturation under colonial rule has led to the disappearance of hundreds of indigenous languages and has created dominant world languages, such as French, Spanish, English, Russian, and Chinese, that are now spoken across the globe. In many of the former colonies of Africa and Latin America, a few European languages have largely replaced diverse native tongues.
   - Christianity, in particular, has benefited from colonialism. In many formerly colonized regions of the world, local, animist religions have been either replaced or reconfigured to accommodate Christian beliefs. Christianity is now the dominant religion throughout the Americas.

- In general, colonialism has led to the homogenization of linguistic and religious geography. However, interesting and diverse new pidgin languages, in places like Southeast Asia and the Caribbean, and hybrid religions, in the American Southwest, have also resulted.

LANGUAGE

1. **Main points:**
   - Language extinction can occur as a result of a variety of different factors. The most common cause is that of colonialism. European powers took control over numerous countries in Africa and South America during the colonization era and imposed their languages on these Native American and native African societies. Many of these societies had well-developed languages but did not have established alphabets; thus, the obliteration of their languages was relatively easy. Other causes of language extinction include diseases that wipe out entire populations and acculturation, which is when one culture dominates another culture and the dominating culture's language prevails.
   - Language provides many insights and helpful clues for understanding both historic cultures and historic migration patterns. When a language becomes extinct, the world essentially loses the means to learn about an entire culture and whatever that culture had to offer the world. It also means the loss of important clues for understanding historic migration patterns.
   - Today, in parts of Scotland, Ireland, and Wales, Celtic is being brought back from near-extinction. Hebrew was revived after World War II when Israel became an independent state, and Native Americans from Alaska to the tip of South America have begun to reestablish their distinct and unique linguistic heritage. The revival of these languages provides these cultures with a means to reestablish a very important aspect of their cultural identities.

RELIGION

1. **Main points:**
   - Christianity is an evangelical religion, meaning that its practitioners have a mandate to spread the gospel. Buddhists have no such mandate.
   - Christians have been extremely successful at spreading their beliefs. Christianity is now the most widespread world religion, with practitioners in every corner of the globe, particularly in the former mission lands and colonized regions of North and South America.
   - Christian missionaries spread the gospel through relocation diffusion; they purposefully moved to new regions of the world to convert native peoples.
   - Until relatively recently, Buddhism was mostly limited to central and eastern Asia. Historically, Buddhism has spread through contagious diffusion, meaning that it was passed on to people because of their proximity to other practitioners.
   - In recent years, Buddhism has gained a foothold with Americans and Europeans, due to their increasing exposure to Eastern cultures and to Buddhism's peaceful and individualistic teachings.

ETHNICITY AND POPULAR CULTURE

1. **Main points:**
   - Cultural imperialism is dominance by one culture over another. For example, American fast food chains, pop music, and films have infiltrated other countries across the world. Traditional British, French, Japanese, Spanish, and Russian cultures have also been widely disseminated.
   - Although many people in these places enjoy their access to Western popular culture, others claim that new ways of life are diluting traditional cultural practices and social systems.
   - Cultural extinction is one potential consequence of imperialism. In cultural extinction, traditional ways of life are lost as new, dominant ways are adopted. Linguistic diversity, in particular, is affected by cultural imperialism, as more and more people abandon native tongues in favor of widespread world languages like English and French.

ADDITIONAL RESOURCES

Conzen, M. (ed.). 2010. *The Making of the American Landscape, 2nd Edition.* Boston, Massachusetts: Unwin Hyman.

This text presents a comprehensive view of the cultural evolution of the American landscape. Written by a team of leading scholars, each essay examines how historical forces of human settlement have shaped the land over the past 10,000 years, focusing most on the past three centuries. With meticulous illustrations, the authors show the reader how to analyze the historical transformations in today's landscapes. They investigate ethnic and cultural movements along with environmental challenges and urbanization trends as the dominant forces behind the shaping of America's landscapes.

Lane, Belden C. 2001. *Landscapes of the Sacred: Geography and Narrative in American Spirituality.* Baltimore, Maryland: Johns Hopkins University Press.

Lane explores the connections between spirituality and place evidenced in Native American, early French and Spanish, Puritan New England, and Catholic worker traditions. He also addresses how to analyze the landscape to understand the symbol-making processes of religious tradition and experience.

Mitchell, D. 2000. *Cultural Geography: A Critical Introduction.* Malden, Massachusetts: Blackwell Publishers.

Mitchell takes a less traditional approach to cultural geography. Instead of focusing on the geography of language, religion, and ethnicity (as was done in this chapter), he looks at cultural change in everyday settings, specifically examining cultural politics. The book is divided into three parts: first, Mitchell discusses the development of cultural geography and examines cultural theory both within the discipline of geography and other disciplines. Second, he explores the landscape, which is the fundamental unit of analysis for the cultural geographer, and what it means and how to understand its production and use. Finally, he explores different aspects of cultural politics by discussing aspects of control and resistance within and across different cultural groups.

Rayburn, Alan. 2001. *Naming Canada: Stories about Place Names from Canadian Geographic.* Toronto, Canada: University of Toronto Press.

This book contains a compilation of 61 articles of a few pages each, published in *Canadian Geographic* from December 1983/January 1984 to November/December 1993. Each article explores such aspects of toponymy as the name Canada itself, names and pronunciations of local places, war commemorations, native names, the borderless land of Acadia, Spanish and Portuguese names, and the trail of names left by the Mackenzie expeditions.

Standage, Tom. 2006. *A History of the World in Six Glasses.* New York: Walker Publishing Company.

Political Patterns and Processes

4

IN THIS CHAPTER

→ WHAT IS POLITICAL GEOGRAPHY?

→ THE GEOGRAPHY OF LOCAL AND REGIONAL POLITICS

→ TERRITORY, BORDERS, AND THE GEOGRAPHY OF NATIONS

→ INTERNATIONAL POLITICAL GEOGRAPHY

→ SPATIAL CONFLICT

 Key Terms

Antecedent boundary	Geometric boundaries	Physical boundaries
Balkanization	Geopolitics	Political geography
Buffer state	Gerrymandering	Popular vote
Centrifugal forces	Heartland theory	Prorupted state
Centripetal forces	Imperialism	Reapportionment
Colonialism	International organization	Rectangular state
Commonwealth of	Irredentism	Redistricting
Independent States	Landlocked state	Relic boundaries
Compact state	Law of the sea	Rimland theory
Confederation	Lebensraum	Self-determination
Democratization	Microstate	Shatterbelt
Devolution	Nation	State
Domino theory	Nationalism	Stateless nation
East/west divide	Nation-state	States' rights
Electoral College	North American Free	Subsequent boundary
Electoral vote	Trade Agreement	Superimposed boundary
Elongated state	North Atlantic Treaty	Supranational organization
Enclaves	Organization	Territorial dispute
European Union	North/south divide	Territorial organization
Exclaves	Organic theory	Theocracy
Federalism	Organization of Petroleum	Unitary state
Fragmented state	Exporting Countries	United Nations
Frontier	Perforated state	

WHAT IS POLITICAL GEOGRAPHY?

Political geography is one of the oldest fields in the discipline of human geography. Political geographers use the spatial perspective to study political systems at all geographic scales, from local governments to international political systems. According to the geographers Paul Knox and Sallie Marston, political geography can be considered within the context of two complementary perspectives. The first perspective focuses on the impacts of economic, cultural, and physical geography on political systems. For example, some Middle Eastern governments owe their organizational structures to the teachings of Islam. Governments controlled through divine guidance or religious leadership such as these are called **theocracies**, and they provide an excellent illustration of the impact of culture on politics. The second perspective flips this around and views political systems as the driving force behind different countries' economic and cultural systems. In Kashmir, a region that lies on the border between India and Pakistan, many Muslims, who have more in common with their Pakistani neighbors, live on the Indian side of the border, while many Hindus, who have more in common with their Indian neighbors, live in Pakistan. The official border between these two countries, drawn by a political agreement in 1972, has been disputed for many years, and outbreaks of violence occur regularly in the region. This example shows how political structures can have important implications for culture. In recent years, political geographers contributed many important insights to the study of issues involving poverty, war, culture, ethnicity, and environmental change.

STRATEGY

Some nations exist as communal identities without the benefit of an internationally recognized, cartographically defined state, such as the Palestinians, Kurds, and Quebecois. They are called "stateless nations." Can you think of other examples? What are some problems associated with stateless nations?

THE GEOGRAPHY OF LOCAL AND REGIONAL POLITICS

Most political geographers focus their studies on one or more of the following geographic scales: local politics, national politics, and international politics. The country is the fundamental unit of political geography. A country, more formally called a **state**, may be composed of more than one nation. A **nation** consists of a group of people with a common political identity, but every nation does not have its own state. For example, Palestinians in the Middle East often consider themselves to be a **stateless nation** because they have been fighting to establish their own state for decades but have yet to achieve their goal. Israel, however, has achieved statehood designated by an internationally recognized government and territorial borders. The term **nation-state** refers to a geographically defined sovereign state composed of citizens with a common heritage, identity, and set of political goals. This situation rarely (if ever) exists in the modern real world.

It is interesting to note that our current notion of the nation-state is, itself, a relatively new idea. Our modern concept of the nation-state arose in 18th-century Europe, demonstrated for the first time in the French and American Revolutions. The modern nation-state differs from older political ideologies in that the citizens of a modern state are members of a country composed of people and their institutions, not subjects to a king or queen. Unfortunately, many states created over the past 200 years contain political boundaries left over from a colonial system that failed to recognize preexisting ethnic and religious boundaries. Because of this, many of today's states must constantly represent themselves as unified nations, despite the tremendous historic and cultural divides they contain. For countries like Afghanistan, Indonesia, and Rwanda, this task has proven extremely difficult, and internal ethnic conflict is an ongoing problem.

STRATEGY

Many groups of people (nations) around the world have engaged in battles to increase their sovereignty. These groups include indigenous peoples, such as Native American first-nations groups, which are subject to the laws of the US federal government, but also have some degree of autonomy. Can you think of some of the benefits and drawbacks of increased sovereignty for indigenous peoples living under the authority of larger states?

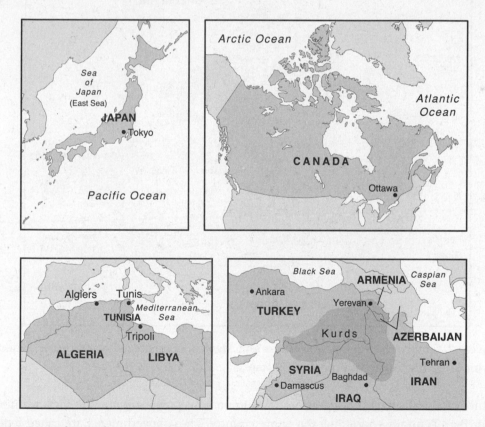

Figure 4.1 Relationships among states and nations. Japan contains a relatively uniform nation within state boundaries; Canada exemplifies a multinational state with two official languages; the Arab nation extends across many states in northern Africa and the Middle East; and the Kurds have no state they can claim as their own—thus exemplifying a stateless nation.

Most countries are organized into a geographically based hierarchy of local government agencies. This division of land, or **territorial organization**, into more easily governable units serves several important functions. First, territorially organized governments have a basis for efficiently delegating administrative functions in what may otherwise be a large and unwieldy area. Second, territorially organized governments can allocate resources through local agencies that may be more in touch with the needs of the people under their jurisdiction. Third, national governments organized by territory usually give their local territories some degree of autonomy, such as the ability to enact laws, police their lands, and tax local citizens. However, the degree to which power is distributed between local and national agencies causes much political debate. In the United States, the issue of **states' rights** arose in the early days of the republic, divided the country during the Civil War, resurfaced during the Civil Rights Movement, and even today surrounds issues of environmental regulation and management of natural resources.

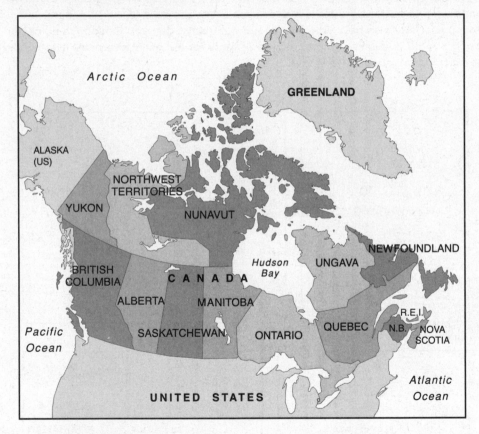

Figure 4.2 Canada, like the United States and Mexico, is organized by territory. Canada's provinces are the rough equivalent of US states.

Under **federalism**, governments bestow autonomous powers upon their local territories rather than centrally controlling the entire country. These governments are called federal states. Federal states vary in the degree of autonomy they give to their local territories. **Devolution** is the process by which central governments delegate statutory powers to lower levels of government, such as the state and county. States that give little or no autonomy to their local territories are called **unitary states**. The United States, Canada, and Mexico, all federal states organized into territories (called states in the United States, *estados* in Mexico, and provinces in Canada), designate a certain level of political power to local areas, allowing voting individuals greater influence in political processes. Territories within states are usually subdivided even further into smaller areas such as counties, cities, school districts, and voting precincts. As a result, local government agencies may have overlapping functions, and several agencies may have jurisdiction over the same geographical areas. In such cases, local government agencies must work to delegate services and authority efficiently. Some cities, such as Toronto, Canada, and Jacksonville, Florida, have worked to eliminate this problem by consolidating their government functions into a limited number of agencies that have authority over the entire metropolitan area.

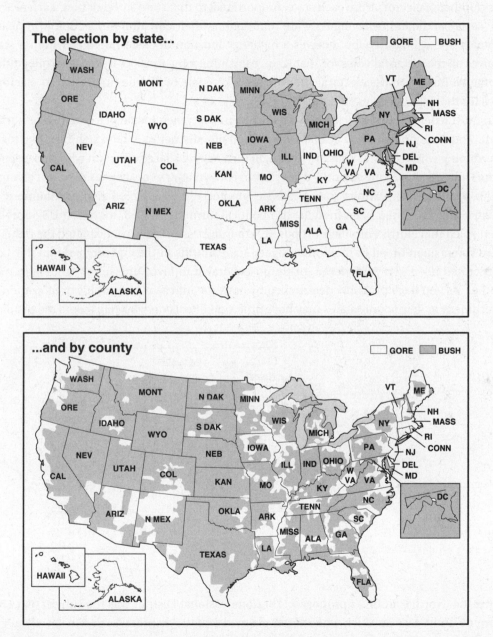

Figure 4.3 In the 2000 US presidential election, George W. Bush won most of the states, but Al Gore picked up a few of the most populous and urbanized states, including New York and California. At the county level the pattern looks different, and we can see that some states were nearly evenly divided. What other interesting patterns can you detect?

TIP

The question of how to delegate authority in a federalist country has crucial implications for everything from the collection of taxes, to the use of natural resources, to the ways that votes are counted in an election. You should be able to construct an essay using examples of how various federal configurations affect contemporary democratic societies.

Geographic organization of the state into locally governed areas has dramatic implications for individual representation. In a democracy like the United States, voters elect officials to posts that are associated with specific geographic areas. Congressional representatives, for instance, serve the people from their own home district. As all Americans learned in the aftermath of the 2000 election, even our presidents are chosen on a state-by-state basis, and the individual who captures the popular vote does not necessarily win the election. In 2000, Al Gore won the **popular vote**, which includes all the votes cast in all the states, by a slim margin but lost the **electoral vote** to George W. Bush. The **Electoral College** consists of a specific number of electors from each state, proportional to that state's population. Each elector chooses a candidate, believing they are representing their constituency's choice. In nearly all states, the candidate who receives a higher proportion of electoral votes within a state receives all the electoral votes for that state, explaining why, in 2000, it was determined that George W. Bush won the electoral vote when the key state of Florida finally cast its electoral votes for the Republican candidate.

In most federal states, officials represent citizens from their locale in a congress or parliament. Thus, the political and ethnic composition of a district may be a significant factor in determining where district lines are drawn, in turn having a large effect on who gets elected. In the United States, congressional districts are redrawn after every census to reflect changes in the population of various states. This geographic exercise, called **reapportionment** or **redistricting**, has often been fraught with political turmoil. In 1993, a divided US Supreme Court ruled that North Carolina's proposed 12th Congressional District violated the rights of white voters guaranteed in the Voting Rights Act. Plaintiffs in the case argued that the 12th District had been **gerrymandered**, or purposely drawn to favor one set of candidates over another. As you have probably determined by now, the inherent geographic organization of government in democratic states may have profound effects on who represents the people.

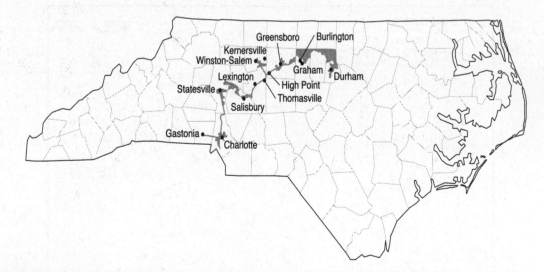

Figure 4.4 North Carolina's proposed 12th Congressional District was the subject of a 1996 Supreme Court case in which white voters claimed that the district was drawn specifically to consolidate the African American vote.

TERRITORY, BORDERS, AND THE GEOGRAPHY OF NATIONS

Chances are that the majority of maps you have looked at of the earth's surface are political maps depicting the boundaries separating unique territories across the globe. In fact, most people's impression of the earth's surface involves a near innate understanding of the political division of space, as almost all globes and world maps contain these designations. One of the main tasks of a political geographer involves understanding the evolution of these bounded territories. The geographic location and designation of political boundaries and the size and shape of the territories they contain play an extremely significant role in a state's economy, its political stability, its relations with other nations, and its culture.

When thinking on the scale of a country, the first thing that should be obvious to any student of political geography is that each country has a unique land base, and a particular set of physical properties and natural resources. The world's largest country, in terms of land area, is Russia, which occupies some 17 million square kilometers, or about 11% of the earth's land surface. Tiny countries, such as Vatican City and San Marino, both of which are located within the larger borders of Italy, are known as **microstates**. In general, larger countries tend to have larger pools of natural resources, but this is not always the case. Despite Canada's immense size and abundance of fresh water, agriculture is limited to the far southern portion of the country, where a reasonably long growing season permits the cultivation of crops. Australia has abundant minerals and much sunshine but is dominated by arid deserts. Brazil's great rainforests, which have tremendous stores of minerals, water, and timber, also tend to have highly leached, infertile soils. Thus, the size of the country does not necessarily guarantee greater levels of natural resources.

Countries also take a wide variety of shapes and sizes. Fiji, with its many small islands, is what geographers call a **fragmented state**. Chile, which is stretched thin along South America's Pacific coast, is an **elongated state**, and Angola, in western Africa, is more or less a **rectangular state**. Countries like Poland that have relatively rounded shapes are **compact states**, and countries like Italy that completely surround other smaller states are referred to as **perforated states**. Nepal is a completely **landlocked state**, meaning that it is completely surrounded by other countries, and Thailand, which has a long, thin arm jutting out from the rest of its territory into the Malay Peninsula, is a **prorupted state**. In addition to these general shapes, many countries have small, outlying holdings that are entirely separated from the bulk of their landmass. Some of these detached pieces, called **exclaves**, lie completely within the boundaries of another country. During the Cold War, West Berlin, at that time owned by West Germany, was an exclave surrounded on all sides by East German territory. In short, the unique shape and size of each country is a product of its physical geography and political history.

Each country's unique shape and location offers it both advantages and disadvantages in the scope of global politics. Countries with extremely large land bases, such as Russia and Canada, must deal with a unique set of problems related to the administration of vast physical landscapes. Countries that have proruptions or exclaves must be able to incorporate effectively areas that are physically detached into the national mainstream. And fragmented states must determine ways to create a cohesive national fabric out of many little swatches of land.

TIP

Landlocked states are at a distinct strategic, military, and economic disadvantage. Can you name three landlocked states and describe how their distance from watercourses has affected their history?

Figure 4.5 The various shapes of countries

Some countries also occupy extremely strategic sites. Israel, the Korean Peninsula, and Panama, all relatively small areas, loom large on the world political scene because of their strategically important locations. Hawaii, at the center of the Pacific Basin, and Istanbul, Turkey, at the crossroads of Europe and Asia, also occupy key positions in the geopolitical landscape. What counts as key, of course, changes through time. For most of American history after European colonization, New Orleans held an extremely important site at the mouth of the Mississippi. Thomas Jefferson once remarked that whoever controls New Orleans controls the entire Mississippi River, which at the time was the main artery for transporting goods and services into the central part of the North American continent. Although New Orleans today remains an important port city, it no longer seems as economically essential as it did in the early 19th century.

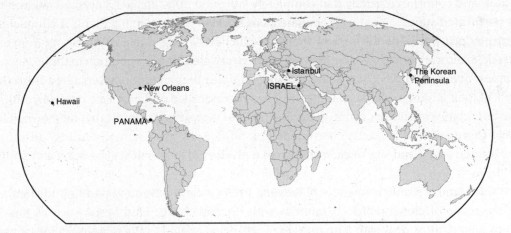

Figure 4.6 Politically and militarily strategic places in the current geopolitical system. New Orleans has lost much of the geopolitical significance it had during the 19th century.

Another important geographic feature of a country is its border. **Physical boundaries** follow important features in the natural landscape such as rivers or mountain ridges. Some borders do not follow any significant features in the landscape, but are purely based on political decisions. However, some political areas are defined by **geometric boundaries**—meaning that they follow straight lines and have little to do with the natural or cultural landscape, and some are marked with well-defined landmarks, such as the wall that separates Tijuana, Mexico, from San Diego County, California. The origin and evolution of particular boundaries provide extremely important clues for understanding current political tensions within certain countries. **Subsequent boundaries** are drawn after a population has established itself and respect existing spatial patterns of certain social, cultural, and ethnic groups. When a boundary is given to a region before it is populated, it is called an **antecedent boundary**, which carries little significance until the area becomes more populated. The western boundary between the United States and Canada was designated by treaty in 1846, when very few people occupied that region, so the boundary did not carry much social or cultural significance. As the area became increasingly populated the boundary became an important division between countries. The opposite of antecedent boundaries, **superimposed boundaries** are drawn after a population has been settled in an area and do not pay much attention to the social, cultural, and ethnic compositions of the populations they divide. These types of boundaries prevailed in Africa during colonialism and still remain, causing much of the political tensions prevalent in this continent. A **relic boundary** is a national border that no longer exists, but has left some imprint on the local cultural or environmental geography.

Within their boundaries, countries must contend with forces that work to pull them apart, while promoting the forces that bind them together. **Centrifugal forces** pull countries apart and include regionalism, ethnic strife, and **territorial disputes**. By the late 1980s, the central government of the Soviet Union had become weak, and the country was being pulled apart by powerful centrifugal forces, including the nationalist aspirations of its many republics. **Balkanization** refers to the contentious political process by which a state may break up into smaller countries. The word comes from the Balkans region of Eastern Europe, which is now composed of a handful of small states. Balkanization can occur when **enclaves** develop with their own ethnic identities, or when central governments increasingly devolve administrative authority to their constituent territories.

STRATEGY

A free-response question on a past APHG exam focused on the tension between centrifugal and centripetal forces in state formation and viability. If given a map of a country, such as India or Canada, could you write an essay about how these forces shape national and international politics?

Centripetal forces bind countries together and include strong national institutions, a sense of common history, and a reliance on strong central government. Sometimes even negative external forces or threats can pull a nation together. The September 11 attacks on the United States, while extremely devastating, joined many Americans in a time of great vulnerability. Certain symbols in the landscape provide evidence of the loyalty of a country's population. After these attacks, many Americans started displaying the American flag prominently, from the windows of their cars and homes to pins placed on lapels or backpacks. Other symbols of strong centripetal forces include good institutions, strong traditions and values, and an effective circulation and communication system connecting all parts of a country.

Centripetal forces can also become destructive. The feeling that one's country should be internally cohesive and should have political autonomy is called **nationalism**. At controlled levels, nationalism can be a healthy centripetal force binding a nation together. However, some nations elevate themselves to such an extent that they adopt doctrines placing their individual nations above all others on the earth's surface. Nationalism has been associated with militaristic regimes, power-hungry leaders, dangerous group mentalities that prevent introspection, and racist ideologies. Nazi Germany is the classic example of nationalism run amok.

Concise political boundaries do not define all geographic spaces. Certain areas of the earth's surface have yet to be officially designated with political boundaries and as such are more accurately described as poorly defined frontiers. Although the meaning of the term **frontier** has been debated for more than 100 years by geographers and historians, it generally connotes an area where borders are shifting and weak, and where peoples of different cultures or nationalities meet and lay claim to the land. One example of a modern frontier is the western Amazon Basin. In this remote region, the national borders between Brazil, Peru, Bolivia, Colombia, and Venezuela mean very little on the ground. Instead, the Amazon is characterized by weak administrative powers, lush rain forests, and rugged terrain. Another interesting example of a frontier is Antarctica. Australia, Norway, France, Argentina, Chile, New Zealand, and the United Kingdom have all made territorial claims on Antarctica; however, their borders have no meaning whatsoever on the ground. In some cases, frontiers are filled by **buffer states**, which occupy the spaces between two larger, potentially oppositional countries.

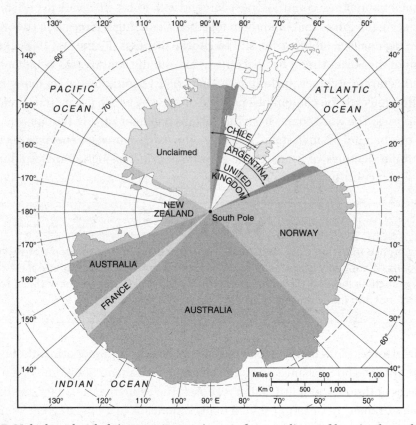

Figure 4.7 Nebulous land claims on Antarctica conform to lines of longitude and have little relationship to physical features or human impacts.

Another important and highly contested frontier, the world's oceans, cause boundary confusion and disputes that have largely been resolved through the **law of the sea**. Countries with coastal access have historically claimed sovereignty over certain strips of adjacent waters. Centuries ago, uneven and very country-specific designations of territorial waters caused little international conflict; however, as discovery of particular resources ranging from fish to oil from continental shelves increased, states across the globe scrambled to claim certain areas of the "high seas" as their own. Legislation controlling the designation of coastal waters quickly became necessary to avoid international conflict. In 1958, the UN Conference on the Law of the Sea (UNCLOS I) met for the first time to decide on uniform laws to govern the ownership of one of the world's most giant frontiers. The first two meetings of this group proved unsuccessful, and it was not until their third meeting, ending in 1983, that they devised a coherent system of laws to govern this vast and resource-rich area. The provisions of this law have been generally adopted across the globe and include two important clauses. The first one describes the restriction of territorial seas to 12 nautical miles (19 km) from a specific shoreline, in which ships of other countries have right of passage. The second designates an exclusive economic zone (EEZ), which recognizes a state's economic rights to 200 nautical miles (370 km) from shore. Within this zone, each state has the right to explore and exploit natural resources in the water, seabed, and subsoil below.

The single most important global geopolitical phenomenon of the past 500 years is **colonialism**. Well before the time of Christopher Columbus, colonialism began shaping world history and human geography in ways that are difficult to overemphasize. An adequate treatment of the role of colonialism in shaping global politics would take many volumes, but a few points are worth mentioning here. Although we usually think of Britain, Spain, Portugal, and France as the great colonial powers, these countries are only a few of the colonizing forces that have vied for land and power in modern history. Sweden, Russia, Austria, China, and Japan were all great colonial powers. In ancient times, Rome, Greece, and even the great Aztec civilization of Mexico were all colonialist states. Although the great European colonial empires largely disbanded their colonies during the 20th century, countries like the United States and China still control the destinies of millions of colonized people throughout the world.

In many cases, imperialism has arisen where colonialism once flourished. While colonialism involves the official government rule of one state over another, **imperialism** describes a situation in which one country exerts cultural or economic dominance over another without the aid of official government institutions. Many South American and African countries that were European colonies during the 18th and 19th centuries are still dominated by Spanish, French, or British culture, even though their former colonizers no longer officially govern them. For example, Kenya, which gained independence from Britain in 1963, still retains a legacy of British language, religion, and administrative systems. The United States is a particularly dominant imperial power in the sense that American popular culture and economics affect the lives of millions of people throughout the world.

TIP

Geographers who work in the area of post-colonial studies think about the legacy of 19th and 20th century colonial arrangements on world culture, politics, social justice, and economic development. Can you describe some ways that the lingering history of colonialism affects contemporary countries in Africa, Asia, and South America?

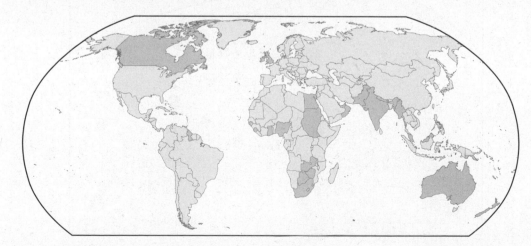

Figure 4.8 The shaded areas represent the British Empire at its colonial apex.

Countries under imperialist domination have one very significant feature that people living under colonial powers do not: self-determination. **Self-determination** is the right of a nation to govern itself autonomously and thus to determine its own destiny. Imperialist powers may dominate the cultural and economic relations of a less-powerful country, but that country still maintains autonomous political power. Although self-determination represents a great advantage for newly independent countries, former colonies frequently remain dominated by their historic colonizers. As a result, imperialism has become an insidious and long-term problem for many African, Asian, and South American countries.

INTERNATIONAL POLITICAL GEOGRAPHY

Three early theories of international political geography are worth mentioning here, before moving on to a discussion of current global politics. First, in the late 19th century, Friedrich Ratzel proposed his **organic theory** of the evolution of nations, which was later developed by the Swedish political scientist Rudolf Kjéllen into the field of **geopolitics**. Geopolitical theorists believed that nations must expand their land base in order to maintain vibrancy. Countries that did not expand eventually disintegrated, like an organism that fails to find food sources on which to grow and thrive. This theory was completely discarded by geographers after it was used by Adolf Hitler to justify his military aggression during World War II. Hitler's expansionist theory of **lebensraum**, which was based on a drive to acquire "living space" for the German people, was based on geopolitical theory.

Another important and related concept is the **Heartland theory**, which was developed by Sir Halford Mackinder in the beginning of the 20th century. According to Mackinder, the great geographical "pivot" point of all human history was in northern and central Asia, the most populous landmass on earth, and he who rules the heartland, rules the world. However, many geographers took exception to this theory. First, most of Mackinder's contemporaries believed that the world's oceans provided the avenue to colonial conquest, not land. Second, history did not hold sufficient evidence to support this area as the geographical basis of world conquest. One dissenter, Nicholas Spykman, in fact argued that the **rimland**, the area surrounding the heartland, was most important for world political power. Although these theories provide political geography with a rich history, they have little importance in the scope of the modern field.

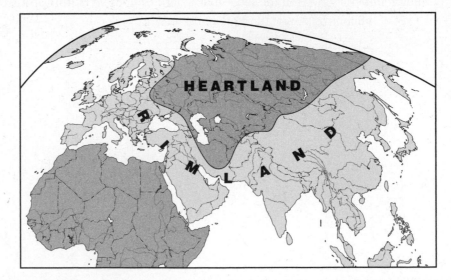

Figure 4.9 The heartland of Sir Halford Mackinder and the rimland of Nicholas Spykman

One of the most important trends in current global politics is the development of international alliances. International alliances take several forms, one of which is the **international organization**. An international organization is an alliance of two or more countries seeking cooperation with each other without giving up either's autonomy or self-determination. The **United Nations** (UN) and the **North American Free Trade Agreement** (NAFTA) are both international organizations, but they encompass much different geographic scales. The UN, a global international organization, includes most of the world's autonomous states and is specifically focused on international peace and security. Because almost every country enjoys membership, the UN presents a powerful global force ensuring internationally approved standards of behavior. With collective action, this organization can enforce certain political decisions such as economic sanctions to isolate misbehaving countries, forcing them to comply with UN standards. On the other hand, USMCA is a regional accord that links the United States, Canada, and Mexico through economic arrangements aimed at opening borders and promoting trade. The **European Union** (EU) represents a good example of a **supranational organization**. These types of organizations are similar to international organizations, but to some extent member nations must relinquish some level of state sovereignty in favor of group interests. The European Union includes over a dozen European states, has a central administrative center in Brussels, Belgium, and a unified currency, the Euro. Joining this union does require giving up some autonomy, but for a greater good. If strength comes in numbers, then the unification of Europe into a single cooperative community has surely produced a great world power.

TIP

The 2005 APHG exam included a free-response question on the concepts of supranationalism and devolution. Devolution is the tendency of national governments to delegate authorities to regional entities, such as state governments and municipalities.

Figure 4.10 The European Union

The **Organization of Petroleum Exporting Countries** (OPEC) is an international economic organization whose member countries all have a single thing in common—they all produce and export oil. This exemplifies an economic alliance. Certain countries also join together for military purposes. The **North Atlantic Treaty Organization** (NATO) is one such union. These types of organizations may require member states to allow other members to establish military bases within their territories and thus can be quite internationally significant. **Confederations** are similar to international organizations in that they bring several autonomous states together for a common purpose. The **Commonwealth of Independent States** (CIS) is a confederacy made up of independent states of the former Soviet Union who have united because of their common economic and administrative needs. The confederation of southern states that seceded from the United States before the Civil War is another classic example of this type of alliance.

An important point to remember about international alliances is that they have changed dramatically since the late 1980s. From 1945 to 1989, the Cold War defined the global geopolitical situation. During this period, an **east/west divide** separated the largely democratic and free-market countries of Western Europe and the Americas from the communist and socialist countries of Eastern Europe and Asia. Some Western leaders were concerned that if communism caught on in a few key countries that were in close proximity to the sphere of Soviet influence, then many others would fall in quick succession, tipping the scales of world power in favor of communism. This idea was called the **domino theory**. Of course, the east/west divide was a gross generalization with many exceptions, capitalist Japan and communist Cuba being the most obvious examples, and the domino effect never occurred. However, these are the ways that many people thought of world politics during the Cold War.

Since the dismantling of the USSR and the European revolutions of the late 1980s, the world's global geopolitical axis has shifted. Now, the most obvious division of world power lies in the disparity between the north and south. The **north/south divide** describes the division between the wealthy countries of Europe and North America (as well as Japan and Australia) and the generally poorer countries of Asia, Africa, and Latin America. Whereas the east/west divide was mainly ideological and political, the north/south divide is mainly economic. This economic inequality was created by a long history of colonization and dominance. Now, the formerly colonized countries of the south are struggling to lift themselves from the crushing poverty and disorder that has characterized the postcolonial period.

A central goal of US diplomacy and international organizations, such as the United Nations, is **democratization**. Democracy can take many forms. Government structures in the United States, for example, differ in many ways from the diversity of forms present in Europe, Latin America, Asia, and elsewhere. Many people believe, however, that some form of democratic representation is a fundamental human right and is essential for promoting social welfare, economic development, and security. In recent decades, democracy has spread to new areas as autocratic regimes have given way to more transparent, popularly elected governments. Yet the transition to democracy has proven extremely turbulent in some areas, and weak or superficial forms of democracy have emerged in other areas. In many parts of the world, people still live in countries where they lack the ability to choose their leaders or to participate meaningfully in political processes.

SPATIAL CONFLICT

It should be quite evident by now that a large component of political geography involves the investigation of historical and current conflicts over territory at all scales, all across the globe. Whether it be the designation of waters to specific countries, the effects of imposed boundaries on many of the colonized countries throughout the developing world, or tensions in choosing voting districts within the United States, space ownership has proven one of the most volatile issues throughout human history. Spatial conflict occurs at all geographic scales. Currently in Los Angeles, the San Fernando Valley is seeking secession from the city. The citizens of the valley think that Los Angeles does an inadequate job supplying their civic needs, and that if given authority to spend their own tax money, they would do a better job providing for their citizens. On a different scale, in 2002 Spain and Morocco disputed ownership of a tiny island off the coast of Morocco called Perejil (and Leila in Morocco). Both countries worked to establish an agreement without using military force. Finally, on a global scale, in recent decades debate and regulation over environmental problems that cross national boundaries have moved to the top of the agenda in international policy discussions. Issues such as ozone depletion, biodiversity loss, and climate change will eventually affect all people, but certain advanced countries quite often contribute more to these global problems, making universal legislation difficult to unilaterally enforce.

Political geography may be one of the most controversial subfields within human geography. The designation and demarcation of space to particular unified populations constantly causes political turmoil. Even as you read this chapter, different counties within your state, different countries within continents, and different continents across the globe discuss, debate, and fight over territory. As such, political geographers contribute great understanding of the evolution and constant changes in political boundaries, as well as an understanding of how individuals and groups within these boundaries successfully interact with other countries across the globe.

KEY TERMS

ANTECEDENT BOUNDARY A boundary line established before an area is populated.

BALKANIZATION The contentious political process by which a state may break up into smaller countries.

BUFFER STATE A relatively small country sandwiched between two larger powers. The existence of buffer states may help to prevent dangerous conflicts between powerful countries.

CENTRIFUGAL FORCES Forces that tend to divide a country.

CENTRIPETAL FORCES Forces that tend to unite or bind a country together.

COLONIALISM The expansion and perpetuation of an empire.

COMMONWEALTH OF INDEPENDENT STATES Confederacy of independent states of the former Soviet Union that have united because of their common economic and administrative needs.

COMPACT STATE A state that possesses a roughly circular, oval, or rectangular territory in which the distance from the geometric center is relatively equal in all directions.

CONFEDERATION A form of an international organization that brings several autonomous states together for a common purpose.

DEMOCRATIZATION The process of establishing representative and accountable forms of government led by popularly elected officials.

DEVOLUTION The delegation of legal authority from a central government to lower levels of political organization, such as a state or country.

DOMINO THEORY The idea that political destabilization in one country can lead to collapse of political stability in neighboring countries, starting a chain reaction of collapse.

EAST/WEST DIVIDE Geographic separation between the largely democratic and free-market countries of Western Europe and the Americas from the communist and socialist countries of Eastern Europe and Asia.

ELECTORAL COLLEGE A certain number of electors from each state proportional to and seemingly representative of that state's population. Each elector chooses a candidate, believing they are representing their constituency's choice.

ELECTORAL VOTE The choice expressed collectively by the electoral college to determine the president and vice-president of the United States.

ELONGATED STATE A state whose territory is long and narrow in shape.

ENCLAVES Any small and relatively homogenous group or region surrounded by another larger and different group or region.

EUROPEAN UNION International organization comprising Western European countries to promote free trade among members.

EXCLAVE A bounded territory that is part of a particular state but is separated from it by the territory of a different state.

FEDERALISM A system of government in which power is distributed among certain geographical territories rather than concentrated within a central government.

FRAGMENTED STATE A state that is not a contiguous whole but rather separated parts.

FRONTIER An area where borders are shifting and weak and where peoples of different cultures or nationalities meet and lay claim to the land.

GEOMETRIC BOUNDARIES Political boundaries that are defined and delimited by straight lines.

GEOPOLITICS The study of the interplay between political relations and the territorial context in which they occur.

GERRYMANDERING The designation of voting districts so as to favor a particular political party or candidate.

HEARTLAND THEORY Hypothesis proposed by Halford Mackinder that held that any political power based in the heart of Eurasia could gain enough strength to eventually dominate the world.

IMPERIALISM The perpetuation of a colonial empire even after it is no longer politically sovereign.

INTERNATIONAL ORGANIZATION An alliance of two or more countries seeking cooperation with each other without giving up either's autonomy or self-determination.

IRREDENTISM A policy of advocating for the return of a territory to a country it formerly belonged to.

LANDLOCKED STATE A state that is completely surrounded by the land of other states, which gives it a disadvantage in terms of accessibility to and from international trade routes.

LAW OF THE SEA Law establishing states' rights and responsibilities concerning the ownership and use of the earth's seas and oceans and their resources.

LEBENSRAUM Hitler's expansionist theory based on a drive to acquire "living space" for the German people.

MICROSTATE A state or territory that is small in both population and area.

NATION Tightly knit group of individuals sharing a common language, ethnicity, religion, and other cultural attributes.

NATIONALISM A sense of national pride to such an extent of exalting one nation above all others.

NATION-STATE A country whose population possesses a substantial degree of cultural homogeneity and unity.

NORTH AMERICAN FREE TRADE AGREEMENT Agreement signed on January 1, 1994, that allows the opening of borders between the United States, Mexico, and Canada.

NORTH ATLANTIC TREATY ORGANIZATION An international organization of member states that have joined together for military purposes.

NORTH/SOUTH DIVIDE The economic division between the wealthy countries of Europe and North America, Japan, and Australia and the generally poorer countries of Asia, Africa, and Latin America.

ORGANIC THEORY The view that states resemble biological organisms with life cycles that include stages of youth, maturity, and old age.

ORGANIZATION OF PETROLEUM EXPORTING COUNTRIES An international economic organization whose member countries all produce and export oil.

PERFORATED STATE A state whose territory completely surrounds that of another state.

PHYSICAL BOUNDARIES Political boundaries that correspond with prominent physical features such as mountain ranges or rivers.

POLITICAL GEOGRAPHY The spatial analysis of political phenomena and processes.

POPULAR VOTE The tally of each individual's vote within a given geographic area.

PRORUPTED STATE A state that exhibits a narrow, elongated land extension leading away from the main territory.

REAPPORTIONMENT The process of a reallocation of electoral seats to defined territories.

RECTANGULAR STATE A state whose territory is rectangular in shape.

REDISTRICTING The drawing of new electoral district boundary lines in response to population changes.

RELIC BOUNDARIES Old political boundaries that no longer exist as international borders, but that have left an enduring mark on the local cultural or environmental geography.

RIMLAND THEORY Nicholas Spykman's theory that the domination of the coastal fringes of Eurasia would provide the base for world conquest.

SELF-DETERMINATION The right of a nation to govern itself autonomously.

SHATTERBELT A region of persistent political fragmentation due to devolution and centrifugal forces.

STATE A politically organized territory that is administered by a sovereign government and is recognized by the international community.

STATELESS NATION A group of people with a common political identity who do not have a territorially defined, sovereign country of their own.

STATES' RIGHTS Rights and powers believed to be in the authority of the states rather than the federal government.

SUBSEQUENT BOUNDARY Boundary line established after an area has been settled that considers the social and cultural characteristics of the area.

SUPERIMPOSED BOUNDARY Boundary line drawn in an area ignoring the existing cultural pattern.

SUPRANATIONAL ORGANIZATION Organization of three or more states to promote shared objectives.

TERRITORIAL DISPUTE Any dispute over land ownership.

TERRITORIAL ORGANIZATION Political organization that distributes political power in more easily governed units of land.

THEOCRACY A state whose government is either believed to be divinely guided or a state under the control of a group of religious leaders.

UNITARY STATE A state governed constitutionally as a unit, without internal divisions or a federalist delegation of powers.

UNITED NATIONS A global supranational organization established at the end of World War II to foster international security and cooperation.

CHAPTER SUMMARY

Political geography is one of the oldest fields in the discipline of geography. Political geographers use the spatial perspective to study political systems from local and regional politics, to national politics, to international politics. At local scales, political geographers study issues like territorial organization, representation, and voting patterns. At national and international scales, geographers study the relationship between physical geography and politics, the historical geography of colonialism and imperialism, the formation of international alliances, and the current political tensions between the wealthy countries of the industrialized north and the poorer countries of the less-developed south.

PRACTICE QUESTIONS AND ANSWERS

The Geography of Local and Regional Politics

MULTIPLE-CHOICE QUESTIONS

1. A _____ is a group of people with a common political identity, and a _____ is a country with recognized borders.

   (A) territory . . . federalism
   (B) nation . . . territory
   (C) state . . . nation
   (D) nation . . . state
   (E) territory . . . state

2. _____ governments are organized into a geographically based hierarchy of local government agencies.

   (A) Federal
   (B) Territorial
   (C) Consolidated
   (D) Electoral
   (E) National

3. With its system of regional provinces, Canada is an example of a(n)

   (A) microstate.
   (B) electoral state.
   (C) reapportioned state.
   (D) federal state.
   (E) nation-state.

4. The drawing of new voting districts is called

   (A) reapportionment.
   (B) gerrymandering.
   (C) reelection.
   (D) redrawing.
   (E) discretization.

5. When voting districts are redrawn in such a way that they purposely favor a political party, they have been

   (A) vetoed.
   (B) reapportioned.
   (C) redistricted.
   (D) gerrymandered.
   (E) reelected.

1. Political systems are often organized geographically.

    (A) What is the difference between territorial organization, federal organization, and electoral voting?
    (B) What are the advantages of organizing political systems by geographic areas?

Territory, Borders, and the Geography of Nations

MULTIPLE-CHOICE QUESTIONS

1. Indonesia is an example of a(n)

    (A) elongated state.
    (B) microstate.
    (C) compact state.
    (D) fragmented state.
    (E) prorupted state.

2. Which of the following is a landlocked country?

    (A) Peru
    (B) Germany
    (C) Burma
    (D) Afghanistan
    (E) Colombia

3. In Antarctica, geometric political borders do little to organize a vast

    (A) frontier.
    (B) borderland.
    (C) wasteland.
    (D) tundra.
    (E) territory.

4. _____ forces work to pull countries apart, while _____ forces work to bind them together.

    (A) Centripetal . . . centrifugal
    (B) Centrifugal . . . centripetal
    (C) Communist . . . democratic
    (D) Capitalist . . . socialist
    (E) Socialist . . . centripetal

5. When one country exerts political, economic, or social influence over another without the aid of official government institutions, it is called

   (A) dominance.
   (B) imperialism.
   (C) colonialism.
   (D) federalism.
   (E) territorialism.

6. For many years, French Canadians from Quebec sought _____, or the right to govern themselves and to establish their own independent state.

   (A) nationalism
   (B) self-determination
   (C) anticolonialism
   (D) reapportionment
   (E) colonization

FREE-RESPONSE QUESTION

1. Many factors contribute to a state's political stability.

   (A) What are some of the forces that bind a state together? (use specific examples)
   (B) What are some of the forces that cause disunity within a state? (use specific examples)

International Political Geography

MULTIPLE-CHOICE QUESTIONS

1. Hitler's nationalist/expansionist philosophies drew in part from

   (A) self-determination.
   (B) sound historical evidence.
   (C) organic geopolitical theory.
   (D) rimland theory.
   (E) heartland theory.

2. When countries come together for a common purpose, somewhat limiting their own individual powers, the resulting body is called a(n)

   (A) international organization.
   (B) confederacy.
   (C) supranational organization.
   (D) union.
   (E) national alliance.

3. OPEC is an example of a(n)

   (A) supranational organization.
   (B) commonwealth.
   (C) confederacy.
   (D) international organization.
   (E) national organization.

4. The _____ was based on control of land, markets, and political ideology, whereas the _____ is based on wealth and poverty.

   (A) east/west divide . . . north/south divide
   (B) domino theory . . . heartland theory
   (C) north/south divide . . . east/west divide
   (D) organic theory . . . rimland theory
   (E) core/periphery . . . east/west divide

5. _____ boundaries characterize much of Africa as they ignore cultural and tribal differences across space.

   (A) Superimposed
   (B) Subsequent
   (C) Colonial
   (D) Antecedent
   (E) Territorial

FREE-RESPONSE QUESTION

1. The European Union and North American Free Trade Agreement are two common examples of international organizations.

   (A) Describe each of these two organizations and their purposes.
   (B) What do each of these two organizations have in common? How are they different?

Answers for Multiple-Choice Questions

THE GEOGRAPHY OF LOCAL AND REGIONAL POLITICS

1. **(D)** Although the terms "nation" and "state" are often used interchangeably, they actually have specific and quite different definitions. Nation connotes a common sense of political identity, while a state has an official government and geographic borders.

2. **(A)** A federal system is a form of territorial, or geographically subdivided, organization in which governments give their constituent territories some degree of autonomy. Federal governments are frequently broken down into states, counties, parishes or other local areas, municipalities, voting districts, and so on, each of which has specific powers within its borders.

3. **(D)** Canada, like the United States and Mexico, has a federally organized government in which power is distributed to local areas, in this case, regional provinces.

4. **(A)** After a census, the demographic data collected is used to form more accurate boundaries around voting districts to ensure fair representation within a political district. This process is called reapportionment or redistricting.

5. **(D)** The purposeful drawing of a political district to favor a political party is called gerrymandering, named both for the first person accused of doing this, and also for the shape of the first gerrymandered district, which resembled the shape of a salamander.

TERRITORY, BORDERS, AND THE GEOGRAPHY OF NATIONS

1. **(D)** Indonesia consists of about 13,000 islands and islets scattered throughout the giant Malay Archipelago. Because the country is separated into so many small pieces, it is referred to as fragmented. Elongated states, like Chile, are stretched long and thin, and compact states, like Poland, are nearly circular.

2. **(D)** Landlocked countries have no outlet to the ocean. Surprisingly few countries are fully landlocked—Afghanistan is one.

3. **(A)** Several countries have claimed portions of Antarctica as their own; however, these territorial claims have little meaning on the ground. In 1959, twelve countries signed an international treaty establishing large portions of the continent as an international commons, protected for conservation and scientific research.

4. **(B)** Forces such as regionalism are centrifugal, meaning that they work to weaken central authority and cohesiveness, while forces such as nationalism generally work to solidify central authority and bind countries together.

5. **(B)** Colonialism involves official institutional domination of one group over another; imperialism usually includes unofficial forms of social, cultural, or economic dominance.

6. **(B)** For many years Canadians of French descent have struggled to maintain their political autonomy and cultural heritage. In recent years, separatist movements have waned, but citizens of Quebec still fiercely defend their unique history, language, and traditions.

INTERNATIONAL POLITICAL GEOGRAPHY

1. **(C)** Organic geopolitical theory, which held that states must expand their land base in order to grow and maintain viability, influenced many 19th and 20th century nationalist politicians, including Adolf Hitler.

2. **(C)** In supranational organizations member states must give up some dimension of their individual autonomy for some greater cause, such as political or economic security. In international organizations, no such sacrifice is called for.

3. **(D)** In OPEC, petroleum-exporting countries have joined together to regulate and stabilize oil markets around the world. This is an example of an international organization based on economic gain.

4. **(A)** During the Cold War, countries were aligned on either side of an east/west divide, based largely on control of land and political ideology. In the post–Cold War era, the most obvious world axis divides the poorer countries of the south from the wealthier countries of North America and Europe.

5. **(A)** During colonialism, many nations imposed political boundaries designating certain territories for themselves without paying attention to any divisions they may have caused between different tribes. Even after decolonization, these boundaries remain, causing much of the bloodshed that has occurred and will continue to occur on this continent.

Answers for Free-Response Questions

THE GEOGRAPHY OF LOCAL AND REGIONAL POLITICS

1. **Main points:**
   - Territorial organization implies a geographically based hierarchy of official duties and organizations.
   - In a federal government, the local territories, states, districts, and so on, each have autonomous powers, such as policing, creating local laws, and representing their citizens at the national level.
   - The United States has an electoral system of representation in which each political party appoints representatives for each state, and then the party that wins the popular vote within the state gets to cast electoral votes for its candidate. The electoral system can have important implications, as it did in the 2000 presidential election when Al Gore narrowly won the overall popular vote, but George W. Bush won the electoral vote.
   - Some of the advantages of organizing by geographic areas include more direct representation and accountability of local elected officials, more efficient administration of remote or expansive areas, increased political attention to local or regional issues, and more efficient allocation of public funds and resources.

TERRITORY, BORDERS, AND THE GEOGRAPHY OF NATIONS

1. **Main points:**
   - Centrifugal forces work to weaken central authority and cohesiveness, while centripetal forces work to solidify central authority and bind countries together.

- One example of a country experiencing powerful centrifugal forces was the Soviet Union in the late 1980s. During this period, the USSR had a weak central government and a depressed economy and was composed of increasingly defiant constituent republics. Russia, the Union's central state, could no longer hold the confederacy together, and the USSR disbanded in 1991.
- During this same period, the Soviet's chief rival, the United States, had a strong central government and a relatively robust economy and was composed of states that were generally deferential to the federal government.

INTERNATIONAL POLITICAL GEOGRAPHY

1. **Main points:**
   - The EU is a supranational organization, whereas NAFTA is an international agreement.
   - In the EU, member states give up a degree of their individual autonomy in areas such as economic and military policy. The EU now has a central administrative capital in Belgium and a single currency, the Euro. For their sacrifice, member states have become part of a greater and more stable economic and military power.
   - In NAFTA (which has been replaced by the USMCA), the United States, Canada, and Mexico have agreed to open their borders to increased trade. In the process, they have made considerable concessions; however, they have not given up their individual powers in the way the European Union members have.
   - Both the EU and NAFTA are attempts by several states to band together for increased freedom and economic prosperity; however, the states of the European Union have gone much further toward creating a single, unified body with a sole political identity than the NAFTA countries have.

 ADDITIONAL RESOURCES

Kaplan, Robert D. 2013. *The Revenge of Geography: What the Map Tells Us About Coming Conflicts and the Battle Against Fate.* New York: Random House.

Marshall, Tim. 2016. *Prisoners of Geography: Ten Maps that Explain Everything About the World.* New York: Scribner.

Wong, Jan. 1997. *Red China Blues: My Long March from Mao to Now.* New York: Anchor.

Agricultural and Rural Land-Use Patterns and Processes

5

IN THIS CHAPTER

→ **HISTORICAL GEOGRAPHY OF AGRICULTURE**

→ **GEOGRAPHY OF MODERN AGRICULTURE**

→ **AGRICULTURE AND THE ENVIRONMENT**

 Key Terms

Agribusiness	Food security	Planned agricultural
Agriculture	Genetically modified	economy
Animal husbandry	organisms	Plantations
Aquaculture	Green Revolution	Salinization
Biotechnology	Horizontal integration	Shifting cultivation
Capital-intensive	Hunting and gathering	Slash-and-burn agriculture
agriculture	Industrial Revolution	Specialty crops
Commercial agricultural	Intensive cultivation	Subsistence agricultural
economy	Labor-intensive	economy
Commodity chains	agriculture	Sustainability
Dairying	Livestock ranching	Swidden
Desertification	Mechanization	Topsoil loss
Domestication	Mediterranean agriculture	Transhumance
Extensive agriculture	Organic agriculture	Urban sprawl
Feedlots	Pastoralism	Vertical integration
Fertile Crescent	Pesticides	von Thunen model

HISTORICAL GEOGRAPHY OF AGRICULTURE

Long ago, when people first began domesticating plants and animals, there was no sign that their actions would dramatically change the face of the earth. However, during the past 10,000 years, **agriculture** has become an endeavor of enormous proportions, with dramatic consequences for Earth's physical and human geography. Agriculture consists of the purposeful planting of crops or raising of livestock for human sustenance. This development has played an important role in the development of human societies, fostering the growth of urban civilizations and initiating global patterns of trade. In fact, it is hard to overestimate the importance of agriculture in human history. Today, changes in agricultural techniques and rural land use are being felt all over the world, as new technologies are introduced, as large corporations gain a greater share of commodity markets, and as traditional areas of **extensive agriculture**, which involve dispersed, widespread ranching and farming, are brought into more **intensive cultivation**, forcing smaller plots to produce greater yields. In some areas, agricultural plots are being lost completely to **urban sprawl**, and in other areas the soil is rapidly eroding, becoming infertile and thereby decreasing the ability of future farmers to reap a harvest from the land. This chapter examines the problems and challenges of providing for Earth's population while still caring for the land.

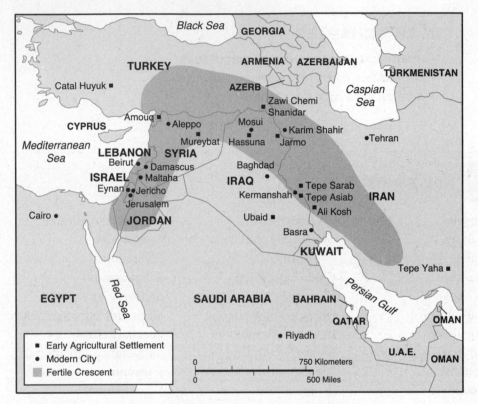

Figure 5.1 The Fertile Crescent

If asked where agriculture originated, most Americans would probably say the **Fertile Crescent**. The name *Fertile Crescent* instantly conjures up images of ancient civilizations and bountiful, sun-drenched valleys—an image starkly different from what most people probably have of today's Fertile Crescent, which includes parts of Iraq, Syria, Lebanon, and Turkey. It is true that the Fertile Crescent was an early hearth of agriculture, but it was actually only one of many places where people independently domesticated plants and animals over the past 10,000 years. **Domestication** initially occurred when humans consciously began to manipulate plant and animal species in order to sustain themselves. Other locations of independent domestication include modern-day Peru, central Mexico, East Africa, India, and China. Agricultural innovations diffused far and wide from these early hearths. Although some native peoples never adopted sedentary agriculture, domesticated plants and animals were widespread long before the age of European exploration began, over 500 years ago. It is also worth noting that westerners probably think of the Fertile Crescent as the birthplace of agriculture because Europeans inherited most of their crops and agricultural practices from there, not because it was the most important site of early farming or animal husbandry.

Agriculture was not invented in the way modern advances have been. It took thousands of years and many false starts for people to suitably domesticate plants and animals so that they became useful and dependable. Before the establishment of domesticated crops, all humans were involved in **hunting and gathering**; their diets consisted of animals they captured and wild plants they collected. Even after agriculture became firmly established, most people still hunted and gathered to round out their diet. Only recently, and only in some parts of the world, have the majority of people ceased hunting and gathering altogether. In the early days of agriculture and **animal husbandry**, domestication was incidental. As hunter-gatherers foraged for their food, they naturally chose the best fruits, nuts, and grains. In doing so they inadvertently spread the seeds of the plants that best suited their purposes, thus selecting for even bigger nuts, even tastier grains, and even juicier fruits. Over thousands of years, some of these plants became domesticated, meaning they had been permanently altered by people for human use. Even when people began to shift slowly to subsistence agriculture, millennia ago, their lives did not necessarily become safer or more secure. It is useful to think of early agriculturalists not as making a conscious decision to shift from hunting and gathering to farming, but as slowly diversifying their survival strategies to provide themselves with a wider range of nutritional sources, and more options if any one strategy failed. (See Figures 5.2 and 5.3.)

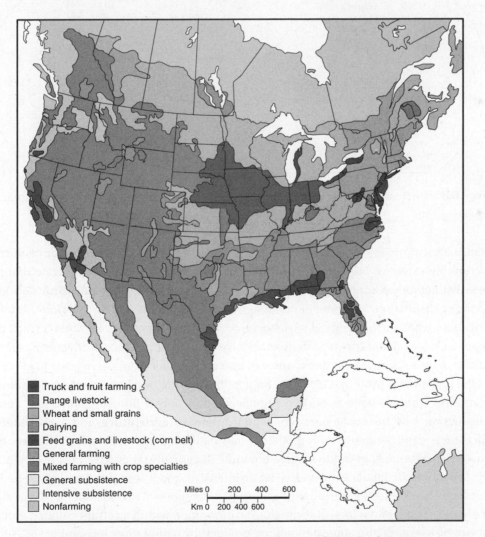

Figure 5.2 Dominant agricultural activities in the United States

Three different types of economies have traditionally governed agricultural production and distribution. In **subsistence agricultural economies**, farmers produce goods to provide for themselves and others in the local community. In **commercial agricultural economies**, a competitive market, where farmers freely market their goods with the goal of making a profit, determines agricultural production. Finally, in **planned agricultural economies**, which are associated with communist-controlled countries, the government controls both the supply and the price of goods that are distributed through government agencies. While these three systems seem quite different from one another, very few countries solely abide by one method. Some subsistence farmers produce excess, which they use to trade for goods for themselves and their families. In the United States, commercial farming generally prevails; however, the government does control price and production to some extent by providing subsidies and incentives to farmers across the country. And in many Latin American countries, governments encourage, and sometimes even demand, that farmers produce export commodities.

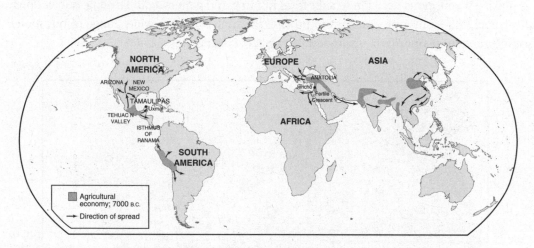

Figure 5.3 The hearths of early agriculture and the initial diffusion of plant and animal domestication

Once agriculture is categorized according to its economic purposes, further distinctions describe the extent of land under cultivation and the methods used to cultivate that land. These distinctions, described in terms of intensive and extensive cultivation, provide another important distinction for understanding agricultural variation across the globe. Intensive cultivation, whether for personal sustenance or commercial activity, involves a small piece of land with large labor inputs to generate a large amount of produce. Conversely, extensive cultivation usually involves large expanses of land and smaller amounts of labor to generate a specific agricultural product. Generally, population densities are high in intensely cultivated agricultural systems, while extensive systems support only a limited population. Further, a distinction can be made between **capital-intensive agriculture** and **labor-intensive agriculture:** capital-intensive methods use mechanical goods such as machinery, tools, vehicles, and facilities to produce large amounts of agricultural goods, a process requiring very little human labor. Conversely, labor-intensive goods use human hands in large abundance to produce a given amount of output. It is important to note that this distinction is not always a result of the level of technological innovation within a certain country, although that can be a large factor. Some agricultural products by nature must be handpicked, such as strawberries, to ensure that the fruit does not get damaged.

The world's first farmers operated under a subsistence economy, producing to support their family and local community. Subsistence farming takes on many different forms in various places across the globe. In **slash-and-burn agriculture**, which is common in the tropics, farmers raze the vegetation in a plot, farm it for a few years, and then move on to another plot with fresh soil. Slash-and-burn agriculture is a form of **shifting cultivation**, and land that has been cleared for farming is called **swidden**. The slash-and-burn system allows fragile tropical soils to recover and rainforest vegetation to quickly reoccupy recently farmed plots. However, this type of agriculture usually cannot support dense human populations, because individual plots rapidly lose their fertility after the first few years of cultivation, and it takes many years for abandoned plots to regenerate themselves. In the Amazon Basin of Brazil, slashing and burning has been going on for thousands of years. However, in the 1970s and 1980s, the Brazilian government encouraged many of its citizens to move to the interior, and the amount of land being converted increased dramatically. Although the cutting has decreased, large areas of the rainforest have now been converted to livestock grazing.

Pastoralism is another type of subsistence agriculture, based on nomadic animal husbandry. Pastoral peoples are found mainly in the dry, mountainous areas of Africa and Asia, where harsh climates render cultivation unfeasible. This type of agricultural activity provides an excellent example of extensive subsistence cultivation as nomadic livestock herders constantly search for forage to feed their livestock and thus cover a wide range of geographic space. The herders rely solely on their livestock to provide food, clothing, and even shelter. This form of agriculture is currently experiencing rapid decline as various economic, physical, and cultural changes force these people to change their livelihood.

Although subsistence farmers frequently participate in small local markets, the majority of their food is grown on site for local consumption. Because subsistence farmers rarely have much land and are almost always lacking in equipment, fertilizers, and other important agricultural inputs, they frequently lead a precarious existence. Subsistence agriculture—including both farming and animal husbandry—is still the dominant lifestyle and source of food in some areas of the world today, but, in general, competitive markets control most of the world's current agricultural activity.

Esther Boserup, an important agricultural geographer, formalized the transition from extensive subsistence forms of agriculture to more intensive cultivation of the land necessary to support greater populations. In contrast to Thomas Malthus, whose model was discussed in Chapter 3, Boserup viewed population growth as a positive force driving agricultural innovations that could support more people. Her model proposes a five-stage progression, in which each stage represents a significant increase in both the intensity of the cultivation system and the number of families it can support. Stage one, called forest-fallow cultivation, involves 20–25 years of letting fields lie fallow after 1–2 years of cultivation. In bush-fallow cultivation, stage two, farmers cultivate the land for 2–8 years followed by a fallow period of 6–10 years. In stage three, the fallow period shortens to just 1–2 years between cultivated periods. In the next stage, farmers begin annual cropping, leaving the land fallow for only several months between plantings. In the final stage, the most intensive system, multicropping the same plot bears several crops a year with little or no fallow period. With each stage, the land can support greater populations, but each transition also involves greater depletion of soil nutrients. While many might argue the unsustainable nature of this system, Boserup argued that the increased levels of productivity would counteract the land's being rendered infertile from overuse.

TIP

A good way to study for the exam is to try to make up multiple-choice questions on your own. Can you write a good multiple-choice question based on the key terms described in bold text on this page?

During the modern era, there have been at least four pivotal periods in the history of agriculture. The first key moment in agriculture's history has been mentioned, and it occurred when humans began modifying plant and animal species to sustain themselves. Seed domestication and technology about as advanced as an animal-driven plow actually describes the majority of agricultural history. Up until the late 18th century, most people throughout the world were farmers or hunter-gatherers of some sort. However, the second pivotal moment in agricultural history arrived with the **Industrial Revolution**, which began in the late 1700s in England and rapidly spread to Western Europe and the United States. The effects of this time period dramatically altered the global geography of agriculture.

Three components of the Industrial Revolution were particularly important for the transformation of agriculture in Western Europe and North America. First, during the Industrial Revolution millions of people migrated from rural areas into the cities of France, England, Germany, and the United States. These new urbanites came to cities, such as London, Manchester, Chicago, and New York, looking for jobs in factories and a better way of life. When they arrived, they created enormous new markets for the agricultural products produced in adjacent rural areas. Second, **mechanization** replaced human hands with agricultural technology, allowing farmers to produce more crops with less work. Finally, increased access to efficient forms of transportation, such as trains and steamboats, allowed farmers to ship their products farther at a lower cost. In fact, increased transportation technology has played a large role in determining which areas of the globe transitioned into commercial agricultural economies. Many isolated spots on the earth's surface remain subsistence economies simply because of their limited access to other parts of the world. Additionally, technological advances like refrigerated boxcars were particularly important because they allowed farmers to ship items great distances to urban consumers. Between 1780 and 1850, these three factors revolutionized farming in the newly industrialized world.

STRATEGY

The Green Revolution provides an excellent topic for an essay question. Can you write—and then answer correctly—a sample essay question based on the Green Revolution?

Another important period in the modern history of agriculture came after World War II. Beginning in the late 1940s, the industrialized countries of the northern hemisphere began transferring a great amount of technology, machinery, fertilizers, and other agricultural inputs to the less-developed countries of Africa, Asia, and Latin America. This episode, called the **Green Revolution**, continued into the 1960s, when developed countries finally realized the detrimental effects these new technologies wreaked on the environment. Instead of alleviating hunger, new machinery, "miracle" seeds, elaborate irrigation systems, and potent fertilizers were devastating the land, destroying traditional modes of agricultural production, and shattering ancient social structures. The Green Revolution encouraged rampant land speculation, vast human migrations, and unsustainable farming practices. Agronomists noticed that techniques developed for farming in the temperate climates were often unsuitable for tropical agriculture. And economists watched as multinational corporations began to steer local economies away from producing food for local consumption and toward producing **specialty crops** for export, such as peanuts and pineapples. As with many other well-meaning development projects, the Green Revolution failed largely because proponents did not consider the potential side effects of their actions. Today, many areas once farmed sustainably by local people for local consumption are now planted with specialty crops grown for export to North America and Western Europe.

Another agricultural transformation, no less contested than the Green Revolution, currently drives much of the world's agricultural activity. High-tech agriculture, which employs computerized irrigation systems, long-term weather predictions, and **genetically modified organisms**, is changing agricultural practices throughout the world. Although most people agree that sophisticated irrigation and satellite-based weather predictions can only help farmers, the issue of **genetically modified organisms** is hotly debated. All crops are genetically modified, in the sense that they have been altered from their original genetic state by selection over time for human use. However, today's genetically modified foods are different. Some of these foods are the products of organisms that have had their genes altered in a laboratory for specific purposes, such as disease resistance, increased productivity, or nutritional value. Many agribusiness corporations have embraced the promise of control, predictability, and efficiency that genetically modified organisms represent. Critics say that such products have not been proven safe for people and that their effects on surrounding ecosystems could be devastating. As the debate rages, genetically modified foods are being produced by large corporations and small farmers throughout the world.

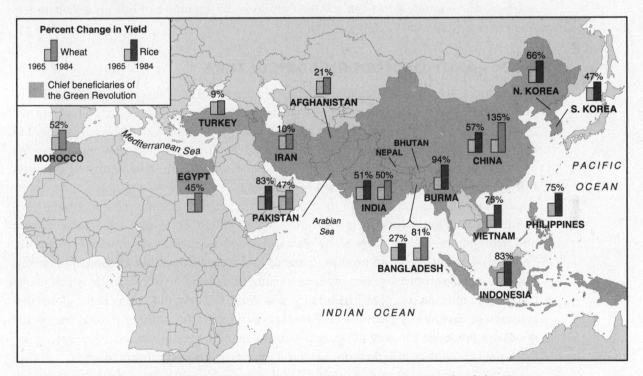

Figure 5.4 This map demonstrates the increases in yields in Asia as a result of the Green Revolution. While increased yields of both wheat and rice certainly helped support expanding populations in these regions, the Green Revolution also forced many local farmers to abandon their farms.

Perhaps the most important trend in modern agriculture has been the development of multinational **agribusinesses**. Today, a handful of giant corporations dominate a significant fraction of the world's agricultural markets through their control of land, technology, machinery, shipping, packaging, and marketing, such that the farm no longer maintains its position as the centerpiece of agricultural activity. California, Florida, and Texas gave birth to many agribusiness conglomerates during the first half of the 20th century, and these corporations have since expanded to Africa, Southeast Asia, and Latin America. Another important aspect of this transformation is the issue of integration. Agribusiness firms have long practiced **horizontal integration**, a form of corporate organization in which several branches of a company or several commonly owned companies work together to sell their products in different markets. More recently, agribusiness firms have also adopted **vertical integration**, in which the same firm controls multiple aspects or phases of a **commodity chain**. A commodity chain is a linked system of processes that gather resources, convert them into goods, package them for distribution, disperse them, and sell them on the market. The same firm may, for example, own a seed company, a fertilizer company, a tractor company, and a grain distribution company—giving it broad powers to control the overall market. Although the transformation from small, local farms to integrated agribusiness operations has had some positive effects, it has also had many negative and unintended consequences. One in particular, the demise of the American family farm, has forced many traditional farmers into unemployment. During the last couple of decades, new movements have arisen to reestablish local food production in both the highly industrialized and the less-developed world. Cooperatives, where money generated benefits the local economy, and local farmer's markets, where family farms can sell their produce, are examples of this grassroots movement to recapture local food production.

GEOGRAPHY OF MODERN AGRICULTURE

Geographers look at both the spatial variation in agricultural activities and agricultural methods currently in use across the globe. The spatial variation in subsistence agriculture has just been discussed; the global distribution of commercial agriculture also varies across space as climate, soil, the availability of material inputs, and the dominant culture system determine agricultural production within a region. For example, oranges and other citrus fruits cannot thrive outside of tropical climates, sandy desert soils cannot produce tomatoes, and Asians depend on rice as a staple within their diet. These variables, in large part, determine agricultural production across the globe; however, certain technologies, such as greenhouses and hydroponics, have allowed many places to overcome their environmental limitations. For example, Icelanders can produce bananas in simulated tropical environments (greenhouses), and desert dwellers can produce tomatoes thanks to hydroponics, in which plants can grow in nutrient solutions instead of in soil. Thus, it is helpful to think of the global distribution of agricultural practices and products as a constantly shifting mosaic, as specific goods are produced in many different regions by many different methods. However, a few key types of commercial agriculture are worth discussing and mapping out here.

- ■ **COMMERCIAL LIVESTOCK** production takes multiple forms and can be divided up into two major categories: **livestock ranching** and **dairying**. Livestock ranching is widespread throughout much of western North America, South America, southern Africa, western Asia, and Australia. One interesting practice in livestock ranching is called **transhumance**. Transhumance is the seasonal movement of livestock between different ranges. In many regions, livestock are moved into the mountains in the summer and then down into the valleys in the winter. Dairying is another important form of animal husbandry that is common in northern Europe and the northern United States.

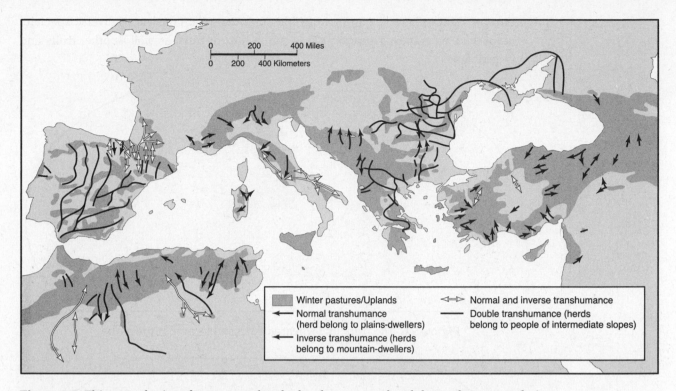

Figure 5.5 This map depicts the routes taken by herders across the globe as they move their flocks from summer to winter pastures for grazing.

- **COMMERCIAL GRAIN FARMING**, which includes wheat and corn, occurs in the North American Great Plains and in southern Russia. A large component of commercial grain farming goes toward feeding livestock. In fact, in Western Europe, three-fourths of cropland is devoted to grain farming specifically for livestock consumption. In general, the market value of meat from livestock surpasses that of grains; thus, many farmers choose to convert their grain into meat by feeding it to livestock.

- **TROPICAL PLANTATIONS** grow crops such as sugarcane and coffee, and are widespread throughout the tropics, in Central and South America, Africa, Asia, and the Caribbean. Plantations generally have some form of foreign control through either investments, management, or marketing, and often employ people not native to the region. Additionally, while many of the crops grown on plantations are suitable to the tropical environments where they exist, they are not usually native to those areas and are almost always exported to other countries rather than consumed locally.

- **AQUACULTURE** is an ancient form of food production, dating back thousands of years in places such as Egypt and China, in which fish or other aquatic organisms are cultivated for human consumption. Aquaculture production has exploded over the past few decades, however, to meet the global demand for seafood, with most production in tropical countries such as Mexico, Ecuador, and Thailand. Today, major aquaculture products include seaweed, shellfish, tilapia, shrimp, and salmon, which is grown mostly in the temperate regions like Canada.

- Finally, **MIXED AND SPECIALTY CROP FARMING** is extremely diverse, and the particular forms it takes depend largely on climate. In the humid, subtropical southeastern United States, citrus fruits, vegetables, and nuts are grown alongside cattle ranches. **Mediterranean agriculture**, practiced in the Mediterranean-style climates of Western

Europe, California, and portions of Chile and Australia, consists of diverse specialty crops such as grapes, avocados, olives, and a host of nuts, as well as other fruits and vegetables.

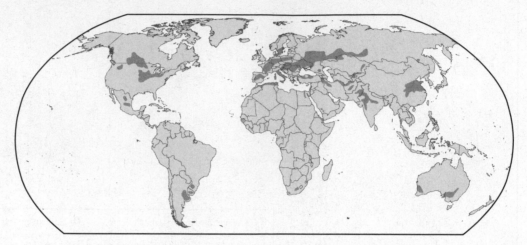

Figure 5.6 The principal wheat-growing areas across the globe

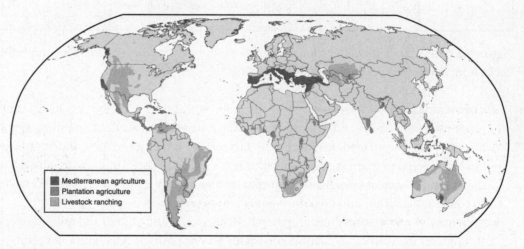

Mediterranean agriculture
Plantation agriculture
Livestock ranching

Figure 5.7 The geographic distribution of Mediterranean agriculture, plantation agriculture, and livestock ranching. Ranching tends to occur in midlatitude climates, while Mediterranean agriculture dominates moderate coastal climates, and plantations flourish in tropical coastal regions.

On a smaller scale, the regional distribution of agricultural practices has also been of interest to many geographers. Johann Heinrich von Thunen described one particularly important

NOTE

The 2008 APHG exam included a free-response question that asked students to compare von Thunen's agricultural land-use model with Burgess's urban land-use model, which also uses concentric rings to describe the organization of space in cities.

model of the regional distribution of agriculture during the 19th century. Von Thunen noticed that lands that appeared to have exactly the same physical geography were actually being used for very different agricultural purposes. He explained this phenomenon through the concept of rent. According to the **von Thunen model**, rent, or land value, will decrease the farther one gets away from central markets. Conversely, rent is highest in close proximity to urban markets. Thus, only the agricultural products that used the land intensively, have high transportation costs, and were in great demand would be located close to urban markets. Products that were in lower demand, required more extensive land use, or were less expensive to ship would be found farther away from the markets, where rent was lower. More specifically, von Thunen speculated that dairying and gardening of fruits and vegetables would be located close to the urban market, while extensive cattle ranching, mixed farming, and orchards would be located farther away. Because fruits, vegetables, and dairy products spoil more quickly, require more sensitive forms of transportation, and in general generate higher prices, farmers can afford to pay the higher price of rent near the market. Although, in real landscapes, this pattern is complicated by many factors, it still describes actual patterns of agricultural land use surrounding many cities.

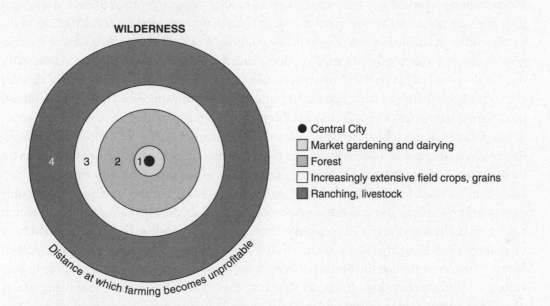

Figure 5.8 Agricultural activity that generates goods that are expensive to transport or perish quickly occurs closer to the market, whereas goods that do not require expensive transport and maintain a longer shelf life are produced on land farther from the central city.

Many other features characteristic of modern agricultural geography were described in the previous section in the discussion of the roots and development of agriculture through history. In the future, it is safe to hypothesize increasing transitions of extensive agricultural systems into intensive ones that can generate greater amounts of goods on smaller amounts of land. Even activities seemingly impossible to transition, such as livestock ranching, have begun traveling along the path toward intensive cultivation in the form of feedlots. **Feedlots** concentrate the raising of livestock in a small geographic space, where they are fed hormones and other fattening grains to prepare them for slaughter at a much more rapid pace and in a much smaller space. Esther Boserup was correct in her observation that increasing population levels necessitate transformations in the world's agricultural systems in order to provide for greater numbers of people. In large part, the technological and biological innovations that have occurred and continue to occur in modern agriculture are responding to the need to produce enough goods to provide for an ever-increasing global population.

For these reasons, **food security** has emerged as a major issue for agricultural geographers. In 1996, the World Food Summit defined food security as the ability of "all people, at all times" to "have access to sufficient, safe, nutritious food to maintain a healthy and active life." Today, however, approximately 2 billion people around the world lack some degree of food security due to individual poverty, regional or national famine, or other problems. Lack of food security can lead to a range of public health maladies, particularly for vulnerable children in countries such as India, where tens of millions are undernourished and underweight. Many scholars are also concerned that climate change and increases in the price of oil may exacerbate these food security problems.

Two of the dominant forces of modern agriculture that work to provide maximum yields for greater populations include biotechnology and agribusiness. Both topics were briefly mentioned earlier, but in order to understand the world's current agricultural system, more time must be spent describing these two dramatic shifts in agricultural production. **Biotechnology** involves any techniques used to modify living organisms in such a way that they improve plant and animal species and, in turn, plant and animal production. As mentioned earlier in the discussion of genetically modified foods, these processes allow farmers greater control over the goods they produce, thereby allowing them greater yields. With the development of plants that can resist certain pests and weeds, as well as plants that are clones from the tissues of other plants, the importance of both space and time in agricultural productivity decreases. These forms of technology allow farmers to grow virtually any product anywhere on the globe. However, biotechnology comes with very serious side effects. Before discussing those, it is important to first note that private firms that have patents on the methods and products they achieve develop most of the biotechnology techniques discussed so far. Thus, both countries in the periphery and small farmers lose out because they cannot afford to transition into these new modes of production. Furthermore, little research has been conducted studying the possible effects genetically modified organisms might have on other nonmodified organisms, such as butterflies and insects, that may pollinate modified plants. Thus, even though the Biotechnologic Revolution may seem to be the answer to providing for future populations, it may be similar to the Green Revolution in that the products are placed on the market without full knowledge of all their various repercussions.

The development of agribusinesses has also radically changed traditional agricultural production. This change evidences itself in several different forms, but the most specific form is a transition from agricultural production to food production, and a transition of the role of the farm as the centerpiece in agricultural production. Food production differs from agricultural production in that it includes an addition of economic value to an agricultural product through canning, refining, packing, or packaging. The modern grocery store provides a perfect example of this process. All the many goods outside of fruits, vegetables, and grains that line the majority of shelves are a result of production processes that process traditional agricultural goods and then package them before they are placed on the shelf in the form of chicken strips or frozen waffles. Additionally, evidence of the rise of the transnational corporation (TNC) within agribusiness also can be seen at most modern grocery stores. The fact that you can get almost any fruit or vegetable from all over the globe at any time of the year implies some level of corporate control over agricultural goods in other parts of the world, usually peripheral countries. In fact, many developing countries encourage TNCs to grow certain goods within their bounds for export in hope that these activities will stimulate their economies. However, this system forces local farmers to abandon traditional methods of agriculture, and a large proportion of the profits generated from these large agribusinesses accrue to the corporation rather than the local economy; thus, they do not always provide the economic benefits the host country hopes for. Additionally, they represent a dramatic transition in the control of agricultural activity. The farm used to be the center controlling

TIP

The 2009 APHG exam included a free-response question that compared the decline in American dairy farming, since 1970, to the rise in organic farming during the same period.

force, but the business or corporation now holds that place as it controls what seeds are grown where, where goods are packaged or processed, and finally where goods are sold. As the world's need for food continues to expand, agricultural geographers and policy makers will face many sticky issues and difficult challenges regarding sustainable and geographically even food production.

AGRICULTURE AND THE ENVIRONMENT

Farmers and agricultural engineers have a very difficult job. How can they protect the environment while ensuring a sustainable harvest, providing safe, reliable, and high-quality food for a growing population? Historically, agriculture has had many adverse impacts on the environment. **Pesticides**, such as DDT, the effects of which were made famous by Rachel Carson in the book *Silent Spring*, have harmed wildlife populations; polluted rivers, lakes, and oceans; and worked their way through the food chain all the way up to human beings. **Topsoil loss**, or erosion, is a tremendous problem in areas with fragile soils, steep slopes, or torrential seasonal rains. Because fertile topsoil tends to accumulate very slowly, this essential resource, once lost, could take thousands of years to replace. Another soil conservation issue is salinization. **Salinization** occurs when soils in arid areas are brought under cultivation through irrigation. In arid climates, water evaporates quickly off the ground surface, leaving salty residues and rendering the soil infertile. As a result of these two soil problems, millions of acres of formerly arable land have become infertile. **Desertification** is the process by which formerly fertile lands become increasingly arid, unproductive, and desert-like. Thus, soil conservation will be essential if the world is to move toward a future of sustainable agricultural production.

STRATEGY

Organic agriculture—farming without the aid of artificial inputs such as pesticides, chemical fertilizers, and genetically engineered seeds—has grown tremendously during the past decade. Where do you think you would be most likely to find organic farms in the United States? Why? If you can develop a plausible hypothesis for the emerging geography of organic farming, then you will be well on your way to understanding many of the key concepts presented in this chapter.

Another important environmental issue having to do with agricultural production is rural land-use change. In many areas of the United States, urban sprawl has overtaken formerly productive agricultural areas, converting fields and orchards to parking lots and subdivisions. Local government planning commissions are attempting to halt this process through agricultural zoning ordinances and tax incentives. However, in places such as California, where land values have risen dramatically over the past quarter century, many farmers find it difficult to pay property taxes and resist the quick financial return associated with developing their land. Unfortunately, some of the lands that are being developed for housing are among the most fertile. Another trend in agricultural land use is the shift from traditional, low-intensity operations, such as ranching, to high-intensity production associated with specialty crops and orchards. These specialty crops frequently offer a greater profit per acre per year; however, they also convert rural landscapes containing diverse habitats to comparatively sterile environments. Interestingly, this pattern has actually reversed itself in New England, where vast areas that were cleared for farms during the 18th and 19th centuries have now reverted back to forest. It is up to local planners, land owners, and voting citizens to decide how to best address the need for housing and services, while preserving traditional landscapes, habitats, and rural lifestyles.

One way to think about agriculture and the environment is in terms of **sustainability**. The word sustainability has many definitions. In its simplest form, it is a set of policies or practices by which societies can ensure that the people of the future have the same access to resources—and thus the same opportunities—as the people of today. Policies or practices that violate this definition may be unsustainable. Yet many questions remain. How do we know if certain practices are truly sustainable? How far into the future should we think? How can we be sure that the people of tomorrow will even want the same resources we have today? Is it better to exploit resources now to promote economic growth or to save them for tomorrow, even if that means some people today are going without? What are appropriate regulations to promote sustainability? Who should decide?

Region	Overgrazing	Deforestation	Agricultural Mismanagement	Other	Total	Degraded Areas as Share of Total Vegetated Land
Asia	197	298	204	47	746	20%
Africa	243	67	121	63	494	22%
South America	68	100	64	12	244	14%
Europe	50	84	64	22	220	23%
North and Central America	38	18	91	71	158	8%
Australia, New Zealand and the South Pacific	83	12	8	0	103	13%
World	679	579	552	155	1,965	17%

(million hectares)

Figure 5.9 The mismanagement of soil has caused massive soil degradation across the globe.

A popular manifestation of the sustainability movement is **organic agriculture**. Organic farming rejects or severely limits the use of artificial fertilizers, pesticides, hormones, antibiotics, additives, and genetically modified organisms. Organic farmers use crop rotation, natural fertilizers such as manure, and biological pest control to promote healthy, vigorous crops. The United States, European Union, and many countries have strong standards for labeling products as "organic," unlike other words such as "natural," which have little meaning when written on a food product label.

Agricultural geography's complexity should be evident by now as you have explored all the many ways humans have devised to sustain themselves from the land. Throughout human history, agriculture has experienced many dramatic transitions in an effort to support ever-expanding human populations more efficiently. Often, the strategies used to accomplish this goal have unexpected and detrimental consequences on the physical environment. As greater levels of technology and scientific improvements are introduced into agricultural activity, it seems almost that a corresponding negative effect occurs on the natural environment. Agricultural geographers face many challenges in the future as they seek to understand all the various repercussions of modern agricultural techniques, while simultaneously devising methods to improve Earth's yield to accommodate an ever-increasing global population.

KEY TERMS

AGRIBUSINESS The set of economic and political relationships that organize food production for commercial purposes. It includes activities ranging from seed production, to retailing, to consumption of agricultural products.

AGRICULTURE The art and science of producing food from the land and tending livestock for the purpose of human consumption.

ANIMAL HUSBANDRY An agricultural activity associated with the raising of domesticated animals, such as cattle, horses, sheep, and goats.

AQUACULTURE The cultivation or farming (in controlled conditions) of aquatic species, such as fish. In contrast to commercial fishing, which involves catching wild fish.

BIOTECHNOLOGY A form of technology that uses living organisms, usually genes, to modify products, to make or modify plants and animals, or to develop other microorganisms for specific purposes.

CAPITAL-INTENSIVE AGRICULTURE Form of agriculture that uses mechanical goods, such as machinery, tools, vehicles, and facilities, to produce large amounts of agricultural goods—a process requiring very little human labor.

COMMERCIAL AGRICULTURAL ECONOMY All agricultural activity generated for the purpose of selling, not necessarily for local consumption.

COMMODITY CHAINS A linked system of processes that gather resources, convert them into goods, package them for distribution, disperse them, and sell them on the market.

DAIRYING An agricultural activity involving the raising of livestock, most commonly cows and goats, for dairy products such as milk, cheese, and butter.

DESERTIFICATION The process by which formerly fertile lands become increasingly arid, unproductive, and desert-like.

DOMESTICATION The conscious manipulation of plant and animal species by humans in order to sustain themselves.

EXTENSIVE AGRICULTURE An agricultural system characterized by low inputs of labor per unit land area.

FEEDLOTS Places where livestock are concentrated in a very small area and raised on hormones and hearty grains that prepare them for slaughter at a much more rapid rate than grazing; often referred to as factory farms.

FERTILE CRESCENT Area located in the crescent-shaped zone near the southeastern Mediterranean coast (including Iraq, Syria, Lebanon, and Turkey), which was once a lush environment and one of the first hearths of domestication and thus agricultural activity.

FOOD SECURITY People's ability to access sufficient safe and nutritious food to maintain a healthy and active life.

GENETICALLY MODIFIED FOODS Foods that are mostly products of organisms that have had their genes altered in a laboratory for specific purposes, such as disease resistance, increased productivity, or nutritional value, allowing growers greater control, predictability, and efficiency.

GREEN REVOLUTION The development of higher-yield and fast-growing crops through increased technology, pesticides, and fertilizers transferred from the developed to developing world to alleviate the problem of food supply in those regions of the globe.

HORIZONTAL INTEGRATION A form of corporate organization in which several branches of a company or several commonly owned companies work together to sell their products in different markets.

HUNTING AND GATHERING The killing of wild animals and fish as well as the gathering of fruits, roots, nuts, and other plants for sustenance.

INDUSTRIAL REVOLUTION The rapid economic changes that occurred in agriculture and manufacturing in England in the late 18th century and that rapidly spread to other parts of the developed world.

INTENSIVE CULTIVATION Any kind of agricultural activity that involves effective and efficient use of labor on small plots of land to maximize crop yield.

LABOR-INTENSIVE AGRICULTURE Type of agriculture that requires large levels of manual labor to be successful.

LIVESTOCK RANCHING An extensive commercial agricultural activity that involves the raising of livestock over vast geographic spaces typically located in semi-arid climates like the American West.

MECHANIZATION In agriculture, the replacement of human labor with technology or machines.

MEDITERRANEAN AGRICULTURE An agricultural system practiced in the Mediterranean-style climates of Western Europe, California, and portions of Chile and Australia, in which diverse specialty crops such as grapes, avocados, olives, and a host of nuts, fruits, and vegetables make up profitable agricultural operations.

ORGANIC AGRICULTURE The use of crop rotation, natural fertilizers such as manure, and biological pest control—as opposed to artificial fertilizers, pesticides, hormones, antibiotics, additives, and genetically modified organisms—to promote healthy, vigorous crops.

PASTORALISM A type of agricultural activity based on nomadic animal husbandry or the raising of livestock to provide food, clothing, and shelter.

PESTICIDES Chemicals used on plants that do not harm the plants, but kill pests and have negative repercussions on other species that ingest the chemicals.

PLANNED AGRICULTURAL ECONOMY An agricultural economy found in communist nations in which the government controls both agricultural production and distribution.

PLANTATION A large, frequently foreign-owned piece of agricultural land devoted to the production of a single export crop.

SALINIZATION Process that occurs when soils in arid areas are brought under cultivation through irrigation. In arid climates, water evaporates quickly off the ground surface, leaving salty residues that render the soil infertile.

SHIFTING CULTIVATION The use of tropical forest clearings for crop production until their fertility is lost. Plots are then abandoned, and farmers move on to new sites.

SLASH-AND-BURN AGRICULTURE System of cultivation that usually exists in tropical areas, where vegetation is cut close to the ground and then ignited. The fire introduces nutrients into the soil, thereby making it productive for a relatively short period of time.

SPECIALTY CROPS Crops, including items like peanuts and pineapples, that are produced, usually in developing countries, for export.

SUBSISTENCE AGRICULTURAL ECONOMY Any farm economy in which most crops are grown for nearly exclusive family or local consumption.

SUSTAINABILITY A set of policies or practices by which societies can ensure that the people of the future have the same access to resources and thus the same economic and environmental opportunities as people living today.

SWIDDEN Land that is prepared for agriculture by using the slash-and-burn method.

TOPSOIL LOSS When the top fertile layer of soil is depleted through erosion. It is a tremendous problem in areas with fragile soils, steep slopes, or torrential seasonal rains.

TRANSHUMANCE The movements of livestock according to seasonal patterns, generally lowland areas in the winter, and highland areas in the summer.

URBAN SPRAWL The process of urban areas expanding outward, usually in the form of suburbs, and developing over fertile agricultural land.

VERTICAL INTEGRATION A form of corporate organization in which one firm controls multiple aspects or phases of a commodity chain.

VON THUNEN MODEL An agricultural model that spatially describes agricultural activity in terms of rent. Activities that require intensive cultivation and cannot be transported over great distances pay higher rent to be close to the market. Conversely, activities that are more extensive, with goods that are easy to transport, are located farther from the market where rent is less.

CHAPTER SUMMARY

During the past 10,000 years, agriculture has become an endeavor of enormous proportions, with dramatic consequences for Earth's physical and human geography. The first agriculturalists were hunter-gatherers who gradually, over thousands of years, adopted farming as another strategy to ensure their survival. By the beginning of the Colonial Period, agriculture was widespread throughout the world. Four important episodes—the first conscious cultivation of plants, the Industrial Revolution, the Green Revolution, and the Biotechnologic Revolution—have dramatically altered the way farmers work and the way people eat. In many countries, subsistence agriculture and animal husbandry still feed most people; however, large-scale commercial agriculture is an increasingly important endeavor throughout the world. Escalating human populations and unsustainable farming techniques have created a host of social and environmental problems that threaten the potential for future populations to reap sustainable harvests from the land.

PRACTICE QUESTIONS AND ANSWERS

Historical Geography of Agriculture

MULTIPLE-CHOICE QUESTIONS

1. The first agriculturalists were

   (A) commercial farmers.

   (B) European entrepreneurs.

   (C) also hunter-gatherers.

   (D) also ranchers.

   (E) most likely males.

2. Slash-and-burn agriculture is

   (A) not sustainable.

   (B) practiced in high, mountainous regions.

   (C) typical for tropical forests.

   (D) a relatively new invention.

   (E) always completely sustainable.

3. The Industrial Revolution transformed Western agriculture

   (A) through mechanization and the creation of new markets.

   (B) with biotechnology.

   (C) through technological and religious change.

   (D) by eliminating agricultural pests.

   (E) by eliminating plant hybridization.

4. The Green Revolution greatly increased crop production in some countries

   (A) without adverse side effects.

   (B) as a replacement for deindustrialization.

   (C) with some adverse side effects.

   (D) by encouraging the cultivation of local crop varieties.

   (E) by introducing organic agricultural methods.

5. Which of the following was NOT a location of independent plant and animal domestication?

   (A) India

   (B) Iraq

   (C) California

   (D) China

   (E) Peru

6. Ranching is a good example of which type of agricultural system?

   (A) Intensive subsistence cultivation
   (B) Extensive commercial cultivation
   (C) Labor-intensive agriculture
   (D) Capital-intensive agriculture
   (E) Controlled agriculture

FREE-RESPONSE QUESTION

1. The Industrial Revolution's impacts extended beyond industry to other areas of the economy.

   (A) Describe how the Industrial Revolution affected modern agricultural production.
   (B) What are some of the positive and negative effects of industrial agriculture?

Geography of Modern Agriculture

MULTIPLE-CHOICE QUESTIONS

1. The modern global geography of agriculture is determined by

   (A) climate.
   (B) soil.
   (C) cultural traditions.
   (D) All of the above
   (E) Only (A) and (B)

2. According to von Thunen, the regional geography of agriculture is determined by

   (A) land area.
   (B) rent.
   (C) urban marketing.
   (D) availability of material inputs.
   (E) climate.

3. _____ is (are) widespread in semiarid climates throughout the world.

   (A) Ranching
   (B) Tropical plantations
   (C) Dairying
   (D) Slash-and-burn agriculture
   (E) Rice paddies

4. The effects of biotechnology

   (A) are positive because it allows for much greater agricultural yields.
   (B) are negative because its expense limits its availability to all farmers across the globe.
   (C) are unknown because very little research has been conducted on them.
   (D) All of the above
   (E) None of the above

5. Agribusiness has had all of the following effects on agriculture, EXCEPT

   (A) the farm is no longer the center of agricultural activity.
   (B) TNCs often control agricultural activity abroad.
   (C) family farmers, through increasing technology, are producing goods for the global economy.
   (D) agriculture has become a multilevel process of production, processing, marketing, and consumption.
   (E) some corporations essentially dictate agricultural production in other countries besides their own.

FREE-RESPONSE QUESTION

1. Current agricultural systems differ dramatically from agricultural production just 50 years ago.

   (A) What are some of the driving forces behind the world's current agricultural system?
   (B) What are some benefits and disadvantages of global-scale agricultural production?

Agriculture and the Environment

MULTIPLE-CHOICE QUESTIONS

1. In arid climates, like southern California and the Middle East, _____ can cause the soil to become salty and infertile.

   (A) erosion
   (B) topsoil loss
   (C) salinization
   (D) saltation
   (E) droughts

2. _____ is a common cause of decreasing farmland in rapidly growing urban areas.

   (A) Urban sprawl
   (B) Topsoil loss
   (C) Loss of material inputs
   (D) Industrialization
   (E) Agribusiness

3. DDT is an example of a _____ that has had negative effects all the way through the food chain.

   (A) herbicide
   (B) pesticide
   (C) bacteria
   (D) fungicide
   (E) genetically modified organism

4. Soil specialists must work to overcome the negative effects of _____ associated with agricultural production.

    (A) fertilization and salinization

    (B) pesticides and fertilization

    (C) salinization and topsoil loss

    (D) topsoil loss and gentrification

    (E) the Green Revolution and agribusiness

FREE-RESPONSE QUESTION

1. Modern agricultural production necessarily involves benefits and disadvantages.

    (A) Describe some of the inputs in the modern agricultural system.

    (B) What are the environmental implications of modern agriculture?

Answers for Multiple-Choice Questions

HISTORICAL GEOGRAPHY OF AGRICULTURE

1. **(C)** Hunter-gatherers were the first individuals to domesticate plants and animals. Even after domesticating certain species, most hunter-gatherers continued hunting and gathering to round out their diet.

2. **(C)** Slash-and-burn agriculture is an agricultural method that introduces nutrients into the soil through burning of organic matter. It occurs mostly in tropical forests, where the soil is rather nutrient poor, as a way to increase soil productivity. Slash and burn can be unsustainable if farmers who practice it do not allow the land enough time to regenerate itself. Generally, the system of shifting cultivation can be quite sustainable if managed sensibly.

3. **(A)** During the Industrial Revolution, many people migrated from rural areas to large urban centers, generating a great need for agricultural goods within those centers. Furthermore, the technology characteristic of the Industrial Revolution transformed agricultural production as mechanization allowed for much more rapid cultivation of greater expanses of land.

4. **(C)** The Green Revolution brought technology, miracle seeds, fertilizer, and other inventions of the developed world into developing nations to stimulate agricultural growth. While certainly these new inputs led to increased agricultural productivity, they carried with them detrimental implications for the local natural environment.

5. **(C)** Although many people associate the Fertile Crescent with the first site of agricultural activity, in reality, several other places across the globe independently began domesticating plants and animals for human sustenance at around the same time. California, although currently an extremely important location of agricultural activity, was not one of these first hearths.

6. **(B)** Ranching is an agricultural activity that takes place over large expanses of land and as such is a good representative of an extensive commercial agricultural activity. Additionally, it does not require either large amounts of human labor or capital inputs.

GEOGRAPHY OF MODERN AGRICULTURE

1. **(D)** Geographers are concerned with what is grown where on the earth's surface. Both currently and historically, climate, soil, and cultural traditions determine this pattern. With increasing technology, these factors are losing their potency, but they still remain the dominant forces in the world's current agricultural mosaic.

2. **(B)** According to the von Thunen model, the land located nearest the market will have the highest rent. Generally, agricultural activities that occur in this region are those that are expensive to transport and require intensive cultivation.

3. **(A)** Ranching takes place in the drier climates across the globe, where wide expanses of land that are not very good for cultivation exist. Both tropical plantations and slash-and-burn agriculture exist in warm, moist climates, and dairying usually exists in cooler climates, such as northern Europe and the northern United States.

4. **(D)** While the effects of biotechnology have not been critically investigated, it still is affecting global agricultural production, allowing for much greater yields of certain products. Because the processes of biotechnology are patented by private companies, only large agribusinesses can afford them, thereby limiting their use.

5. **(C)** Agribusiness has largely contributed to the demise of the family farm. As agriculture becomes increasingly controlled by large corporations that have the technology to mass produce goods without much human capital, it loses its need for human labor. Furthermore, the agricultural activities that require large amounts of labor have been relocated to parts of the globe where human labor is cheaper.

AGRICULTURE AND THE ENVIRONMENT

1. **(C)** Salinization occurs when arid environments use irrigation to provide enough moisture for plant production. When the water evaporates, it leaves a salty residue, which eventually causes the soil's infertility.

2. **(A)** As urban areas continue to expand outward, usually in the form of suburbs, the development often takes over agricultural land.

3. **(B)** The effects of DDT, made known by Rachel Carson in *Silent Spring*, demonstrate the various detriments pesticide use has all the way through the food chain. All the other options may too have effects through the food chain, but DDT is a pesticide specifically designed to eliminate insects, or pests, that threaten certain agricultural products.

4. **(C)** Chemical fertilizers may cause negative effects on the soil, but some forms of fertilizers are organic, meaning that they use natural products rather than chemicals to stimulate plant growth and, therefore, do not harm the soil. Both salinization and topsoil loss destroy the soil's properties and must be overcome by soil scientists if they want the land to remain fertile.

Answers for Free-Response Questions

HISTORICAL GEOGRAPHY OF AGRICULTURE

1. **Main points:**
   - The impacts on agriculture as a result of the Industrial Revolution can be categorized into three main areas: rural-to-urban migrations, mechanization, and transportation.
   - With the Industrial Revolution came many new factory jobs in large urban areas. This stimulated a mass rural-to-urban migration, specifically in France, Germany, the United Kingdom, and the United States. These large population centers generated a great need for agricultural products from the surrounding rural areas. Thus, the Industrial Revolution created a giant market for agricultural products.
   - Second, mechanization of certain agricultural activities radically transformed the amount of agricultural output in relation to input of human labor. The introduction of tractors and combines slowly began to transform agricultural activity in the developed regions of the globe.
   - Finally, increased transportation technology allowed farmers to ship their goods greater distances for smaller costs, allowing them to grow more than they could traditionally as they had greater access to distant markets. Additionally, refrigerated transportation technology allowed the more time-sensitive products to extend their geographical market.

GEOGRAPHY OF MODERN AGRICULTURE

1. **Main points:**
   - First, the driving forces behind the current agricultural system are responding to an increasing need to provide enough food for an ever-expanding global population. As such, scientists, economists, business people, and policy makers have had to devise products, processes, and plans to accommodate such a large population.
   - Two of the main forces in use to provide greater agricultural yields are biotechnology and the rise of agribusinesses.
   - Biotechnology is the use of organisms to improve other organisms. These technological processes have improved agricultural production through the development of pest- and disease-resistant plants and the technology for cloning specific plant species. However, biotechnology is expensive and patented within the private sector, limiting its use to agribusinesses that can afford to profit from it. Furthermore, the environmental impacts have not been thoroughly investigated. Most of the European community has banned the sale of genetically modified foods because they have not been proven safe, and many are suspicious of their effects on other species, including humans.
   - Agribusiness has transformed agricultural production into a global food chain of production, processing, and consumption. Corporations have replaced farms as the driving force behind agriculture and have expanded production capabilities dramatically such that almost any individual in the developed world can obtain any agricultural good at any time of the year. Because corporations have overtaken local agricultural production, they have displaced many family farmers and disrupted many agricultural traditions. Furthermore, often large-scale agricultural production involves the use of chemicals to ensure specific yields, in turn dramatically affecting the natural environment.

AGRICULTURE AND THE ENVIRONMENT

1. **Main points:**
   - Modern forms of agriculture wreak many forms of destruction on the land. Common destroying forces include pesticide use, salinization, loss of topsoil, and urban sprawl.
   - Pesticide use introduces certain chemicals into soil mixtures that are ingested into plants and other organisms and eventually make their way up the food chain, into humans. These chemicals have varying effects on different species. For example, when birds ingest too much DDT, a pesticide popular not too long ago, their egg shells no longer harden, killing many birds before they even have a chance to hatch.
   - Both salinization and topsoil loss, if not counteracted, can render the soil infertile. Salinization occurs in arid areas that use irrigation to water plants. When water evaporates in these environments, it leaves a salty residue, which over time destroys the soil's chemical and biological makeup. Topsoil loss occurs through erosion, which may be the result of overcultivation, or when steep slopes are cultivated. When topsoil loss occurs, the most nutrient-rich layer of soil disappears, dramatically affecting the soil's ability to be productive.
   - With urban sprawl, certain fertile lands are being paved over; thus, arable land is not being used for its best purpose. When this process occurs, it necessitates the need for intensive cultivation systems elsewhere, and these levels of intensity can have very negative effects on the environment.

ADDITIONAL RESOURCES

Kenner, Robert. 2009. *Food, Inc.* [video recording]. Magnolia Pictures.

Pollan, Michael. 2007. *The Omnivore's Dilemma: A Natural History of Four Meals.* New York: Penguin.

Schlosser, Eric. 2012. *Fast Food Nation: The Downside of the All-American Meal.* New York: Mariner Books.

Cities and Urban Land-Use Patterns and Processes

6

IN THIS CHAPTER

→ **HISTORICAL GEOGRAPHY OF URBAN ENVIRONMENTS**

→ **CULTURE AND URBAN FORM**

→ **THE SPATIAL ORGANIZATION OF URBAN ENVIRONMENTS**

→ **URBAN PLANNING**

 Key Terms

Action space	Galactic City Model	Multiple-nuclei model
Beaux Arts	Gateway cities	New urbanism
Blockbusting	Gentrification	Node
Boomburb	Ghettoization	Postmodern architecture
Borchert's epochs	Great Migration	Primate city
Central business district	Hinterland	Rank-size rule
Central-place theory	Industrial Revolution	Sector model
City Beautiful movement	Inner-city decay	Segregation
Colonial cities	Islamic cities	Squatter settlements
Concentric-zone model	Latin American cities	Suburbs
Edge cities	Medieval cities	Urban-growth boundaries
Environmental justice	Megacities	Urban morphology
European cities	Megalopolis	Urban revitalization
Exurbanite	Metacities	Urban sprawl
Feudal cities	Metropolitan areas	White flight
Forward capital	Modern architecture	World Cities

HISTORICAL GEOGRAPHY OF URBAN ENVIRONMENTS

Where do you think the world's first cities arose? If you said "in the same places that were hearths of early agriculture," then you were right. In fact, the growth of early cities was only possible after people had developed sedentary agriculture to the point when farmers began producing surplus crops, or more food than their families alone could eat. When this happened—in places like the Middle East, China, Peru, and the Mississippi Valley—some people were able to quit farming and take up other occupations, such as carpenter, merchant, artisan, scholar, priest, and doctor. The flowering of these different occupations fueled the growth of cities and increased demand for more agricultural products. Thus, early cities were closely linked to their adjacent agricultural regions.

Although the first urban settlements date back several thousand years, urbanism actually spread very slowly, with many fits and starts. In Europe, a few Mediterranean cities, such as Athens and Rome, grew markedly during the Classical Period. However, the Middle Ages, or "dark ages," interrupted the blossoming of urban life in Europe, and Western culture lapsed into a stagnant period of few intellectual advances or cultural endeavors. The feudal system, which dominated most of Europe during this time, discouraged urbanism and confined most people to lives as uneducated peasant-farmers. Most European **feudal cities** lacked diversity, cultural vibrancy, and active trade and served mainly as centers of military or religious power. Interestingly, during the period we normally think of as the Dark Ages, great cities outside of Europe, in the Middle East, Far East, Indian subcontinent, Mesoamerica, and South America, were prospering.

Figure 6.1 The towns and cities of Europe in 1350

During the Renaissance Period, which lasted from about 1350 to 1650, European culture was reborn and cities became vibrant centers of learning. Urban growth accelerated dramatically after the Hundred Years' War, which ended in 1453, and dozens of new towns began to spring up out of the countryside. The great European cities that first emerged during this period include Dublin, Madrid, Prague, Vienna, Amsterdam, and Barcelona. It was only during this period that the cities of Northern and Western Europe began to compete, in size, wealth, and complexity, with those of the Middle East, Asia, and the Americas.

The Colonial Period, which began during the Renaissance and lasted through the 19th century, represented a new time for the world's cities. During this period, European colonial powers sent their explorers to every corner of the globe in the name of God and gold. At the beginning of the Colonial Period, some of the world's greatest cities—and most sophisticated cultures—were located in the Americas. When Cortez and his ragtag troops first stumbled into the Aztec capital of Tenochtitlan, it was probably the largest and richest city in the world. But when the local leaders fell victim to Cortez's fiendish trickery, superior armaments, and alien diseases, the city was reduced to a shadow of its former glory, smaller communities broke off from the larger alliance, and the people of central Mexico became subjects of the Spanish crown. Such was the fate of many native peoples in Africa and the Americas that eventually fell under some sort of European colonial rule. Many of the old Native American cities later became **colonial cities**, and served as regional administrative centers for the European powers. Current-day Mexico City, with its picturesque colonial architecture, sprawls outward from the ancient site of Tenochtitlan. Consequently, although the Colonial Period was a time of vigorous trade, diversification, and growth for European cities, it was also a time of calamitous decline and chaotic transformation for cities in many other parts of the world.

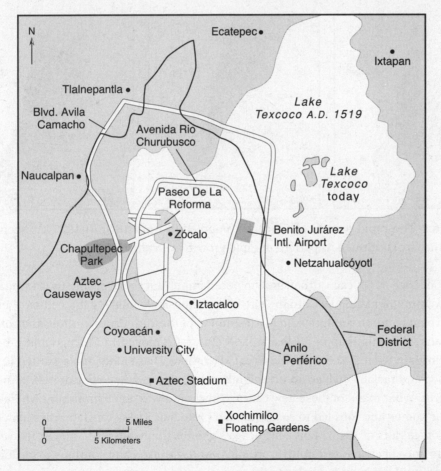

Figure 6.2 The ancient city of Tenochtitlan once existed as a world center of wealth and culture. It deteriorated after colonization and now houses one of the world's most populous and polluted cities—Mexico City. The ancient city was built on an island in Lake Texcoco, which has become a landfill, currently housing most of the poor population of Mexico City.

It was not until the 18th century that urbanism really exploded on a global scale. The Industrial Revolution, which, as discussed earlier, began in England during the 18th century and spread rapidly to Western Europe and North America, propelled much urban growth during this period. The **Industrial Revolution** stimulated tremendous population growth in cities such as Manchester and Chicago, which became centers for processing, manufacturing, shipping, and finance. Chicago, which became a central hub for railroads carrying wheat, beef, timber, and other commodities, grew from a small village in 1840 to a city of 1 million in 1900. By 1930 Chicago had over 3 million inhabitants, qualifying it as one of the world's fastest-growing industrial centers.

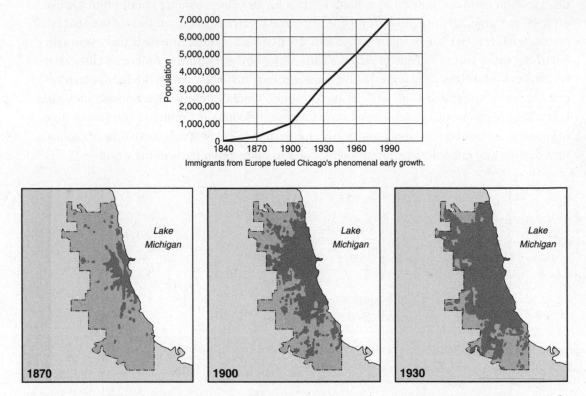

Immigrants from Europe fueled Chicago's phenomenal early growth.

Figure 6.3 The rapid growth of Chicago's population from 1870 to 1930. As seen on the graph, the city continued its pattern of rapid growth even after 1930.

In cities such as Chicago, urban growth posed important social problems. During the 19th century, immigrants flooded into the city by the thousands, making it a center of both cultural diversity and urban unrest. In the beginning of the 20th century, thousands of African Americans from Mississippi, Louisiana, and elsewhere moved to Chicago and other cities in the North in what is known as the **Great Migration**. These individuals wanted to flee the depressed and racially divided South and take advantage of new opportunities in the industrial North. What many of these new Chicagoans found was an environment where competition for homes and jobs led to suspicion and prejudice. These tensions culminated in the Chicago race riots of 1919. Many of Chicago's ghettos that were formed over 100 years ago, and were hubs of civil unrest in 1919, are still poverty stricken and continue to be dominated by recent immigrants and people of other races.

During this period, some cities also grew rapidly because of the strategic economic advantages owing to their locations. For example, New York and San Francisco both grew tremendously in the 19th century, in large part because of their proximity to raw materials and markets, their excellent natural harbors, and their position as gateway cities. **Gateway cities** act as ports of entry and distribution centers for large geographic areas. San Francisco, which, like St. Louis, has frequently been called "the gateway to the West," experienced a population explosion after gold was discovered in California's Sierra Nevada Mountains in 1849. Similarly, for the millions of European immigrants who entered the United States through Ellis Island between 1892 and 1954, the New York harbor represented an open door to a new life with endless opportunities. New York's highly symbolic urban landscape, which includes monuments to freedom, such as the Statue of Liberty, emphasizes the values and opportunities that so many immigrants sought when they came to America.

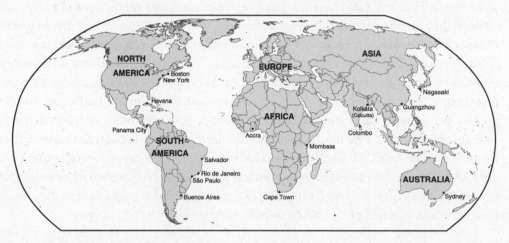

Figure 6.4 Gateway cities almost exclusively occupy key coastal positions, providing tremendous ports of entry for immigrants, as well as corridors for trade and distribution.

CULTURE AND URBAN FORM

From Brussels to Beijing and from London to Lima, the urban geography of today's cities is incredibly diverse. Geographers, architects, historians, and planners have devised a variety of ways of describing, categorizing, and comparing these varied urban environments. This section provides a basis for understanding the geography of contemporary urban environments. However, it is worth noting that all cities are, to some extent, products of their own unique histories and, as a result, each has a distinctive cultural landscape.

Comparing cities in different regions provides a great way for understanding the structures of urban environments. On a regional or even continental scale, cities are frequently products of similar cultures and related histories, and this provides a basis for insightful comparisons. One example of a region characterized by comparable urban environments is Europe. **European cities** take many forms; however, many of Europe's great cities matured during the Medieval Period and still retain characteristics that were typical of cities of that time. Typical **medieval cities** are extremely densely packed, with narrow buildings and winding streets, contain an ornate church that prominently marks the city center, and are surrounded by high walls that provided defense against attack. This type of urban organization harkens back to a time when rival city-states vied for regional dominance, religious leaders held the balance of power, and transportation was limited to horses and foot traffic. In the centuries since, many such cities have jumped their walls and spread out into the surrounding countryside. However, some are still intact, and many large European cities still contain a central medieval core. Excellent examples of intact medieval cities include Montepulciano, Assisi, and the dozens of other quaint hilltop towns that dot the Italian countryside.

Islamic cities, such as Mecca in Saudi Arabia, owe their distinctive urban geographies to the teachings of the Muslim faith. As in medieval European cities, Islamic cities contain places of worship, mosques at their center, and walls guarding their perimeter. However, Islamic cities have distinctive features, such as bustling open-air markets, courtyards surrounded by high walls, and dead-end streets, which limit foot traffic in residential neighborhoods. Although many features of the traditional Middle Eastern Islamic city are related to Muslim values and religious practices, some are also adaptations to the hot and dry desert climate in which they exist. Light-colored surfaces reflect sunlight, and roofs are designed to capture and recycle rainwater efficiently.

In Asia, Africa, and Latin America, cities represent complex expressions of native culture, colonial dominance, and industrial aspirations, as well as widespread poverty. In places like China, cities frequently contain cultural monuments related to Buddhism and communism. Other common features include colonial buildings, factories, and **squatter settlements** surrounding a symbolic city center or port. African cities, such as Nairobi, Kenya, owe most of their urban form to colonialism, 20th-century industrial expansion, and rapid, unplanned population growth. **Latin American cities**, though generally more developed than their African counterparts, owe much of their urban form to the same sorts of causes. In many Latin American cities, distinctive sectors of industrial or residential development radiate out like the spokes of a wheel from the **central business district**, where most industrial and financial activity occurs. Mexico City, with its grand boulevards originating in the city center, is an excellent example of the Latin American metropolis.

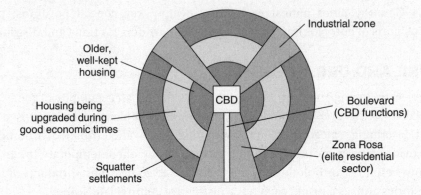

Figure 6.5 Typical Latin American cities contain a central business district at their core, with both rings and sectors segregating a variety of residential and commercial activities.

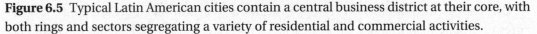

Environmental design can have profound impacts on the forms that urban areas take and on the various arrangements of cities. Although an in-depth discussion of urban design would require many volumes, a few key concepts are worth mentioning here because they have so profoundly affected many of today's largest and most important cities. In the 19th century, European city planners struggled to respond to the lessons of the French and American Revolutions within the context of their new industrial societies. The **Beaux Arts** school, centered in Paris and Vienna, had profound implications for planning in a number of European cities, by stressing the marriage of older, classical forms with newer, industrial ones. Beaux Arts planners designed wide thoroughfares, spacious parks, and civic monuments that stressed progress, freedom, and national unity. The **City Beautiful movement**, which has found important expressions in the United States, drew directly from the Beaux Arts school. Architects from this movement strove to impart order on hectic, industrial centers by creating urban spaces that conveyed a sense of morality and civic pride, which many feared was absent from the frenzied new industrial world. The classic example of City Beautiful design is Chicago, where, following the fire that ravaged the city in 1871, architects created a new urban environment with expansive parks, extravagant monuments, and an orderly street plan. Their goal was primarily to impose a sense of piety and organization on what had previously been a chaotic urban mass.

In the mid-20th century, **modern architecture** reigned supreme. From the modernist's perspective, cities and buildings should act like well-oiled machines, with little energy spent on frivolous details or ornate designs. Instead, efficient, geometrical structures made of concrete and glass dominated urban forms for half a century. Perhaps the world's greatest expression of modernist planning and architecture is Brasilia, the capital of Brazil, which was comprehensively planned and designed by Oscar Niemeyer. Brasilia was meant to be a **forward capital**. It was built in a symbolic location far away from Brazil's population centers, where it would provide a springboard for the development of the country's vast and sparsely populated interior. Brasilia's fantastic modern architecture conveys a sense of futuristic order, scientific progress, and industry. Although many consider modernist buildings, such as those found in Brasilia, to be stark and impersonal, they represent just one expression of the trust in scientific efficiency that captivated the world for much of the 20th century. Modernist structures also dominate many of the American cities that grew so quickly during the mid-20th century.

Postmodern architecture is quickly becoming familiar to many Americans, Europeans, and Australians. Largely a reaction to the feeling of sterile alienation that many people get from modern architecture, postmodernism uses older, historical styles and a sense of light-heartedness and eclecticism. Post-modern buildings combine pleasant-looking forms and playful colors, although they may feel a bit artificial; these architectural forms appropriate old architectural themes in order to convey new ideas, and to create spaces that are more people-friendly than their modernist predecessors.

One important current movement in urban planning and design is known as **new urbanism**. New urbanism began in the United States in the 1980s as a way of changing the suburban, automobile-centered cities of the past into the more sustainable, pedestrian-friendly cities of the future. New urbanism promotes mixed-use developments that include commercial and residential tenants, increased housing density, building and landscape designs that encourage a sense of community integration, and green transportation options such as light rail and bicycle lanes. Portland, Oregon, is an example of an American city known for its efforts to incorporate the principles of new urbanism into its developments.

No two cities on the globe look the same or feel the same or smell the same. Cities are results of a wide variety of political decisions, economic possibilities, geographic opportunities, design initiatives, and a multitude of other factors that contribute to the dynamic processes occurring in every city across the globe, every day. As already discussed, some cities display relics of their history, some of their religious beliefs, and some of their economic activities. Additionally, throughout history, cities have served as palettes for architects and urban designers and display all the various trends and philosophies fading in and out of these fields over time. As such, cities always represent eclectic environments serving all sorts of different functions in different buildings to different kinds of people. Beyond understanding the various dynamics that make cities unique, geographers also strive to explain the layout of urban environments in terms of the spatial organization of all the many activities taking place within a city on a daily basis.

TIP

Geography is in part about observing the things around you— in the places where you live, work, go to school, and vacation. Can you describe the physical form, or urban morphology, of the city or town where you live, based on the concepts defined in this chapter?

THE SPATIAL ORGANIZATION OF URBAN ENVIRONMENTS

One of the fundamental aims of urban geography is to understand how cities are organized. To do this, geographers have developed models that explain and categorize cities in terms of their internal spatial organization. Three main models of urban environments—the concentric-zone model, the multiple-nuclei model, and the sector model—are particularly worthy of consideration. The **concentric-zone model** applies to cities that have rings of development emanating outward from a core, or central business district. In theory, each ring contains different types of development and economic activities. This is because the value of land decreases as you go farther out from the central core. Many geographers cite Chicago as an excellent example of the concentric-zone model. The **multiple-nuclei model** applies to a city that lacks a strong central core, but instead has numerous **nodes** of business and cultural activity. A classic example of this model is Los Angeles, which has a relatively small downtown business district but contains many independent nodes of high land value and vigorous business activity, such as Santa Monica, Hollywood, Westwood, and Pasadena. Cities that conform more to the **sector model** tend to have corridors of different types of development that radiate outward like spokes from the central business district. These corridors often follow long-standing transportation routes, such as wide boulevards, train tracks, or waterways, which affect land values and lend themselves to certain types of development. Some Latin American cities, such as Mexico City, conform to this model. It is worth pointing out, however, that no city matches any model perfectly and that each city's urban organization is tied to both its unique history and its physical geography.

Inner cities frequently surround the central business district and contain dynamic urban geographies. During much of the 20th century, the trend in the United States and Europe was toward the **ghettoization** of inner cities. Many inner cities became dilapidated centers of poverty as affluent whites moved out to the **suburbs**, or residential communities on the outskirts of urban areas, and immigrants and people of color vied for scarce jobs and resources in the declining urban center. This process has often been called **white flight**. It was exacerbated by a practice known as **blockbusting**, which dates to as early as 1900, in which real estate agents and developers encouraged affluent white property owners to sell their homes and businesses at a loss by stoking fears that their neighborhoods were being overtaken by racial or ethnic minorities such as blacks, Hispanics, and Jews. Blockbusting hastened **inner-city decay**, which was particularly extreme in northern cities that had formerly been centers of heavy industry, such as Detroit, Michigan, and Pittsburgh, Pennsylvania. Interestingly, in places like Pittsburgh and Baltimore this pattern has now reversed itself; the inner cities have revived much of their former vibrancy and economic opportunity. **Urban revitalization**, which usually includes the construction of new shopping districts, entertainment venues, and cultural attractions, has enticed young urban professionals, or "yuppies," back into the cities where nightlife and culture are more accessible. The process by which inner-city neighborhoods turn into expensive and fashionable urban districts is called **gentrification**. Although urban revitalization may seem like a great idea, higher-cost living frequently squeezes out the low-income residents who have made these places their homes for many years.

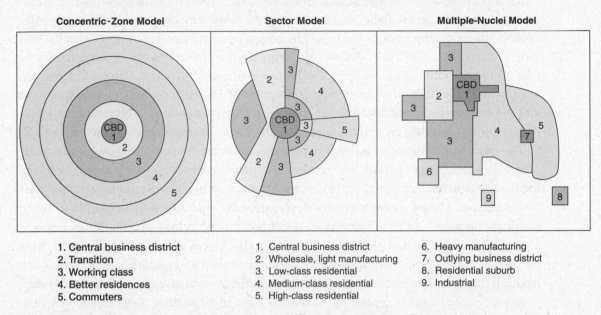

Concentric-Zone Model	Sector Model	Multiple-Nuclei Model

1. Central business district
2. Transition
3. Working class
4. Better residences
5. Commuters

1. Central business district
2. Wholesale, light manufacturing
3. Low-class residential
4. Medium-class residential
5. High-class residential

6. Heavy manufacturing
7. Outlying business district
8. Residential suburb
9. Industrial

Figure 6.6 Three different models urban geographers have developed to explain the variety of forms urban environments can take in terms of the locations of different urban activities.

A related issue is **environmental justice**, which the US Environmental Protection Agency (EPA) defines as the "fair treatment and meaningful involvement of all people regardless of race, color, national origin, or income with respect to the development, implementation, and enforcement of environmental laws, regulations, and policies." Concerns about potential environmental injustices, such as disproportionate exposures to pollution in poor communities and communities of color, have existed for many decades. Formal research on environmental justice only began, however, in the 1980s, with studies of toxic chemicals in the American South. In recent years, studies of environmental justice have expanded to include more work on rural and suburban populations, as well as a growing focus on the uneven distribution of negative impacts associated with global climate change.

STRATEGY

A past AP Human Geography exam included an essay question on the geography of suburbanization. You should be able to describe how factors such as transportation, housing, and social and demographic trends have contributed to the growth of suburbs in the years since World War II.

Perhaps the most important factor affecting the development of contemporary cities is transportation. According to the geographer **John R. Borchert**, American cities have undergone five major **epochs**, or periods, of development shaped by the dominant forms of transportation and communication at the time. These include the sail-wagon epoch (1790–1830), iron horse epoch (1830–1870), steel rail epoch (1870–1920), auto-air-amenity epoch (1920–1970), and satellite-electronic-jet propulsion and high-technology epoch (1970–present).

People's mobility was limited by their transportation options; as a result, Manhattan came to be characterized by densely packed high-rise buildings and narrow cobblestone streets. Unlike New York, Chicago's development was closely related to its role as a railroad hub. Even today, most people who commute into downtown Chicago's central business district, known as "the Loop," do so via trains that branch out from the city center, and Chicago's many distinct urban neighborhoods are linked by commuter rail lines. Los Angeles was planned around the automobile. Today, LA is a sprawling suburban metropolis, with clogged freeways that connect distant nodes of activity, and far-flung suburbs that operate largely independent of each other. Relatively few Angelinos, or residents of Los Angeles, walk to school or work, and the city has little unifying identity; and despite efforts to establish commuter rail lines, public transportation is generally limited to buses that must negotiate congested city streets.

The exact type of urban sprawl that Los Angeles has come to symbolize now plagues many American cities. **Urban sprawl**, which refers to expansive suburban development over large areas in which the automobile provides the primary source of transportation, has many consequences. Urban sprawl increases **segregation** by enabling affluent people to live in ethnically homogenous suburbs, increases pollution caused by long car commutes, degrades a sense of community that people who live close to the places where they work tend to have, and gobbles up open spaces important for public recreation and wildlife habitat. People who have left the inner city and moved to outlying suburbs or rural areas are called **exurbanites**.

In some places, urban sprawl has taken the form of edge cities. **Edge cities**, which are located on the outskirts of larger cities, serve many of the same functions of traditional urban areas, but in a sprawling and decentralized environment often dominated by technology firms and service industries. The classic example of the edge city is Tyson's Corner, Virginia, located near the intersection of several highways outside of Washington, D.C. In this case, Americans would probably do well to learn from Europeans and Canadians who have managed to limit urban sprawl through long-range planning, efficient public transportation, and the establishment of **urban-growth boundaries**. Toronto, Canada, is one example of a city that has used smart growth policies, efficient transportation networks, and a consolidated regional government to create a livable and sustainable urban environment. Many cities in the United States, such as Portland, Oregon, Boulder, Colorado, and Chicago, Illinois, have adopted this model, through the development of pedestrian-friendly urban cores surrounded by open space areas known as greenbelts.

STRATEGY

In 2004, the APHG exam included a free-response question that asked students to compare and interpret two graphs: one depicted population density from the inner city outward to the suburbs, and the other showed a population pyramid for the same area. Upon closer inspection, the two graphs had similar shapes. This revealed that age is an important factor in determining where people live in North American cities. You should be able to complete similar exercises that ask you to compare two sets of geographic data.

Urban geographers also study the roles that cities play in their larger regions. It is clear that cities have profound influences on the communities surrounding them. In fact, the US Census Bureau defines cities in large part by their **metropolitan areas**, which include the central city and all the surrounding communities that have "a high degree of social and economic integration with that core." According to this definition, the city is not solely contained within its spatially defined city limits; it is more accurately conceived of in terms of social and economic relationships within its region. For example, the Tulsa, Oklahoma, metropolitan area is composed of five counties: Creek, Osage, Rogers, Tulsa, and Wagoner.

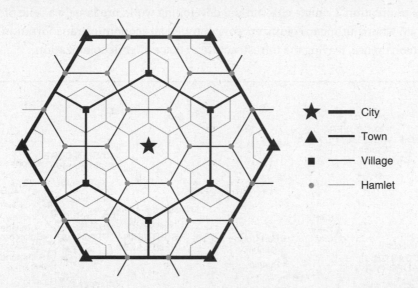

Figure 6.7 The star in the middle represents the center of a city, which has the greatest market area or hexagon encompassing its bounds. Moving down in geographic scale, each type of center requires a smaller population to be economically sustainable. Thus, at the level of the hamlet, an economic function such as a gas station may by profitable, whereas the city's market area can support higher-order activities such as professional sporting events.

Metropolitan areas are in part based on **central-place theory**, which provides a more explicit framework for looking at the relationship between cities and their surrounding communities, based on people's demand for goods and services. According to the central-place theory, large cities serve as the economic hubs of their regions because they provide a great variety of goods and services that are not available in smaller communities. In this view, a region is defined as an area with one central place, or large city, surrounded by increasingly smaller towns and hamlets, each of which contain fewer goods and services than the central place. Thus, people in small towns must occasionally travel to the central place to take advantage of big-city amenities such as professional sports events, museums, and a diversity of stores, music, and food that is simply not available in the **hinterlands** or rural areas. Geographers have also noticed that, in any given region, there should be many small hamlets, some towns, and a few small cities, but only one central city.

The proportion of small towns to large cities is called the **rank-size rule**, and it applies both to regions and to the world as a whole. The rank-size rule says that there is a specific relationship between the relative abundance of settlements of different sizes, and that the smallest settlements should always be the most abundant. More specifically, the rank-size rule states that the population of any given town should be inversely proportional to its rank in the country's or world's hierarchy of cities. Thus, the second largest city should be half the population of the largest city within a certain country. In a global perspective, there are very few cities as big as Tokyo or Sao Paulo, but there are hundreds of cities the size of Cincinnati or Nashville, and there are literally thousands of small cities and little towns throughout the world. Many countries, especially in the developed world, do not display this kind of pattern in terms of their cities' populations. Many of these countries have one **primate city** that overwhelmingly dominates the urban concentration within a country. Seoul, South Korea, contains over one-third of the country's urban population and over one-quarter of the entire country's population! Primate cities in the developing world are largely a relic of their colonial history, when European colonizers concentrated all economic, transportation, and trade activity in one place, leaving the infrastructure in place after decolonization.

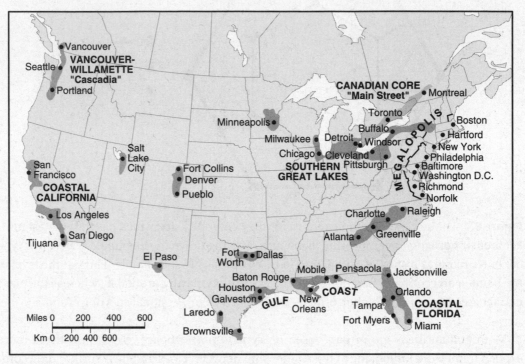

Figure 6.8 The United States demonstrates the megalopolis phenomenon in several areas as neighboring metropolitan areas continually spread out and form an even larger urban complex.

> **TIP**
>
> The world's population is becoming increasingly urbanized. You should be able to describe the causes and consequences of this great migration from country to city in different parts of the world.

In some regions, many towns and cities have grown together and merged over time, creating a **megalopolis**. A megalopolis is an entire region that has become highly urbanized. In North America, megalopolises include the Boston–New York–Philadelphia–Baltimore–Washington, D.C. urban mass along the eastern seaboard and the Los Angeles–Orange County–San Diego–Tijuana megalopolis in southern California and Baja, Mexico. Other, global examples of megalopolises include Sao Paulo–Rio de Janeiro in Brazil and Tokyo–Osaka in Japan.

Megacities are different from megalopolises. Megacities are increasingly a phenomenon of the developing world, where high population growth and migration have caused some urban areas to explode in population since World War II. All megacities are plagued by chaotic, unplanned growth, terrible pollution, and widespread poverty. A ubiquitous feature of these megacities is the squatter settlements that cluster on hillsides and near waterways, and that crop up, seemingly, overnight. Usually constructed of scrap materials, such as aluminum siding and plywood, these makeshift neighborhoods exemplify the tragedy of urban squalor in the less-developed world.

The counterpoint to the chaotic megacity is the **world city**. World cities are centers of economic, cultural, or political activity, and thus have an influence felt across the globe. World cities are commonly categorized into several tiers indicating the extent of their influence, but all world cities enjoy a significant amount of economic influence and prosperity. Top-tier world cities include the economic and cultural powerhouses of New York, London, and Tokyo. Second-tier world cities include seats of government, such as Moscow, Paris, Washington, D.C., and Brussels, Belgium; centers of popular culture, such as Los Angeles and Mumbai, India; and centers of industry, such as Mexico City and Sao Paulo, Brazil. It is interesting to note that, in economic terms, the world's most populous cities are not its most powerful. Indeed, all the most prominent world cities are located in the wealthy and highly industrialized countries.

The 2009 APHG exam included a free-response question on the causes and consequences of squatter settlements in the megacities of the developing world.

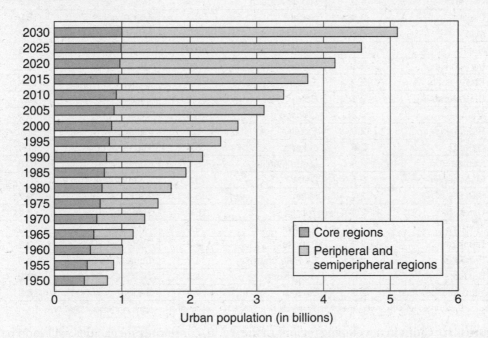

Figure 6.9 Since World War II, most of the world's urban growth can be attributed to peripheral and semiperipheral regions where the majority of the world's megacities expand their numbers on a daily basis.

URBAN PLANNING

A discussion on urban planning provides a unifying medium for understanding how the many aspects of cities discussed so far fit together, and it also provides an avenue for considering the future of the world's cities. Earlier in the chapter, the topic of urban planning was touched on in the discussion of urban form. This section mostly emphasized urban design and how certain overarching architectural trends and philosophies have historically dictated the many forms a city may take. However, the job of the urban planner is usually more complex. Many human geographers enter the field of urban planning, as it provides them with a way to apply their understanding of different human spatial activities within a specific job field. Urban planners consider individuals' perceptions and feelings within urban environments, both specifically and generally, to ensure that the cities they work for provide a comfortable and enjoyable living environment for their citizens.

1950	Population	1980	Population	2010	Population
New York	12.3	Tokyo	21.9	Tokyo	28.8
London	8.7	New York	15.6	Mumbai	23.7
Tokyo	6.9	Mexico City	13.9	Lagos	21.0
Paris	5.4	São Paulo	12.1	São Paulo	19.7
Moscow	5.4	Shanghai	11.7	Mexico City	18.7
Shanghai	5.3	Osaka	10.0	New York	17.2
Essen	5.3	Buenos Aires	9.9	Karachi	16.7
Buenos Aires	5.0	Los Angeles	9.5	Dhaka	16.7
Chicago	4.9	Kolkata	9.0	Shanghai	16.6
Kolkata	4.4	Beijing	9.0	Kolkata	15.6
Osaka	4.1	Paris	8.7	Delhi	15.2
Los Angeles	4.0	Rio de Janeiro	8.7	Beijing	14.3
Beijing	3.9	Seoul	8.3	Los Angeles	13.9
Milan	3.6	Moscow	8.2	Manila	13.7
Berlin	3.3	Mumbai	8.0	Buenos Aires	13.5
Mexico City	3.1	London	7.8	Cairo	13.2
Philadelphia	2.9	Tianjin	7.7	Seoul	12.9
St. Petersburg	2.9	Cairo	6.9	Jakarta	12.7
Mumbai	2.9	Chicago	6.8	Tianjin	12.4
Rio de Janeiro	2.9	Essen	6.7	Istanbul	11.7
Detroit	2.8	Jakarta	6.4	Rio de Janeiro	11.4
Naples	2.8	Metro Manila	6.0	Osaka	10.6
Manchester	2.5	Delhi	5.5	Guangzhou	10.3
São Paulo	2.4	Milan	5.4	Paris	9.7
Cairo	2.4	Tehran	5.4	Hyderabad	9.4
Tianjin	2.4	Karachi	5.0	Moscow	9.3
Birmingham	2.3	Bangkok	4.8	Teheran	9.2
Frankfurt	2.3	St. Petersburg	4.7	Lima	8.8
Boston	2.2	Hong Kong	4.5	Bangkok	8.8
Hamburg	2.2	Lima	4.4	Lahore	8.6

Source: Data, United Nations, *World Urbanization Prospects*. New York: U.N. Department of Economic and Social Affairs, 1998.

Figure 6.10 Cities in developing regions of the world, commonly megacities, will soon be the most highly populated cities on the globe, with the exception of Tokyo.

Urban planners work with behavioral geographers to understand and predict human behavior in urban environments. Together, they have discovered that urban spatial behavior is, in large part, determined by individual perceptions of the specific urban environments they interact with. Additionally, these perceptions, when generalized, often determine how cities are rated in terms of whether or not they are good places to live. Kevin Lynch in his famous book, *Image of the City*, published in 1960, studied individuals' perceptions of urban environments to determine the legibility of urban spaces. How well an individual could, in effect, read a landscape depended on both the city's physical structure as well as the individual's orientation. Both of these factors contribute to an individual's comfort level and confidence in successfully navigating through that city. Lynch classified the geographical contents of cities into five main elements: paths, edges, districts, nodes, and landmarks. Paths consist of channels of movement such as highways, sidewalks, or transit lines, with dominant paths ideally containing concentrations of activities. Edges mark the boundaries between two areas and come in the form of rivers, rail lines, large walls, or forested spaces. Areas that the individual both enters and leaves within a city are called districts, and they can be both mental and physical. Physical districts include the central business district, or industrial areas, whereas mental or emotional districts include spaces such as dangerous neighborhoods or wealthy areas that may have more well-defined boundaries in the individual's mind than in the actual landscape. Nodes are any focal points within the city, such as the junction of paths, and, similar to landmarks, highly visible structures enhance their image. Finally, landmarks are any point references external to the observer, usually used for directional cues within a city. All of these factors, according to Lynch, work together to form a legible or illegible city. If these factors do not cohesively form a legible landscape, individuals do not believe that they can orient themselves within that city, and their mobility, in turn, is severely limited.

Another important collaboration between planners and behavioral geographers comes through the investigation of individual **action spaces**. These spaces comprise all the parts of a city in which daily movement occurs. In small towns, an individual's action space could cover the entire town, but in large cities, it's more likely that an individual's action space covers just a portion of the city's geographic space. Generally, five rules apply to people's daily spatial behavior:

1. **PEOPLE TEND TO MAKE MANY MORE SHORT TRIPS THAN LONG TRIPS.** For example, although you probably go to school or to work every day, you go on longer trips to see distant relatives much less frequently.

2. **PEOPLE TEND TO WORRY LITTLE ABOUT DISTANCE WHEN MOVING AROUND VERY CLOSE TO HOME, BUT WHEN THE TRIP REQUIRES YOU TO VENTURE FARTHER AWAY FROM HOME, DISTANCE BECOMES AN IMPORTANT FACTOR IN DECIDING WHETHER OR NOT TO GO.** Running down to the corner store is not something you give very much thought, but going on a trip to Europe might require many months of planning.

3. **THE LONGEST TRIPS TEND TO BE WORK-RELATED.** Although this probably does not apply to you, it might apply to your parents if their work requires them to make business trips to other states or countries.

4. **TRANSPORTATION LIMITS SOME SPATIAL BEHAVIORS.** Are you old enough to drive? If you are, can you remember how hard it was to get around when your transportation options were more limited? If you are not old enough to have a license, or if you do not have access to a car, then you know exactly what this rule is all about.

5. **PEOPLE TEND TO AVOID PERCEIVED HAZARDS, WHETHER OR NOT THOSE HAZARDS POSE A REAL THREAT.** Are there certain parts of town that you avoid? Do you hate going over tall bridges? These are both perceived hazards that could affect your daily travel.

While all of these rules do not explicitly apply to urban spatial behavior, they provide important insights for planners in terms of how people conceive of space and time. Thus, they can help planners make informed decisions when thinking about transportation planning, urban design, and the location of various facilities such as schools, grocery stores, and banks.

Urban planning can be an extremely difficult job because planners must deal with the highly dynamic nature of the city in working to provide for all the city's diverse inhabitants. Additionally, planners must always keep future growth in mind, and consider all the implications increasing populations might have on sewage-treatment needs, health services, education services, utilities, social-welfare programming, and many other services cities provide their inhabitants. Furthermore, as planners devise ways for cities to better serve community needs, they often meet with opposition from many different forces. As they formulate new plans, they must consider all possible avenues and arguments and seek development initiatives that accommodate all different kinds of needs and opinions. At times, it seems almost impossible!

TIP

A free-response question on the 2003 APHG exam asked students to consider the distribution of urban areas in core and periphery countries. What could you say about the differences in the distribution of urban areas, and the connections between them, in countries such as Argentina and Germany?

Cities, with all their many dynamic forces and opportunities, house the majority of the world's population. Every day, they draw tourists, immigrants, young and opportunistic college graduates, and people of different ethnic groups into their bounds. Urban geographers investigate all the many forces at work within a city contributing to its unique character, unique set of opportunities, and unique problems and social ills. By generalizing some city characteristics across the globe, geographers have developed models to generally explain both the spatial distribution of cities across the globe and the spatial distribution of activities within a city. Furthermore, by collaborating with behavioral geographers, urban planners have a greater understanding of how individuals perceive and behave within urban environments. All of these observations and tools will become increasingly necessary for urban geographers and urban planners as they strive to design policies to provide services for ever-increasing urban populations.

KEY TERMS

ACTION SPACE The geographical area that contains the space an individual interacts with on a daily basis.

BEAUX ARTS This movement within city planning and urban design that stressed the marriage of older, classical forms with newer, industrial ones. Common characteristics of this period include wide thoroughfares, spacious parks, and civic monuments that stressed progress, freedom, and national unity.

BLOCKBUSTING As early as 1900, real estate agents and developers encouraged affluent white property owners to sell their homes and businesses at a loss by stoking fears that their neighborhoods were being overtaken by racial or ethnic minorities.

BOOMBURB A large, rapidly growing city that is suburban in character but resembles population totals or large urban cores.

BORCHERT'S EPOCHS According to the geographer John R. Borchert, American cities have undergone five major epochs, or periods, of development shaped by the dominant forms of transportation and communication at the time. These include the sail-wagon epoch (1790–1830), iron horse epoch (1830–1870), steel rail epoch (1870–1920), auto-air-amenity epoch (1920–1970), and satellite-electronic-jet propulsion and high-technology epoch (1970–present).

CENTRAL BUSINESS DISTRICT The downtown or nucleus of a city where retail stores, offices, and cultural activities are concentrated; building densities are usually quite high; and transportation systems converge.

CENTRAL-PLACE THEORY A theory formulated by Walter Christaller in the early 1900s that explains the size and distribution of cities in terms of a competitive supply of goods and services to dispersed populations.

CITY BEAUTIFUL MOVEMENT Movement in environmental design that drew directly from the Beaux Arts school. Architects from this movement strove to impart order on hectic, industrial centers by creating urban spaces that conveyed a sense of morality and civic pride, which many feared was absent from the frenzied new industrial world.

COLONIAL CITIES Cities established by colonizing empires as administrative centers. Often they were established on already existing native cities, completely overtaking their infrastructures.

CONCENTRIC-ZONE MODEL Model that describes urban environments as a series of rings of distinct land uses radiating out from a central core, or central business district.

EDGE CITIES Cities that are located on the outskirts of larger cities and serve many of the same functions of urban areas, but in a sprawling, decentralized suburban environment.

ENVIRONMENTAL JUSTICE According to the US Environmental Protection Agency, "the fair treatment and meaningful involvement of all people regardless of race, color, national origin, or income with respect to the development, implementation, and enforcement of environmental laws, regulations, and policies."

EUROPEAN CITIES Cities in Europe that were mostly developed during the Medieval Period and that retain many of the same characteristics, such as extreme density of development with narrow buildings and winding streets, an ornate church that prominently marks the city center, and high walls surrounding the city center that provided defense against attack.

EXURBANITE Person who has left the inner city and moved to outlying suburbs or rural areas.

FEUDAL CITIES Cities that arose during the Middle Ages and that actually represent a time of relative stagnation in urban growth. This system fostered a dependent relationship between wealthy landowners and peasants who worked their land, providing very little alternative economic opportunities.

FORWARD CAPITAL A capital city placed in a remote or peripheral area for economic, strategic, or symbolic reasons.

GALACTIC CITY MODEL A circular-city model that characterizes the role of the automobile in the post-industrial era.

GATEWAY CITIES Cities that, because of their geographic location, act as ports of entry and distribution centers for large geographic areas.

GENTRIFICATION The trend of middle- and upper-income Americans moving into city centers and rehabilitating much of the architecture but also replacing low-income populations, and changing the social character of certain neighborhoods.

GHETTOIZATION A process occurring in many inner cities in which they become dilapidated centers of poverty, as affluent whites move out to the suburbs and immigrants and people of color vie for scarce jobs and resources.

GREAT MIGRATION An early 20th-century mass movement of African Americans from the Deep South to the industrial North, particularly Chicago.

HINTERLAND The market area surrounding an urban center, which that urban center serves.

INDUSTRIAL REVOLUTION Period characterized by the rapid social and economic changes in manufacturing and agriculture that occurred in England during the late 18th century and rapidly diffused to other parts of the developed world.

INNER-CITY DECAY Those parts of large urban areas that lose significant portions of their populations as a result of change in industry or migration to suburbs. Because of these changes, the inner city loses its tax base and becomes a center of poverty.

ISLAMIC CITIES Cities in Muslim countries that owe their structure to their religious beliefs. Islamic cities contain mosques at their center and walls guarding their perimeter. Open-air markets, courtyards surrounded by high walls, and dead-end streets, which limit foot traffic in residential neighborhoods, also characterize Islamic cities.

LATIN AMERICAN CITIES Cities in Latin America that owe much of their structure to colonialism, the rapid rise of industrialization, and continual rapid increases in population. Similar to other colonial cities, they also demonstrate distinctive sectors of industrial or residential development radiating out from the central business district, where most industrial and financial activity occurs.

MEDIEVAL CITIES Cities that developed in Europe during the Medieval Period and that contain such unique features as extreme density of development with narrow buildings and winding streets, an ornate church that prominently marks the city center, and high walls surrounding the city center that provided defense against attack.

MEGACITIES Cities, mostly characteristic of the developing world, where high population growth and migration have caused them to explode in population since World War II. All megacities are plagued by chaotic and unplanned growth, terrible pollution, and widespread poverty.

MEGALOPOLIS Several metropolitan areas that were originally separate but that have joined together to form a large, sprawling urban complex.

METACITIES Larger than megacities, metacities describe an urban region where multiple dense areas/cores are interspersed with suburbs and green spaces (and squatter settlements in the case of developing countries).

METROPOLITAN AREA Within the United States, an urban area consisting of one or more whole county units, usually containing several urbanized areas, or suburbs, that all act together as a coherent economic whole.

MODERN ARCHITECTURE Point of view, wherein cities and buildings are thought to act like well-oiled machines, with little energy spent on frivolous details or ornate designs. Efficient, geometrical structures made of concrete and glass dominated urban forms for half a century while this view prevailed.

MULTIPLE-NUCLEI MODEL Type of urban form wherein cities have numerous centers of business and cultural activity instead of one central place.

NEW URBANISM A movement in urban planning to promote mixed-use commercial and residential development and pedestrian-friendly, community-oriented cities. New urbanism is a reaction to the sprawling, automobile-centered cities of the mid-twentieth century.

NODE Geographical centers of activity. A large city, such as Los Angeles, has numerous nodes.

POSTMODERN ARCHITECTURE A reaction in architectural design to the feeling of sterile alienation that many people get from modern architecture. Postmodernism uses older, historical styles and a sense of lightheartedness and eclecticism. Buildings combine pleasant-looking forms and playful colors to convey new ideas and to create spaces that are more people-friendly than their modernist predecessors.

PRIMATE CITY A country's leading city, with a population that is disproportionately greater than other urban areas within the same country.

RANK-SIZE RULE Rule that states that the population of any given town should be inversely proportional to its rank in the country's hierarchy when the distribution of cities according to their sizes follows a certain pattern.

SECTOR MODEL A model or urban land use that places the central business district in the middle, with wedge-shaped sectors radiating outward from the center along transportation corridors.

SEGREGATION The process that results from suburbanization when affluent individuals leave the city center for homogenous suburban neighborhoods. This process isolates those individuals who cannot afford to consider relocating to suburban neighborhoods and must remain in certain pockets of the central city.

SQUATTER SETTLEMENTS Residential developments characterized by extreme poverty that usually exist on land just outside of cities that is neither owned nor rented by its occupants.

SUBURBS Residential communities, located outside of city centers, that are usually relatively homogenous in terms of population.

URBAN-GROWTH BOUNDARIES Geographical boundaries placed around a city to limit suburban growth within that city.

URBAN MORPHOLOGY The physical form of a city or urban region.

URBAN REVITALIZATION The process occurring in some urban areas experiencing inner-city decay that usually involves the construction of new shopping districts, entertainment venues, and cultural attractions to entice young urban professionals back into the cities, where nightlife and culture are more accessible.

URBAN SPRAWL The process of expansive suburban development over large areas spreading out from a city, in which the automobile provides the primary source of transportation.

WHITE FLIGHT The abandonment of cities by affluent or middle-class white residents. White flight was particularly problematic during the mid-twentieth century because it resulted in the loss of tax revenues to cities, which led to inner-city decay. This process reversed itself somewhat during the 1990s and 2000s with urban revitalization projects.

WORLD CITIES Centers of economic, cultural, and political activity that are strongly interconnected and together control the global systems of finance and commerce.

CHAPTER SUMMARY

Urban geographers study all aspects of the world's cities, from their historical development, to their spatial organization, to the ways that they interact with the regions surrounding them, to their importance in the world economic system. The first cities arose thousands of years ago in regions where agriculture had gained an early foothold. By the beginning of the Colonial Period, large, prosperous cities existed on every continent except Australia. During the 19th century, the Industrial Revolution fundamentally changed both the way that cities were built and their basic economic and cultural functions. Since then, a variety of architectural movements and transportation innovations have profoundly shaped the urban environment. Many of today's cities still retain much of their historical architecture and layout, but they also represent dynamic products of constantly changing social and cultural forces. A few such places have managed to establish themselves as world centers of economic, cultural, or administrative power. The impoverished and rapidly expanding megacities of the developing world provide a stark contrast to these prosperous world centers.

PRACTICE QUESTIONS AND ANSWERS

Historical Geography of Urban Environments

MULTIPLE-CHOICE QUESTIONS

1. The first cities arose in

   (A) ancient Greece.
   (B) hearths of early agriculture.
   (C) the Indian subcontinent.
   (D) central Mexico.
   (E) near the equator.

2. Some prominent Native American cities later became

   (A) manufacturing hubs.
   (B) agricultural distribution centers.
   (C) gateway cities.
   (D) colonial cities.
   (E) export processing zones.

3. The Industrial Revolution

   (A) had little impact on urban areas.
   (B) spawned vast manufacturing centers.
   (C) began in the Great Lakes region.
   (D) made factory workers obsolete.
   (E) caused an urban-to-rural migration.

4. _____ is an important gateway city.

   (A) Oslo, Norway,
   (B) Perth, Australia,
   (C) Nairobi, Kenya,
   (D) Honolulu, Hawaii,
   (E) Denver, Colorado,

5. During the Middle Ages, _____ dramatically slowed the growth of urban areas.

   (A) feudalism
   (B) colonialism
   (C) the Black Death
   (D) the Renaissance
   (E) the Industrial Revolution

1. The city has evolved dramatically from its original form many centuries ago.

   (A) Describe the historical evolution of the city.

   (B) How has the city varied in form and function across the globe throughout history?

Culture and Urban Form

MULTIPLE-CHOICE QUESTIONS

1. Classic _____ cities have narrow, winding streets, open-air markets, many dead-ends, and courtyards surrounded by high walls.

   (A) medieval European
   (B) Hindu
   (C) Latin American
   (D) Islamic
   (E) colonial

2. Architects and planners from the _____ strove to introduce beauty and impose order on chaotic industrial cities.

   (A) postmodern school
   (B) modernist tradition
   (C) City Beautiful movement
   (D) Beaux Arts school
   (E) classical movement

3. Modernist architecture

   (A) stressed efficiency and geometrical order.
   (B) uses eclectic and classic forms.
   (C) stressed the ornate.
   (D) is limited to newer American cities.
   (E) is characterized by skyscrapers.

4. Asian, African, and South American cities

   (A) contain dominant centers, usually surrounding something of religious significance.
   (B) contain strong manufacturing and industrial sectors within the city.
   (C) display mostly modern forms of architecture as they are recently developing themselves after colonialism.
   (D) contain many structural relics from colonialism.
   (E) usually have a church at the center of the city.

5. Medieval European cities usually contain all the following characteristics EXCEPT

   (A) winding streets and tall, narrow buildings.
   (B) large, ornate cathedrals.
   (C) walls surrounding the city for defense purposes.
   (D) wide streets to accommodate large military troops.
   (E) a high density of buildings.

1. Imagine that you are a visitor in a new city. As you walk the city streets, you take careful notice of the urban landscape.

   (A) What kind of evidence would you look for to see the influence of different urban-design movements?

   (B) What specifically would you look for to symbolize each of the following trends in urban design: Beaux Arts, the City Beautiful movement, modernism, and postmodernism?

The Spatial Organization of Urban Environments

MULTIPLE-CHOICE QUESTIONS

1. Los Angeles provides an excellent example of

   (A) the beaux arts tradition.
   (B) a central business district.
   (C) the multinucleated metropolis.
   (D) the concentric zone model.
   (E) disagglomeration.

2. Many Latin American cities conform more or less to the

   (A) theory of ghettoization.
   (B) the sector model.
   (C) the multinode model.
   (D) inner-city decay theory.
   (E) the concentric-zone model.

3. In cities such as Baltimore, inner-city revitalization has transformed _____ into gentrified urban neighborhoods.

   (A) suburbs
   (B) central business districts
   (C) edge cities
   (D) ghettos
   (E) agglomerations

4. Which of the following cities exemplifies an urban geography defined by railroads?

   (A) Boston
   (B) Mexico City
   (C) Chicago
   (D) San Francisco
   (E) Los Angeles

5. Which of the following best describes edge cities?

   (A) They are located along freeways on the outskirts of major cities.
   (B) They are usually found in Europe and Asia.
   (C) They are small, isolated communities.
   (D) They are designed in the City Beautiful tradition.
   (E) They are gentrified communities.

6. According to the central-place theory,

   (A) small communities bind regions together.
   (B) most people live in mid-sized cities.
   (C) large cities serve as economic hubs.
   (D) regions are impossible to define.
   (E) there are more large cities than small cities.

7. The coastal southern California and northern Baja, Mexico region can be described as a

   (A) central place.
   (B) artificial construction.
   (C) megacity.
   (D) megalopolis.
   (E) agglomeration.

FREE-RESPONSE QUESTION

1. Modern American cities have been highly influenced by transportation.

   (A) Describe how different forms of transportation have affected different American cities.
   (B) How has the automobile transformed modern American cities? Describe any differences between American cities that evolved pre-automobile with those that began after the automobile.

Urban Planning

MULTIPLE-CHOICE QUESTIONS

1. Which of the following was NOT one of the main elements contributing to a city's legibility according to Kevin Lynch?

   (A) Landmarks
   (B) Nodes
   (C) Links
   (D) Edges
   (E) Districts

2. Action space consists of

(A) recreational facilities in an urban area.

(B) the space in which individual daily activity occurs.

(C) spaces within a city designated for transportation.

(D) a diagrammatic representation of the amount of time it takes to travel between activities on a particular day.

(E) the area surrounding the interactions a central place has with the surrounding community.

3. Individual spatial behavior on a daily basis

(A) generally involves more shorter trips than longer trips.

(B) can be described as that individual's action space.

(C) can be limited by transportation possibilities.

(D) mostly involves work-related travel.

(E) All of the above

FREE-RESPONSE QUESTION

1. Behavioral geographers contribute valuable insight and research to many aspects of human geography.

(A) What is urban planning?

(B) Discuss why it might be important for urban planners to consult behavioral geographers when making planning decisions.

Answers for Multiple-Choice Questions

HISTORICAL GEOGRAPHY OF URBAN ENVIRONMENTS

1. **(B)** The first cities only developed after sedentary agriculture advanced to the point at which crop surpluses allowed some people to take up professions other than farming, such as brick laying and carpentry. When and where this occurred, the first cities arose.

2. **(D)** Mexico City is just one example of a great Native American city (Tenochtitlan) that eventually became a center of colonial government administration and military might. Colonial powers used these cities as bases from which to dominate people in the surrounding countryside.

3. **(B)** Chicago is the classic example of a city that was born during the Industrial Revolution and then experienced tremendous growth, becoming a great center of manufacturing, processing, and transportation.

4. **(D)** Honolulu's extremely strategic location, isolated in the center of the Pacific Ocean, has made it an important gateway for travelers heading both east and west. As a result of its position, Honolulu has become an important shipping hub and military base. The city has also attracted an extremely diverse population of people from around the Pacific Rim.

5. **(A)** Feudalism fostered a system of dependence between landholders and lowly peasants that worked the land. During the Middle Ages, this system prevailed, and because of the lack of opportunity imbedded within it, urban areas experienced a period of stagnation.

CULTURE AND URBAN FORM

1. **(D)** Islamic cities are complex landscapes that incorporate both symbolic expressions of the Muslim faith and adaptations to hot desert climates. Muslim cities outside the Middle East—in places like Indonesia—lack desert adaptations but retain traditional Muslim design and architecture.

2. **(C)** Both the Beaux Arts and City Beautiful movements were attempts to create urban spaces that reflected changing social values while harkening back to classical forms. The City Beautiful movement, which is well represented in cities like Chicago and Washington, D.C., was, more specifically, an effort to give new industrial spaces—which lacked the ancient history and sense of place of the older European cities—beauty, order, and a sense of civic identity.

3. **(A)** Boxy, geometrical structures built of concrete and glass typify the modernist movement. Modernist architects tried to convey a sense of futuristic order and scientific rationality on urban spaces. Many such spaces would later be thought of as sterile and impersonal.

4. **(D)** The majority of cities in each of these three areas, at one time, were colonial cities. As such, this extremely devastating historical process has largely determined their spatial organization.

5. **(D)** Medieval streets are generally narrow and winding. The only forms of transportation that traversed them were people on foot or horses, neither of which necessitated wide streets.

THE SPATIAL ORGANIZATION OF URBAN ENVIRONMENTS

1. **(C)** Although Los Angeles has a central industrial core, its downtown area is not nearly as dominant of a city center as, say, the Loop in Chicago. Instead, Los Angeles is a vast metropolis with many "centers," or nodes of commercial and industrial activity.

2. **(B)** The sector model of urban geography describes cities in which different types of development—residential, commercial, industrial—radiate out from the city center like the spokes of a wheel. These spokes often follow avenues of transportation, such as grand boulevards, riverways, or railroads. Mexico City's Paseo de la Reforma is one example.

3. **(D)** Ghettos are economically depressed inner-city neighborhoods often populated by ethnic minorities. During the past twenty years, many American cities have successfully revitalized their inner cities by luring young urban professionals back into funky urban neighborhoods. However, many long-time residents of these former ghettos have been forced to move owing to increasing costs of living. Other cities, such as Detroit, have had much less success at transforming their urban cores.

4. **(C)** Historically, Chicago has been a city of railroads. Chicago has served as the railroad hub of the West since the 1870s and, within the city itself, many people travel by light rail. The elevated train system or "el" moves hundreds of thousands of people per day around the tightly packed urban area.

5. **(A)** Edge cities serve the same functions as many urban areas but, for many, are conveniently located just outside of urban areas, usually on important corridors or freeway systems.

6. **(C)** Large cities, because they generally have a large population, have a much greater economic base and thus can support and provide a greater variety of economic and cultural opportunities.

7. **(D)** This region consists of a conurbation of linked cities such that, as you drive north or south along the coast in this region, you will always be driving through urban areas. These cities, at one time, were probably very geographically separate from one another but, because they have been increasingly expanded, have molded together into a megalopolis.

URBAN PLANNING

1. **(C)** According to Kevin Lynch, the five main elements contributing to a city's legibility include paths, edges, districts, landmarks, and nodes. Links may exist in cities in the form of transportation systems, but they do not exist as one of the fundamental units that make cities more easily readable for their inhabitants and first-time visitors.

2. **(B)** Action spaces are used by behavioral geographers to analyze individuals' daily spatial activities. They describe all the typical interactions individuals have with their environment on any given day, and thus can provide insightful information for urban planners in making decisions on transportation routes and locations of various facilities within a city.

3. **(E)** Behavioral geographers have analyzed numerous people's action spaces to draw general conclusions about human spatial behavior on a daily basis. In general, this kind of behavior conforms to five rules: people take many more shorter trips than longer trips, most trips tend to be work-related, trips of greater distances require more thought and planning, travel decisions can be quite limited by transportation, and people tend to avoid perceived hazards when going about their daily activities.

Answers for Free-Response Questions

HISTORICAL GEOGRAPHY OF URBAN ENVIRONMENTS

1. **Main points:**
   - The first cities arose in areas of early agriculture. As agricultural systems became increasingly developed, farmers were able to grow more than they needed of certain products. These excesses generated a trading system between farmers. Soon, not as many farmers were needed to produce the same amount of goods, and some began specializing in other functions that also served the community, such as carpentry, metalwork, and education.
   - These first urban areas experienced a period of relative stagnation during the Middle Ages when a feudal system prevailed, and most cities remained centers of religious and military power rather than centers of culture and education.
   - During the Renaissance, European cities once again flourished and became cultural and learning centers. Also, during this period, many European countries began

colonizing various parts of Africa and South America, establishing colonial cities of administrative powers, usually overtaking the already existing infrastructure of established native cities.

■ The Industrial Revolution sparked dramatic changes in urban areas across the developed world. The job opportunities in manufacturing, processing, and finance drew thousands of people in from rural areas.

CULTURE AND URBAN FORM

1. **Main points:**

■ Evidence of the Beaux Arts influence and the City Beautiful movement might be difficult to distinguish from each other as both stressed a sense of order and placed an emphasis on civic pride. Clues such as wide thoroughfares and spacious parks give evidence to Beaux Arts, and monuments and spaces dedicated to fostering civic identity and pride exemplify the influence of the City Beautiful movement.

■ Details of modern architecture might manifest themselves in buildings that appear to be extremely functional rather than beautiful. Buildings that are extremely geometrical and made of concrete and glass would provide evidence of modernist influence on a city.

■ Finally, postmodernism would show itself in landscapes of eclecticism where different agricultural forms and colors are combined to be more inviting and unexclusive.

THE SPATIAL ORGANIZATION OF URBAN ENVIRONMENTS

1. **Main points:**

■ The layout of cities across the United States can largely be attributed to the dominant form of transportation available to that city at its period of most-rapid development.

■ For example, comparing two of the major cities in California—Los Angeles and San Francisco—provides an example of the effects of transportation on both the organization and geographical spread of the city. San Francisco developed before the automobile, and most individuals traversed the city on its famous form of transportation, the trolley car. As such, San Francisco is much more compactly developed, with narrow, windy streets that now cause tremendous congestion for automobile traffic. Conversely, Los Angeles developed after the invention of the automobile, and the mobility offered by the car allowed the area to extend geographically into an extremely large metropolitan area. As such, Los Angeles has no real center, just numerous nodes that are quite geographically spread out from one another.

■ Other examples that illustrate the effects of transportation on city design include two other great American cities: Chicago and New York. New York, as one of the first American cities, developed before the automobile, thus limiting its boundaries. As such, Manhattan is an extremely dense city with small streets, also now extremely congested, because they weren't initially planned for automobile traffic. Chicago was a city dominated by rail. Today, Chicago's center, or the Loop, is located just at the edge of Lake Michigan, and spokes, or rail lines, radiate outside of that hub, bringing people in and out of the city on a daily basis.

URBAN PLANNING

1. **Main points:**
   - Behavioral geographers are very interested in how individuals perceive different environments and, in turn, how those perceptions affect both their feelings toward those environments and behavior within them. As such, urban planners can gain much insightful information regarding how individuals perceive urban environments and the types of factors that contribute to greater comfort levels within a city.
   - One such tool urban planners and behavioral geographers can use to understand individual behavior within urban environments is a record of their action spaces. These records generally depict all of the various interactions an individual has within the urban environment on a daily basis. They can provide insights on commonly used places, commonly avoided places, dominant forms of transportation, and utility of other services within the city.
   - Additionally, urban planners have benefited from other research by behavioral geographers on factors in the city that contribute to a city's legibility. If a city is well organized, containing a diversity of unique landmarks, individuals generally feel a greater sense of orientation, contributing to their overall comfort level and confidence to venture out into places that may not be extremely familiar.

ADDITIONAL RESOURCES

Garreau, Joel. 1991. *Edge City: Life on the New Frontier*. New York: Doubleday.

In this book, Garreau traces the genesis of edge cities, discussing both suburbanization and the rise of the American shopping mall, and how these phenomena lead to the development of areas just outside of cities that end up being larger than the central cities they surround. He looks at several specific examples of edge cities within the United States and some of the social and economic effects this type of location has on the surrounding communities.

Kunstler, James Howard. 1994. *The Geography of Nowhere: The Rise and Decline of America's Man-Made Landscape*. New York: Free Press.

Industrial and Economic Development Patterns and Processes

7

IN THIS CHAPTER

→ **INDUSTRIALIZATION**

→ **MODELS OF DEVELOPMENT AND MEASURES OF PRODUCTIVITY**

→ **GLOBAL ECONOMIC PATTERNS**

→ **LOCATION PRINCIPLES**

→ **DEVELOPMENT, EQUALITY, AND SUSTAINABILITY**

→ **GLOBALIZATION**

 Key Terms

Agglomeration

Ancillary activities

Backwash effects

Break-bulk point

Brick-and-mortar businesses

Bulk-gaining industries

Bulk-reducing industries

Commodity dependence

Conglomerate corporation

Core

Core-periphery model

Cottage industry

Deglomeration

Deindustrialization

Development

E-commerce

Economic backwaters

Ecotourism

Export-processing zone

Fast world

Footloose firms

Fordism

Foreign investments

Gender equity

Globalization

Gross Domestic Product

Gross National Product

Human Development Index

Industrial Revolution

Industrialization

Industrialized countries

Least-cost theory

Least-developed countries

Manufacturing region

Maquiladoras

Microlending

Net National Product

Nonrenewable resources

Offshore financial centers

Outsourcing

Periphery

Primary economic activities

Productivity

Purchasing-Power Parity

Quaternary economic activities

Quinary economic activities

Regionalization

Renewable resources

Rostow's stages of development (economic growth)

Rust Belt

Secondary economic activities

Semiperiphery

Service-based economies

Slow world

Spatially fixed costs

Spatially variable costs

Specialty goods

Sustainable development

Tertiary economic activities

Transnational corporation

World cities

World-system theory

INDUSTRIALIZATION

Industrialization has always been a major theme in economic geography, and any discussion of the geography of industry must start with the **Industrial Revolution**. When most people talk about the Industrial Revolution, they are usually referring to the profound technological and economic changes that arose in England during the late 18th century and then rapidly spread to other parts of Europe and North America. Modern factories, mass-produced goods, and modern forms of capital investment are all products of this period. Before the Industrial Revolution, most goods were either produced in small shops or out of the home. Many **specialty goods**, which are assembled individually or in small quantities, often in home-based **cottage industries**, are still produced to this day. But they are not nearly as dominant as they were prior to large-scale industrialization. By the early 20th century, mass production and assembly lines, first championed by Henry Ford, had replaced many specialty goods. The perfection of standardized mass production, which is in part attributed to Ford, is called **Fordism**. In addition to issues of production and investment, geographers and historians also associate the Industrial Revolution with a set of wide-ranging social changes, which began sometime after 1750 and have been unfolding to this day. The rise of wage labor and large-scale urbanization are particularly important examples of the social changes associated with the Industrial Revolution.

For almost 200 years, heavy industry was mostly limited to northern Europe, East Asia, and North America. Britain, France, the United States, Russia, Germany, and Japan remained at the forefront of industrial production and innovation through the middle of the 20th century. These places, which came to be known as the **industrialized countries**, still account for a large portion of the world's total industrial output. However, in recent years, the geography of industrial production has shifted radically. Although the wealthiest countries, where most industrial goods are consumed, still house most major corporations, many firms have relocated their facilities to less-developed countries, where it is cheaper to produce goods. By the 1970s the most highly developed countries, which were once home to the lion's share of industrial facilities, had already started shifting to information and **service-based economies**, which are focused on research and development, marketing, tourism, sales, and telecommunications. This shift has had enormous implications for global economic systems and for the lives of people around the world.

In industrialized countries, this has frequently been a painful transition. The high-tech and service industry jobs that began to dominate these countries' job markets during the 1970s and 1980s generally provided better pay, safer working conditions, less pollution, and a higher standard of living than factory jobs. However, they also required more education and extensive, specialized training from their employees. When companies began to move their production facilities from the older industrialized countries to the newly industrializing countries, such as Mexico, China, and Malaysia, many former factory workers found themselves jobless and with few economic options.

When industrial facilities leave an area, taking that region's economic base with them, it is referred to as **deindustrialization**. Deindustrialization has been particularly extreme in places like the American Midwest and central Britain, where entire regions' economies had previously relied on heavy industry. The Great Lakes is an economic region that has been hit particularly hard by deindustrialization, and Flint, Michigan, is a particularly disheartening example. For decades, Flint's economy revolved around large General Motors (GM) automobile plants located in the town. In the early 1980s, GM announced its plans to move the bulk

TIP

Deindustrialization is a popular topic for AP Exam essay questions. You should familiarize yourself with the history of heavy industry in the United States, and the causes and effects of the shifting economic geography of American manufacturing since the 1970s.

of its production out of Flint to Mexico, which offered cheap labor, flexible environmental regulations, inexpensive land, and enticing tax breaks. This was a good move for GM's shareholders, but it was tragic for the town of Flint and for thousands of factory workers who lost their jobs.

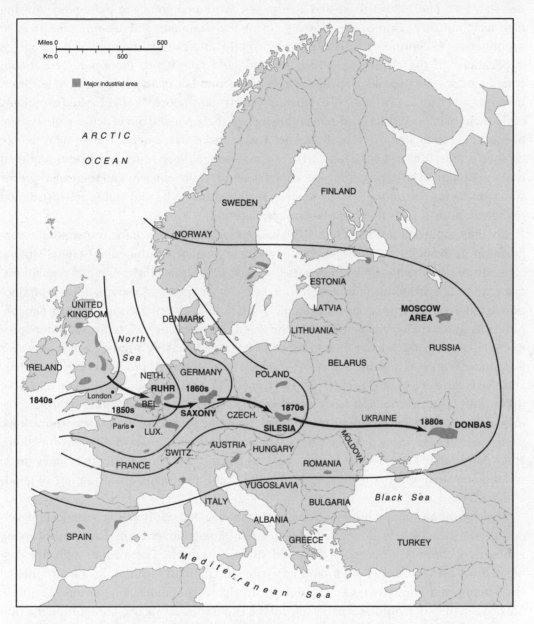

Figure 7.1 Diffusion of the Industrial Revolution from its hearth in central Britain

Another interesting aspect of deindustrialization is that it has highly regionalized effects. The entire United States did not suffer directly as a result of the layoffs in Flint, but the Great Lakes region, which is now commonly referred to as the **Rust Belt** for all its idle factories and aging machinery, was debilitated. In the early 1980s, so many people were moving out of Flint, to areas of the country where their job prospects were brighter, that moving companies like U-Haul could not keep enough trucks available to supply outgoing demand in the region. When one region's economic gain translates into another's economic loss, it is called a **backwash effect**. Although manufacturing is still a significant part of the economies of most historically industrial regions, many such places have not yet been able to regain their former economic vibrancy and sense of civic pride.

While deindustrialization was debilitating some regions of the developed world, other regions were undergoing a different type of economic revolution—the high-tech boom. Between 1960 and 1990, defense and aerospace contributed greatly to the rapidly growing high-tech industry. During the 1990s, software development and e-commerce rose to prominence. **E-commerce** provides a particularly interesting example of geography's implications in the economy. During the mid- and late 1990s, thousands of dot-com companies opened their doors, peddling everything from books, to housewares, to fertilizer. Investors had high hopes for these companies, which many people thought would eventually replace old-fashioned **brick-and-mortar businesses** that operated out of actual stores where people could shop. In the future, all commerce would take place online. Today, many people do shop online, but the transformation to e-commerce has been much less spectacular than many expected. In large part, this is because it is simply more efficient and less expensive for some products to be sold out of individual local outlets than to be sold online and distributed to individual customers from some centralized location.

Another important feature of today's global economy is the rise of the **transnational corporation**. Transnationals, also referred to as multinational corporations, like General Motors, take advantage of geographic differences in wages, labor laws, environmental regulations, taxes, and the distribution of natural resources by locating various aspects of their production in different countries. A classic example, Nike, has its headquarters in Portland, Oregon, but its factories are located in newly industrialized countries such as Indonesia, where production costs are much lower than in the United States. By taking advantage of Indonesia's low production costs and the relatively low cost of transporting its products back to US markets, Nike can net a greater profit. Nike and General Motors are not unique; in fact, most of the world's large corporations have become transnationals, taking advantage of geographic disparities in the global economy. Most transnationals are also **conglomerate corporations**. Conglomerates are firms that comprise many smaller firms that serve different functions. Huge corporations like General Motors, General Electric, and Mitsubishi are actually made up of many smaller firms, operating all over the world, producing a wide variety of goods and services.

Noticing the advantages of joining the global economy, leaders of many economically developing countries have developed schemes to attract foreign investment. **Export-processing zones**, now common throughout the world, are one example. These zones, officially designated for manufacturing, often have accessible distribution facilities, lax environmental restrictions, and attractive tax exemptions for foreign corporations. One prime example is Mexico's system of **maquiladoras**, which dot the United States–Mexico border from the Gulf of Mexico to the Pacific Ocean. Maquiladoras are home to dozens of US firms, such as General Motors, which are able to compete better on the global market as a result. In maquiladora cities, such as Ciudad Juarez, near El Paso, and Tijuana, near San Diego, American companies own large factories that produce and assemble goods for export back to the United States. Maquiladoras have grown tremendously during the past decade, resulting in large part from the North American Free Trade Agreement (NAFTA), which removed many of the obstacles that prevented free trade between Mexico, Canada, and the United States. Although maquiladoras provide desperately needed jobs for thousands of Mexicans, they are also frequently plagued by extremely high crime rates, corrupt government agencies, and terrible pollution.

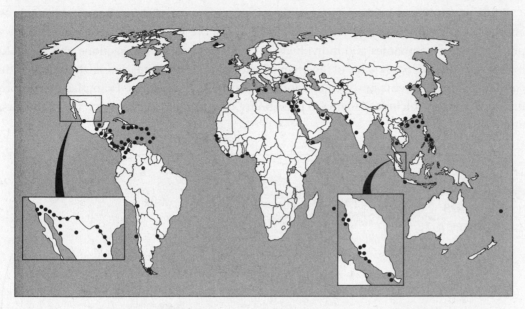

Figure 7.2 Export-processing zones

The process of moving industrial production or service industries to external facilities or organizations often out of the country is called **outsourcing**. Some products and services are more amenable to outsourcing than others. **Bulk-gaining industries** assemble products whose weight is greater after assembly than it was in its constituent parts. Bulk-gaining industries include soda bottling and tend to have production centers close to their markets. In **bulk-reducing industries**, the final product weighs less than its constituent parts. These often include industries based on large amounts of raw materials, such as sawmills that process timber and oil refineries that produce gasoline. A **break-bulk point** is a place, such as a port, where large shipments of goods are broken up into smaller containers to ship to local markets.

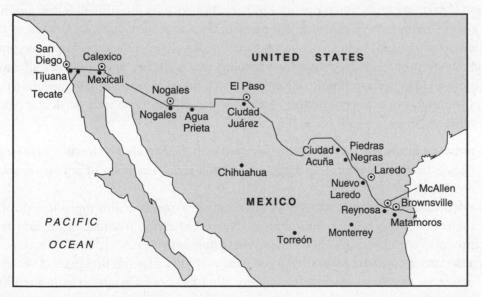

Figure 7.3 Mexico's maquiladoras, or industrial export-processing zones, are located along the border with the United States.

Some governments have encouraged the establishment of **offshore financial centers** as another strategy for initiating economic growth. Offshore financial centers provide a low-profile way for companies and individuals to conduct financial transactions and to avoid high taxes. Although only some of these centers are actually offshore islands, they all provide conducive environments in which to conduct international business. Examples of offshore financial centers include Panama, Luxembourg, Switzerland, Singapore, the Bahamas, and Kuwait.

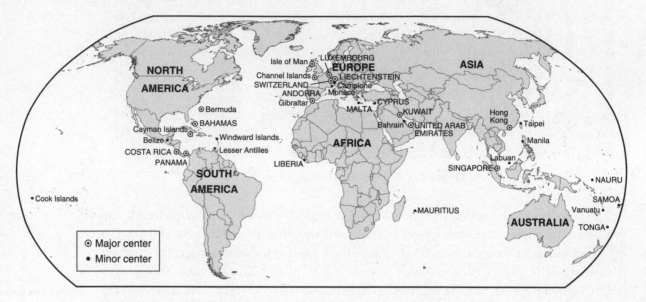

Figure 7.4 Not all offshore financial centers are actually offshore. However, they all provide a variety of incentives for companies to conduct banking and other financial transactions within their borders.

MODELS OF DEVELOPMENT AND MEASURES OF PRODUCTIVITY

Over the past 200 years, the economies of countries such as the United States, Germany, and Great Britain have followed similar paths of economic development. As each of these economies progressed from low levels of economic development and per capita income to higher levels, they tended to rely less on farming, raw materials, and heavy manufacturing and more on light industry, information, and service-based companies. This evolution traces a path through five economic sectors, each of which is associated with particular types of economic activities.

- **PRIMARY ECONOMIC ACTIVITIES** are involved with the harvest or extraction of raw materials. Fishing, agriculture, ranching, and mining are all examples of primary economic activities.
- **SECONDARY ECONOMIC ACTIVITIES** are generally associated with the assembly of raw materials into goods for consumption. Heavy industries, manufacturing, and textile products are all examples of secondary economic activities.
- **TERTIARY ECONOMIC ACTIVITIES** involve the exchange of goods produced in secondary activities. Retailing, restaurants, and any other basic service job occur in the tertiary sector of the economy.

- **QUATERNARY ECONOMIC ACTIVITIES** include research and development, teaching, tourism, and other endeavors having to do with generating or exchanging knowledge.
- **QUINARY ECONOMIC ACTIVITIES** are generally considered a subset of quaternary activities and are those that involve high-level decision making and scientific research.

All countries contain all five types of economic activities. Wealthy countries have extensive tertiary and quaternary sectors, while less-developed countries are dominated by tertiary and secondary activities, and the world's least-developed countries' economies are based almost entirely on primary activities. Switzerland's economy relies on banking, research and development, tourism, and other quaternary activities. Indonesia is extensively industrialized but far less developed than Switzerland; it is currently the site of a vast secondary activity economy, yet it has relatively few advanced quaternary activities. Madagascar, which is one of the world's **least-developed countries**, is in a different situation. Primary economic activities, including farming, livestock ranching, nomadic herding, mining, and fishing, dominate its economy. The shift from primary and secondary economic sectors to tertiary and quaternary will never achieve absolute completion within a country. Even the most highly developed countries still produce agricultural products and extract raw materials from their lands. Nonetheless, the transition in economically dominant activities within a country indicates the important role technology plays in determining a country's level of development.

In 1960, W. W. Rostow, an American economist and historian, developed these observations into a formal theory called **Rostow's stages of development (economic growth)**. In his research, Rostow argued that countries undergo five stages of economic development. During the first stage, the country's economy is dominated by primary activities—productivity, technological innovation, and per-capita incomes remain low. In the second stage, preconditions for economic development arise, including the commercialization of agriculture and increased exploitation of raw materials. In the third stage, **foreign investment** pours in, jumpstarting an economy that was already prepped for growth. An important aspect of the third stage is that a large proportion of foreign investment goes to infrastructure improvements, such as building roads and canals. In the fourth stage, the country develops a broad manufacturing and commercial base. High per-capita incomes and high levels of mass consumption characterize the fifth, and final, stage.

This model has proven to be a useful tool for students of economic development because it seems to follow the path of European and American history and because we can readily identify countries that, right now, seem to be at each of Rostow's stages. You could say, for instance, that Nepal is at the first stage of economic development, while Denmark is at the fifth. However, Rostow's model has also been criticized for assuming that economies will naturally pass through each of the five stages consecutively. Rostow's model did not explicitly account for factors such as global politics, colonialism, physical geography, war, culture, and ethnic conflict, which may cause countries to follow quite different economic trajectories. Nepal's economy may never look like Denmark's, because the two countries' cultures and histories are inherently different. In Saudi Arabia, the presence of oil has created an economic situation that is fundamentally different from either Nepal's or Denmark's. Another problem with Rostow's original model is that it defines the fifth, final, and most highly developed economic stage as being characterized by high mass consumption. Environmentalists and others have criticized Rostow's description of the relationship between development

TIP

Rostow's stages-of-development model has appeared on past AP Human Geography exams in the form of an essay question. It may come up again, so be prepared to discuss its assumptions, predictions, strengths, and weaknesses.

and consumption, claiming that development does not necessarily equal high consumption. For some of these critics, development may mean other things like increased social welfare or ecological sustainability. Finally, the Rostow's stages-of-development model does not account for deindustrialization. In spite of its many criticisms, Rostow's model is an excellent example of a geographic concept that has made an extremely important contribution, partly because of what it says and partly because of the questions and criticisms that arise from it.

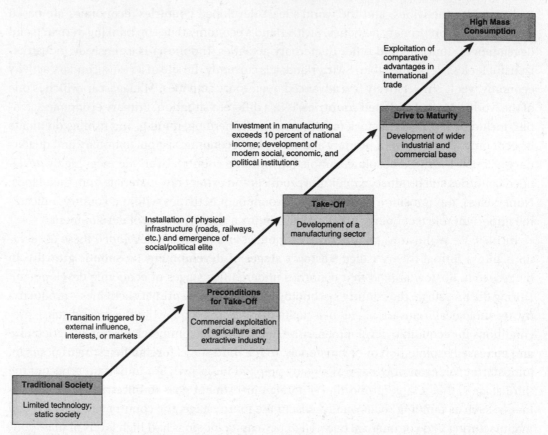

Figure 7.5 Rostow's stages-of-development model caricatures the development of national economies by depicting them as passing through five stages, from traditional societies to high mass consumption.

While Rostow was busy trying to model the stages of development, others were busy developing different indices to try to measure development. As it turns out, **development** is a difficult thing to measure because it means so many different things to so many different people. Determining achievable and acceptable levels of development also involves the issue of standard of living. You should be aware of at least a few measures of development, the way different measures characterize standard of living, and the problems associated with each. Generally, indices of development fall into one of two categories: economic measures or noneconomic measures, which usually consist of a specific measure of social welfare. One important economic measure of development that you have probably heard of is the **Gross National Product** (GNP). The GNP is a measure of all the goods and services produced by a country in a year, including those generated from its investments abroad. Though widely used by economists and reported in the media, the GNP provides a rather broad vision of

productivity. It assumes that development can be measured simply in monetary terms and that any productivity is good productivity, both of which are highly debatable ideas. First, this value, stated in US dollar terms, does not account for the money value of all the goods produced by subsistence economies characteristic of many developing countries. Thus, the GNP systematically reports lower values for productivity in the developing world. To solve this problem, economists have developed a new measure, called **Purchasing-Power Parity**, or PPP, which accounts for what money actually buys within different countries. Second, GNP fails to account for capital that is lost through the exploitation of natural resources. GNP is similar to **Gross Domestic Product** (GDP) except that it omits investments abroad. For a country like Japan, which has extensive investments abroad, the GNP would be significantly larger than the GDP.

STRATEGY

One effective way to study for the exam is to practice comparing, contrasting, and defining related terms. The various measures of economic productivity described in this section are one example of such related concepts. Can you compare, contrast, and define them?

TIP

The 2008 APHG exam included a free-response question that asked students to explain why, in seven developing countries, female enrollment in secondary schools had increased between 1995 and 2003.

Net National Product, or NNP, is a measure of all goods and services produced by a country in a year, including production from its investments abroad *minus* the loss or degradation of natural resource capital as a result of productivity. Let us say that a timber company clear cuts a swath of forest and then sells the timber, earning the company $2 million. Using the GNP, this would result in a $2 million input to the national economy. But under the NNP, the country would add $2 million and then subtract for the loss of that timber, which could have been used for a variety of environmental amenities such as open space, water filtration, or wildlife habitat. With NNP, you must consider not only the profit of the sale but also the loss of the forest. The problem with NNP is that the dollar value of standing timber is notoriously hard to calculate. How much money is it worth to have a healthy forest on yonder hillside? This example should give you an idea of how difficult it is for countries to equitably measure their level of development and economic growth with purely monetary measures.

When noneconomic aspects of development are considered, the equation becomes even more complex. The **Human Development Index** (HDI) is one example of an alternate measure of development. Conceived by the UN Development Program, the HDI calculates development not in terms of money or productivity, but in terms of human welfare. The HDI evaluates human welfare based on three parameters: life expectancy, education, and income. On a global scale, the pattern of human development closely matches that of the GNP. One interesting discrepancy between the two is that, in southern Europe, countries such as Greece, Portugal, and Spain are all rated highly in terms of human development but are slightly less high in terms of GNP.

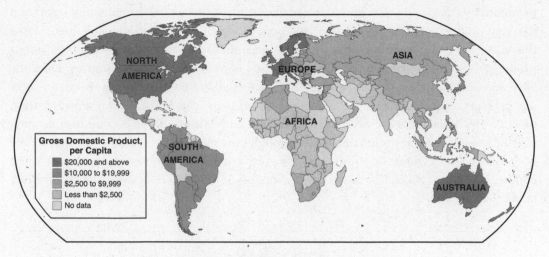

Figure 7.6 Gross Domestic Product by country

Gender equity is an important measure of human welfare that is not necessarily correlated with GNP. In some of the countries with the highest per-capita GNPs, gender equity, or women's welfare, lags far behind. Examples of this include Italy, Japan, and Kuwait, where cultural traditions have long discouraged women's achievement in education, government, and business. Fortunately, in countries such as Japan, women have gained much ground in recent years. One fascinating example of this is the growing acceptance of women as sushi chefs. In Japan, women have long been prevented from becoming master sushi chefs, a lucrative, high-status profession that has traditionally been dominated by men. In recent years, more and more women have been accepted as apprentices in the profession, and male customers are even starting to become accustomed to female chefs. This is one of many examples of the complex interface between economics and culture that defines studies of human welfare and economic development.

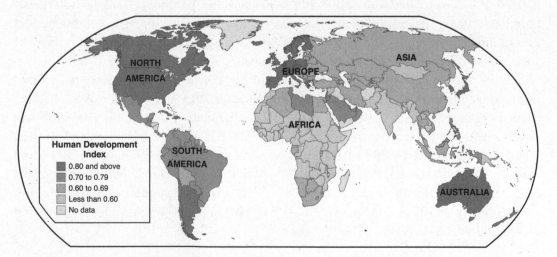

Figure 7.7 Human Development Index by country

Other economic measures of development include indicators such as per-capita energy consumption, the percentage of a country's workforce engaged in certain sectors of the economy, and caloric intake or nourishment levels. While these additional measures may not provide an entirely accurate continuum for judging a country's level of development, they do provide other important clues to a country's economic situation and its people's quality of life.

GLOBAL ECONOMIC PATTERNS

One pattern evident in all measures of economic development is the division of the world's countries into a global economic **core**, **semiperiphery**, and **periphery**. This is called the **core-periphery model**. The core—which includes most of Europe, Japan, the United States, Canada, Australia, and New Zealand—is made up of countries with relatively high per-capita incomes and high standards of living. These highly developed countries also contain the great **world cities**, such as London, Tokyo, and New York, which serve as global centers of economic activity. On the semiperiphery are the newly industrialized countries with median standards of living, such as Chile, Brazil, India, China, and Indonesia. Semiperipheral countries offer their citizens relatively diverse economic opportunities but also have extreme gaps between rich and poor.

The periphery, the world's less-developed countries, includes Africa (except for South Africa), parts of South America, and Asia. Peripheral states have low levels of economic productivity, low per-capita incomes, and generally low standards of living. They also usually lack the infrastructure—paved highways, railroads, sanitation facilities, and telecommunications towers—that would attract foreign investors. Peripheral countries with rapidly growing populations face daunting environmental problems of pollution, deforestation, and topsoil loss, particularly worrisome to subsistence farmers who rely on local timber, water, and crops for their daily needs. Few economic opportunities exist in the peripheral countries, and these places have not benefited significantly from globalization or from the information age. The **slow world** of the periphery is often compared to the **fast world** of the core, where rapid transit, telecommunications, mass media, and computers have cranked up the pace of life. Surprisingly, the living conditions in the poorest of countries are often better than those of the urban poor in the great, semiperipheral metropolises, such as Sao Paulo, Brazil; Mexico City, Mexico; Jakarta, Indonesia; and Mumbai, India. These people live in some of the worst conditions on Earth.

Immanuel Wallerstein's **world-system theory** describes the earth as an interdependent system of countries linked by political and economic competition, similar to the core-periphery model. According to Wallerstein, this network emerged in the late 15th century when European exploration of outside "worlds" began. Increased expeditions fostered the development of shipbuilding and navigation techniques that began to bind certain places closer together in the 16th century. As these political and economic relations strengthened, the areas connected to each other became increasingly exposed to new technologies and innovations. This facilitated a more rapid path to development for certain regions of the globe, while other areas experienced little to no penetration of the benefits of economic competition. Eventually, the pattern between core, peripheral, and semiperipheral countries emerged as the network between the highly industrialized European countries continued to dominate global economic activity. Peripheral countries had little access to the technologies that would facilitate development, and semiperipheral countries, through exploitation by developed countries, were allowed partial entrance into the system. These areas have some dominance over peripheral states but are, in turn, dominated by the core. Consequently, the core-periphery model, according to Wallerstein, began to emerge in the late 15th century, when Europeans began to control global economic and political connections through exploration and colonization.

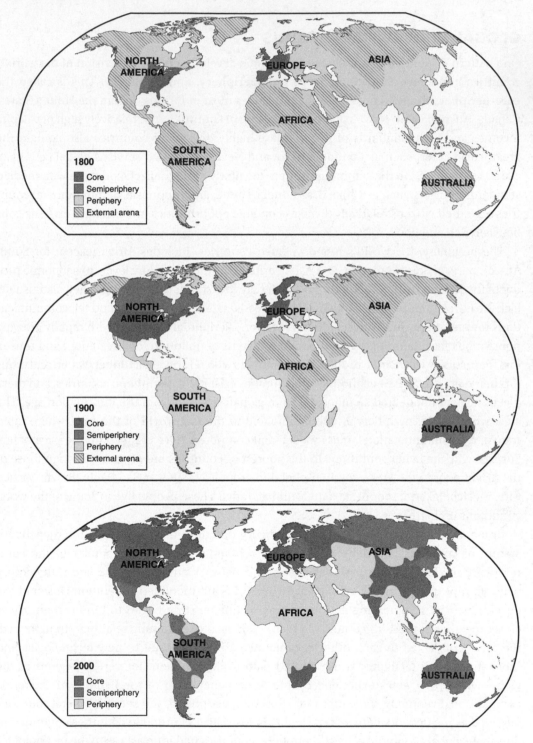

Figure 7.8 The global core, periphery, and semiperiphery by country, 1800 to 2000

LOCATION PRINCIPLES

On smaller scales, economic geographers also study the factors that determine where specific economic activities take place. All industries locate their production facilities based on the following geographic factors.

1. The location a company chooses must provide easy *access* to the materials necessary for production.
2. The location must have an adequate *supply of labor.* For some industries, inexpensive, unskilled labor is best, but for others, such as information technology, an abundance of skilled labor is necessary.
3. *Proximity to shipping and markets* is also a key factor, especially for industries producing items that are either bulky or perishable. These items are either expensive to ship or, by their nature, time-sensitive.
4. The site should be chosen to minimize *production costs.* Firms can minimize production costs by locating in a place with cheap land and labor. Government policies can also have an important impact on production costs. States like Nevada have attracted many firms during the past couple of decades by providing tax incentives for relocating there.
5. *Natural factors,* such as climate, may limit the geographical distribution of certain types of firms, such as agribusiness corporations.
6. The firm's *history* and its leaders' *personal inclinations* may also influence the final choice.

Although no company can hope to maximize all these factors, an optimal balance can give a firm a distinct advantage over their competition.

Another way of looking at these optimization efforts is **least-cost theory**. According to least-cost theory, which was developed by the economist Alfred Weber, firms locate their production facilities in the place that minimizes transportation costs, agglomeration costs, and labor costs. In terms of transportation costs, Weber believed that companies must take into account the cost of transporting both raw materials and finished products. If the raw materials weigh more than the finished products, then facilities should be closer to the materials and are said to have a material orientation. If the finished products weigh more, then facilities exhibit a market orientation by locating closer to the market than to the sources of their raw materials. For example, paper mills transform heavy timber into lightweight paper, making the cost to transport raw materials much greater than that to transport the finished product. Thus, paper mills generally have a material orientation. Tire manufacturers choose a market orientation, since it is significantly more expensive to ship tires a great distance than to ship the various resources used to make tires. An important exception to Weber's theory is that some industries have no real inclination to be located close to either raw materials or primary markets, since their products are so lightweight and valuable. Companies like this are often called **footloose firms**—the diamond and computer chip markets are excellent examples. For some products, the location of a production facility can determine the cost of products. Some products have **spatially fixed costs**, which do not change despite where the product is assembled, while others have **spatially variable costs**, which change depending on where the products are produced.

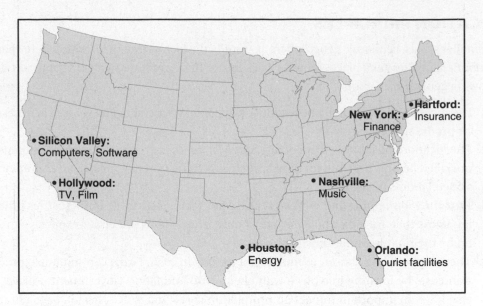

Figure 7.9 Agglomerations of various economic activities in the United States

Another important component of least-cost theory is **agglomeration**. An agglomeration effect occurs when many companies from the same industry cluster together in a relatively small area to draw from the same set of collective resources. For example, computer companies frequently cluster together in places such as California and Austin, Texas, to take advantage of a highly trained labor force. The same is true of fashion designers in Milan, Italy, and Paris, France, and of motion picture studios in Los Angeles, California, and India. Interestingly, as more firms from the same industry locate in particular areas, even more resources become available. This is called a multiplier effect, and it helps to cement particular regions as centers of certain types of industry. As the Silicon Valley became increasingly known for its high-tech firms, it attracted more and more computer experts. The fact that such a talented labor force existed in this area also encouraged other high-tech firms to locate there. By the late 1990s, the San Francisco Bay Area was poised to take the lead in the information revolution and the dot-com boom in particular. The agglomeration of firms in the Bay Area also spawned a number of **ancillary activities**, which are economic activities that surround and support large-scale industries. Ancillary activities can include everything from shipping to food service. The opposite of agglomeration, **deglomeration** occurs when firms leave an agglomerated region to start up in a distant, new place. After the dot-com bust, some high-tech firms left San Francisco because the costs of living were so high.

Agglomeration is actually a part of the larger pattern of regionalization that occurs in every nation's economy. **Regionalization** is the process by which specific geographical areas acquire characteristics that differentiate them from others within the same country. In economic geography, regionalization involves the development of dominant economic activities in particular regions. The primary **manufacturing region** in the United States has historically been the Great Lakes region, which includes Michigan, Illinois, Indiana, Ohio, New York, and Pennsylvania. Although deindustrialization has eroded the manufacturing core of the Rust Belt, it is still the home of much of the United States' remaining industrial base. Similarly, industrial regions exist in southeastern Brazil and central England. In the United States, certain cities have experienced a regionalization process and, as a result, they are known for particular economic activities. In New York, financial markets are king. Hartford, Connecticut, houses a disproportionate number of insurance companies. San Francisco dominates in shipping and technology, and Houston is home to a number of energy firms.

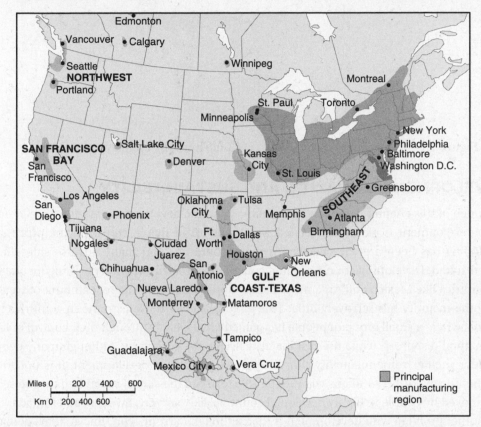

Figure 7.10 Principal manufacturing regions of the United States. Since the 1970s, the Great Lakes region, now known as the Rust Belt, has lost much of its former economic vitality.

While some regions experience economic gains from regionalization, others may become **economic backwaters**. As stated earlier, backwash effects occur when one region experiences tremendous economic growth, while others lag behind or even recede. This process can lead to higher levels of out-migration, exodus of investment capital, and shrinking of the local tax base. In China, the vast majority of industry and commerce is located on the eastern coast, while most of the west has little economic activity. In France, a disproportionate amount of economic activity is centered in Paris. Most of Argentina is relatively undeveloped; however, the city of Buenos Aires is a great, albeit somewhat unstable, seat of industry. In the United States, regional economic backwaters include the upper Great Plains, the lower Mississippi Valley, and parts of the Southwest. While the process of regional industrial development creates vibrant centers of production and commerce, it also tends to leave many other areas out of the loop in terms of development and prosperity.

In some economic backwater areas, governments have developed strategies to enhance local economic development. These initiatives can be undertaken by the central government, as in the case of Mexico's maquiladoras and China's Special Economic Zones, or by local municipal governments. Examples of local development initiatives include urban renewal projects, such as Baltimore's Inner Harbor. Many local development initiatives incorporate both public agencies and private firms.

Figure 7.11 The economic core region of China hugs the country's eastern coast.

DEVELOPMENT, EQUALITY, AND SUSTAINABILITY

For much of this chapter, you have read about economic development. As you have already seen, development is closely correlated with standard of living. However, the process of developing has been a great struggle for many countries, and many adverse side effects have resulted. Developing countries often contain extreme social and economic inequality. In countries like Mexico and Argentina, a small upper class holds a great amount of wealth, while the majority has relatively little. This pattern is even more extreme in countries like Nigeria, where a small group of people has gained great wealth through their control of land and natural resources, while the vast majority live in abject poverty. Although many people consider such extreme inequality simply a "growing pain" of development, it is uncertain whether the majority of Indians will ever be able to attain the sort of social welfare and security enjoyed by people in highly developed countries like Norway, Sweden, and Canada.

Another problem with development is that, although current economic systems demand growth in order to raise the level of productivity and standard of living, this type of economic growth is probably not sustainable into the distant future. Raw materials, which fuel industry, are generally limited either to nonrenewable stocks or renewable stocks that are frequently harvested at rates far in excess of their rate of natural replacement. This poses a serious problem for firms such as oil companies, which must find new ways to make profits through sustainable energy in the future. If no new oil reserves are tapped, our global supply will probably be depleted in less than 100 years. Other adverse side effects of production, such as air and water pollution, must be contained for people's health and well-being. In some encouraging cases, new technologies have enabled us to increase production while actually decreasing pollution, but truly sustainable development still remains elusive.

Sustainable development is an attempt to address the issues of social welfare and environmental protection within the context of capitalism and economic growth. Sustainable development, which has many definitions, is basically the idea that people living today should be able to meet their needs without prohibiting the ability of future generations to do the same. Implicit in this definition is the idea that you should be able to provide for people today without irreparably harming the environment and thus compromising its ability to provide the things that might be needed in the future. This could include a focus on the long-term use of **renewable** (as opposed to **nonrenewable**) **natural resources**. Some advocates of sustainable development have also promoted **ecotourism**, which has proven popular in places such as Costa Rica.

TIP

A free-response question on the 2007 APHG exam asked test takers to examine the concept of economic restructuring, which involves shifts in regional or national economies from one sector to another over time—for example, from manufacturing to service industries.

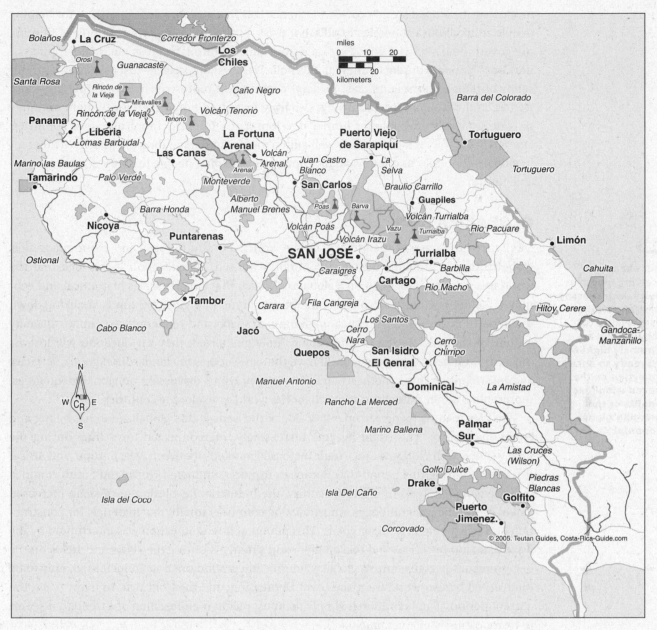

Figure 7.12 National parks of Costa Rica

Unfortunately, there are many problems with the idea of sustainable development. Many people have criticized it for being too vague and not attentive enough to the needs and wishes of local people. Moreover, there is no guarantee that sustainable development will actually result in the conservation of natural resources and biological diversity. In large areas of Africa, safaris have supported wildlife conservation for more than a century, but the results of this long history—for local people and ecosystems—continue to be the subject of vigorous debate.

In today's world, where we have not yet reached the absolute limits of Earth's ability to provide for us, economic inequality arises primarily from the inequitable distribution of wealth and resources. In less-developed countries throughout the world, millions of people suffer from malnutrition, lack of clean water and air, and no access to even basic medical care. However, many experiments and initiatives are currently underway for helping people in the less-developed world provide for themselves in a sustainable fashion and for helping people think about economics in different ways. In the tiny Himalayan country of Bhutan, the government's official policy is to measure success through "gross national happiness," not Gross National Product. Although this policy is less than perfect—just ask Bhutan's thousands of refugees or second-class lowland peasants—it represents a way of thinking about welfare and development that diverges sharply from traditional Western thought and may provide important insights for further development efforts in chronically poor countries. Human geography is a particularly well-suited tool for understanding the injustices inherent in global economics and for evaluating options for helping people to create a truly sustainable future.

GLOBALIZATION

TIP

Your chances of receiving multiple-choice and/or essay questions on the exam related to globalization are extremely high! Are you ready to answer a question on the uneven economic benefits of the emerging global economy?

Of all the catch phrases used to explain the world's current political and economic trends, **globalization** is probably the most commonly abused. Globalization is the idea that the world is becoming integrated on a global scale such that smaller scales of political and economic life are becoming obsolete. Some people argue that globalization is knocking down social barriers, decreasing the meaningfulness of space, and rendering geographic diversity a thing of the past. As the world becomes fully globalized, they say, location will lose its meaning, and people everywhere will have the same access to standardized goods, services, and information. This hypothesis, in light of the previous discussion on global inequalities, represents a rather unrealistic scenario for the world's developing countries.

What few people recognize on either side of the issue is that globalization has a long and circuitous history. The world became increasingly interconnected for a time during the Renaissance, when long-distance trade increased markedly between Asia, Europe, and Africa. It was also during this period that European explorers inducted North and South America into the global system. In the 19th century, the Industrial Revolution once again increased global economic integration, as industrializing countries sought raw materials for construction and new markets for their goods. This period of interconnection was interrupted by the wave of economic crises that rocked the world's markets during the 1880s and 1890s. In the beginning of the 20th century, global economic integration once again increased, only to be interrupted by two world wars, the Great Depression, and the Cold War. In many ways, the current period of globalization is simply the most recent manifestation of a trend that seems to be repeating itself over time.

One extremely important factor sets the current time period of increased global interconnectivity apart from similar patterns in world history. The near instantaneous connections that occur across globalized locations resulting from increased telecommunications technology, specifically the internet, have transformed both the nature of international connections and the space in which these interactions occur. One could argue that cyberspace represents both a new frontier and a revolutionary medium for globalization. The increasingly rapid flow of innovations, information, and capital may have profound implications for regulating economic flows across borders and across the world. This argument lends credence to the idea that geographic national boundaries are becoming more permeable, while technology enables the world economy to become increasingly intertwined. But it is also important to note that even the internet requires a material infrastructure that exists in actual geographic space. This fact alone gives all national governments a certain level of regulation over the activities that occur between globally interconnected nodes.

Others oppose the idea of globalization on the basis of its exclusivity. Although chains such as McDonald's and brand names like Nike may seem omnipresent to the international traveler, globalization has not fulfilled its promise of providing standardized, high-quality goods and services, nor has it improved the lives of the world's poorest people, many of whom still remain isolated from economic growth. For them, the benefits of globalization are nothing more than an abstraction. Proponents of antiglobalization movements complain that the forces of globalization, particularly multinational corporations, are tearing at the fabric of local communities and time-honored cultural practices. Many antiglobalization activists also allege that, because multinational corporations have little stake in local communities and ecosystems, globalization also causes environmental destruction. Globalization proponents counter that expansion of the global economic system will give previously marginalized people more economic power, raise the overall standard of living, increase the accountability of governments, and enable greater access to information. Ironically, the awareness that individuals have of the negative effects of this process can often be attributed to the forces they so radically oppose. Protestors use the World Wide Web, one of the biggest features of globalization, and the connection capabilities it allows to spread their messages across the globe. Regardless of whether you are for or against globalization, it is impossible to ignore its effects.

It should be obvious by now that economic geographers have a tremendously difficult yet extremely important job. Understanding spatial patterns of development may be more simple than determining the explanations behind those patterns, but both present crucial challenges. Patterns, trends, regionalizations, and transitions occur continually at all geographic scales, making it extremely difficult to quantitatively measure development levels across the globe. In the face of these challenges, economic geographers must also make viable suggestions for implementing policies to help initiate sustainable economic growth and to improve the lives of people throughout the world.

KEY TERMS

AGGLOMERATION Grouping together of many firms from the same industry in a single area for collective or cooperative use of infrastructure and sharing of labor resources.

ANCILLARY ACTIVITIES Economic activities that surround and support large-scale industries such as shipping and food service.

BACKWASH EFFECTS The negative effects on one region that result from economic growth within another region.

BREAK-BULK POINT A location where large shipments of goods are broken up into smaller containers for delivery to local markets.

BRICK-AND-MORTAR BUSINESSES Traditional businesses with actual stores in which trade or retail occurs; they do not exist solely on the internet.

BULK-GAINING INDUSTRIES Industries whose products weigh more after assembly than they did previously in their constituent parts. Such industries tend to have production facilities close to their markets.

BULK-REDUCING INDUSTRIES Industries whose final products weigh less than their constituent parts, and whose processing facilities tend to be located close to sources of raw materials.

COMMODITY DEPENDENCE When peripheral economies rely too heavily on the export of raw materials, which places them on unequal terms of exchange with more-developed countries that export higher-value goods.

CONGLOMERATE CORPORATION A firm comprising many smaller firms that serve several different functions.

CORE National or global regions where economic power, in terms of wealth, innovation, and advanced technology, is concentrated.

CORE-PERIPHERY MODEL A model of the spatial structure of development in which underdeveloped countries are defined by their dependence on a developed core region.

COTTAGE INDUSTRY An industry in which the production of goods and services is based in homes, as opposed to factories.

DEGLOMERATION The dispersal of an industry that formerly existed in an established agglomeration.

DEINDUSTRIALIZATION Loss of industrial activity in a region.

DEVELOPMENT The process of economic growth, expansion, or realization of regional resource potential.

E-COMMERCE Web-based economic activities.

ECONOMIC BACKWATERS Regions that fail to gain from national economic development.

ECOTOURISM A form of tourism, based on the enjoyment of scenic areas or natural wonders, that aims to provide an experience of nature or culture in an environmentally sustainable way.

EXPORT-PROCESSING ZONE Area where governments create favorable investment and trading conditions to attract export-oriented industries.

FAST WORLD Areas of the world, usually the economic core, that experience greater levels of connection due to high-speed telecommunications and transportation technologies.

FOOTLOOSE FIRMS Manufacturing activities in which the cost of transporting both raw materials and finished product is not important for determining the location of the firm.

FORDISM System of standardized mass production attributed to Henry Ford.

FOREIGN INVESTMENTS Overseas business investments made by private companies.

GENDER EQUITY A measure of the opportunities given to women compared to men within a given country.

GLOBALIZATION The idea that the world is becoming increasingly interconnected on a global scale such that smaller scales of political and economic life are becoming obsolete.

GROSS DOMESTIC PRODUCT The total value of goods and services produced within the borders of a country during a specific time period, usually one year.

GROSS NATIONAL PRODUCT The total value of goods and services, including income received from abroad, produced by the residents of a country within a specific time period, usually one year.

HUMAN DEVELOPMENT INDEX Measure used by the United Nations that calculates development not in terms of money or productivity but in terms of human welfare. The HDI evaluates human welfare based on three parameters: life expectancy, education, and income.

INDUSTRIAL REVOLUTION The rapid economic and social changes in manufacturing that resulted after the introduction of the factory system to the textile industry in England at the end of the 18th century.

INDUSTRIALIZATION Process of industrial development in which countries evolve economically, from producing basic, primary goods to using modern factories for mass-producing goods. At the highest levels of development, national economies are geared mainly toward the delivery of services and exchange of information.

INDUSTRIALIZED COUNTRIES Those countries, including Britain, France, the United States, Russia, Germany, and Japan, that were all at the forefront of industrial production and innovation through the middle of the 20th century. While industry is currently shifting to other countries to take advantage of cheaper labor and more relaxed environmental standards, these countries still account for a large portion of the world's total industrial output.

LEAST-COST THEORY A concept developed by Alfred Weber to describe the optimal location of a manufacturing establishment in relation to the costs of transport and labor, and the relative advantages of agglomeration or deglomeration.

LEAST-DEVELOPED COUNTRIES Those countries, including countries in Africa (except for South Africa), parts of South America, and Asia, that usually have low levels of economic productivity, low per-capita incomes, and generally low standards of living.

MANUFACTURING REGION A region in which manufacturing activities have clustered together. The major US industrial region has historically been in the Great Lakes, which includes the states of Michigan, Illinois, Indiana, Ohio, New York, and Pennsylvania. Industrial regions also exist in southeastern Brazil, central England, around Tokyo, Japan, and elsewhere.

MAQUILADORAS Cities where US firms have factories just outside the United States–Mexican border in areas that have been specially designated by the Mexican government. In such areas, factories cheaply assemble goods for export back into the United States.

MICROLENDING A provision of small loans to poorer people, typically women, to encourage the development of small businesses that are often community-oriented.

NET NATIONAL PRODUCT A measure of all goods and services produced by a country in a year, including production from its investments abroad, *minus* the loss or degradation of natural resource capital as a result of productivity.

NONRENEWABLE RESOURCES Natural resources, such as fossil fuels, that do not replenish themselves in a timeframe that is relevant for human consumption.

OFFSHORE FINANCIAL CENTERS Areas that have been specially designed to promote business transactions, and thus have become centers for banking and finance.

OUTSOURCING Sending industrial processes out for external production. The term outsourcing increasingly applies not only to traditional industrial functions but also to the contracting of service industry functions to companies to overseas locations, where operating costs remain relatively low.

PERIPHERY Countries that usually have low levels of economic productivity, low per-capita incomes, and generally low standards of living. The world economic periphery includes Africa (except for South Africa), parts of South America, and Asia.

PRIMARY ECONOMIC ACTIVITIES Economic activities in which natural resources are made available for use or further processing, including mining, agriculture, forestry, and fishing.

PRODUCTIVITY A measure of the goods and services produced within a particular country.

PURCHASING-POWER PARITY A monetary measurement of development that takes into account what money buys in different countries.

QUATERNARY ECONOMIC ACTIVITIES Economic activities concerned with research, information gathering, and administration.

QUINARY ECONOMIC ACTIVITIES The most advanced form of quaternary activities consisting of high-level decision-making for large corporations or high-level scientific research.

REGIONALIZATION The process by which specific regions acquire characteristics that differentiate them from others within the same country. In economic geography, regionalization involves the development of dominant economic activities in particular regions.

RENEWABLE RESOURCES Any natural resource that can replenish itself in a relatively short period of time, usually no longer than the length of a human life.

ROSTOW'S STAGES OF DEVELOPMENT (ECONOMIC GROWTH) A model of economic development that describes a country's progression, which occurs in five stages, transforming them from least-developed to most-developed countries.

SECONDARY ECONOMIC ACTIVITIES Economic activities concerned with the processing of raw materials, such as manufacturing, construction, and power generation.

SEMIPERIPHERY Those newly industrialized countries with median standards of living, such as Chile, Brazil, India, China, and Indonesia. Semiperipheral countries offer their citizens relatively diverse economic opportunities but also have extreme gaps between rich and poor.

SERVICE-BASED ECONOMIES Highly developed economies that focus on research and development, marketing, tourism, sales, and telecommunications.

SLOW WORLD The developing world that does not experience the benefits of high-speed telecommunications and transportation technology.

SPATIALLY FIXED COSTS An input cost in manufacturing that remains constant wherever production is located.

SPATIALLY VARIABLE COSTS An input cost in manufacturing that changes significantly from place to place in its total amount and in its relative share of total costs.

SPECIALTY GOODS Goods that are not mass-produced but rather assembled individually or in small quantities.

SUSTAINABLE DEVELOPMENT The idea that people living today should be able to meet their needs without prohibiting the ability of future generations to do the same.

TERTIARY ECONOMIC ACTIVITIES Activities that provide the market exchange of goods and that bring together consumers and providers of services, such as retail, transportation, government, personal, and professional services.

TRANSNATIONAL CORPORATION A firm that conducts business in at least two separate countries; also known as multinational corporations.

WORLD CITIES A group of cities that form an interconnected, internationally dominant system of global control of finance and commerce.

WORLD-SYSTEMS THEORY Theory developed by Immanuel Wallerstein that explains the emergence of a core, periphery, and semiperiphery in terms of economic and political connections first established at the beginning of exploration in the late 15th century and maintained through increased economic access up until the present.

CHAPTER SUMMARY

Economic geography is the study of the flow of goods and services through space. Economic geographers also study the ways in which people provide for themselves in different places and geographic patterns of inequality at all scales of economic organization. Historically, economic geographers have been profoundly influenced by classical economic theory and capitalism. More recently, the opening of markets and the international character of economic flows, in general, have caused many economic geographers to focus more on international economic alliances, cycles of industrialization, poverty, globalization, and development. In this chapter, you looked at multiple geographic scales in an attempt to better understand this extremely diverse and wide-ranging field.

PRACTICE QUESTIONS AND ANSWERS

Industrialization

MULTIPLE-CHOICE QUESTIONS

1. The Industrial Revolution

   (A) began in Germany in the 16th century.
   (B) was initiated by Henry Ford.
   (C) began in England in the 18th century.
   (D) reached its peak in the 1970s.
   (E) began in the United States in the early 20th century.

2. Which of the following are commonly associated with the Industrial Revolution?

   (A) Specialty goods
   (B) Cottage industries
   (C) New forms of capital investment
   (D) The printing press
   (E) Guild industries

3. Deindustrialization has had a dramatic impact on which of the following regions?

   (A) The lower Mississippi Valley
   (B) The Great Plains
   (C) The Great Lakes
   (D) The Pacific Northwest
   (E) The Cotton Belt

4. _____ take advantage of geographic differences in wages, labor laws, environmental regulations, taxes, and the distribution of natural resources by locating various aspects of their production in different countries.

   (A) Conglomerate corporations
   (B) E-businesses
   (C) Transnationals
   (D) Service industries
   (E) Footloose industries

5. Mexico's maquiladoras are examples of

   (A) offshore financial centers.
   (B) transnationals.
   (C) brick-and-mortar businesses.
   (D) export-processing zones.
   (E) ancillary activities.

FREE-RESPONSE QUESTION

1. Economic activities are often categorized as being either primary, secondary, tertiary, or quaternary.

   (A) What is deindustrialization? How does it fit into this categorization of economic activities from a geographic and development perspective?

   (B) What are the effects of deindustrialization?

   (C) How does deindustrialization fit into the global geography of production that has emerged in the last 30 years? What are its backwash effects?

Models of Development and Measures of Productivity and Global Economic Patterns

MULTIPLE-CHOICE QUESTIONS

1. Niger's economy is mostly limited to

   (A) service industries.

   (B) primary economic activities.

   (C) export-processing activities.

   (D) quaternary economic activities.

   (E) nonbasic industry.

2. Rostow's stages-of-development model predicts that each country's economy will progress from

   (A) high consumption to ecological sustainability.

   (B) low output to high input.

   (C) low per-capita incomes to high per-capita incomes and high consumption.

   (D) high levels of pollution to efficient resource use.

   (E) low employment in tertiary activities to high employment in primary activities.

3. The _____ is a measure of all goods and services produced by a country in a year, including production from its investments abroad, *minus* the loss or degradation of natural resource capital as a result of productivity.

   (A) Net National Product

   (B) Gross National Product

   (C) Human Development Index

   (D) Intrinsic-Productivity Index

   (E) Purchasing-Power Parity

4. Gender equity is related to

   (A) Gross National Product.

   (B) cultural traditions.

   (C) education.

   (D) All of the above

   (E) Only (A) and (B)

5. First-tier world cities include

   (A) Tokyo, Mexico City, and Sao Paulo.
   (B) Tokyo, London, and New York.
   (C) Paris, Brussels, and Moscow.
   (D) Washington, Moscow, and London.
   (E) Los Angeles, London, and Paris.

FREE-RESPONSE QUESTION

1. Consider Rostow's stages-of-development model.

   (A) What are its assumptions? What evidence would you use to support it?
   (B) What are its shortcomings? Will all countries eventually conform to this model?

Location Principles

MULTIPLE-CHOICE QUESTIONS

1. Firms try to locate their production facilities to

   (A) maximize spatial accessibility.
   (B) maximize visibility and minimize transportation.
   (C) maximize agglomeration.
   (D) minimize costs and maximize profits.
   (E) minimize competition.

2. The clustering of financial firms on Wall Street in New York is an example of

   (A) least-cost theory.
   (B) agglomeration.
   (C) deindustrialization.
   (D) ancillary industry.
   (E) central-place theory.

3. Mr. Jemstone located his jewelry shop in a place near his home so that he can eat lunch with Mrs. Jemstone every afternoon.

   (A) His locational decision represents a market orientation.
   (B) Since Mr. Jemstone owns his own business, he has chosen an optimal site as he minimizes transport costs to and from work.
   (C) It doesn't matter too much where Mr. Jemstone put his jewelry shop because it is a footloose industry and as long as a viable market exists, he can locate pretty much anywhere.
   (D) Because most jewelry shopping occurs online, the actual location of the shop is unimportant.
   (E) Mr. and Mrs. Jemstone probably participate in a cottage industry.

4. Economic activities that increase and thereby benefit from agglomerations in particular regions are called

   (A) ancillary activities.
   (B) tertiary activities.
   (C) basic sector services.
   (D) quinary activities.
   (E) footloose industries.

5. Which of the following regions is NOT an economic backwater?

   (A) Buenos Aires, Argentina
   (B) Western China
   (C) Sao Paolo, Brazil
   (D) Lower Mississippi Valley
   (E) Upper Great Plains

Development, Equality, and Sustainability and Globalization

MULTIPLE-CHOICE QUESTIONS

1. The idea that resources should be conserved so that people living today can meet their needs without limiting the ability of future generations to do the same is called

   (A) globalization.
   (B) gross national happiness.
   (C) sustainable development.
   (D) environmental conservation.
   (E) subsistence economics.

2. Globalization

   (A) is a new and unique phenomenon.
   (B) has penetrated the entire world.
   (C) is always good for people in the poorest countries.
   (D) has a long and circuitous history.
   (E) does not cause a countermovement of localization.

FREE-RESPONSE QUESTION

1. Globalization is a fairly popular concept used in Human Geography.

   (A) What is globalization? Discuss how the spatial patterns of globalization might coincide with levels of development.
   (B) What are possible explanations for these patterns?

Answers for Multiple-Choice Questions

INDUSTRIALIZATION

1. **(C)** The Industrial Revolution began in England in the 18th century and then spread to continental Europe, particularly France and Germany, and North America. The Industrial Revolution reached its peak in the 19th and early 20th centuries, but by the 1970s many formerly industrialized regions had lost their manufacturing bases to developing countries such as Mexico and China.

2. **(C)** When most geographers, economists, and historians think about the Industrial Revolution, massive factories, standardized goods, and new forms of capital investment are among the first things to come to mind.

3. **(C)** During the first half of the 20th century, the Great Lakes region, including Ohio, Michigan, and portions of Illinois, Pennsylvania, Indiana, and Wisconsin, became vibrant centers of manufacturing and heavy industry. By the 1970s, that same region had slipped into economic depression, as corporations moved their factories to developing countries and relocated their headquarters to Sun Belt states.

4. **(C)** Transnational corporations attempt to maximize their profits by taking advantage of different regulatory and economic situations in different countries. One transnational corporation may have its headquarters in the United States, locate its factories in Mexico and Thailand, and keep its finances at banks in Switzerland and the Cayman Islands.

5. **(D)** Maquiladoras are manufacturing areas located along the US–Mexico border. Products are assembled at maquiladora factories and then exported across the border to US markets.

MODELS OF DEVELOPMENT AND MEASURES OF PRODUCTIVITY AND GLOBAL ECONOMIC PATTERNS

1. **(B)** Countries at low levels of economic development, like Niger, have economies that are mostly oriented toward primary activities, such as fishing, farming, and mining.

2. **(C)** Rostow's stages-of-development model provides a teleological view of economics in which the economies of all nations are headed toward a single ultimate purpose—the attainment of high per-capita incomes and high levels of material consumption.

3. **(A)** The Gross National Product computes the value of all goods and services produced by a country in a year, but the Net National Product also subtracts capital lost through the degradation of natural resources.

4. **(D)** Gender equity, or women's welfare, is related to a wide range of cultural and economic issues, including, but not limited to, economic productivity, cultural traditions, and education.

5. **(B)** Tokyo, London, and New York are the world's three most important centers of economic activity and are thus categorized as first-tier world cities. Other important centers of government, finance, and popular culture, such as Paris, Washington, Brussels, and Los Angeles, are considered to be second-tier world cities.

LOCATION PRINCIPLES

1. **(D)** Firms try to locate their production and distribution facilities in ways that maximize their profit through cutting down on transportation costs.

2. **(B)** Agglomeration occurs when firms find that it is to their advantage to locate close to other firms in the same industry. Advantages include being able to use the same pool of employees and being close to important infrastructure components, such as transportation depots or movie sets.

3. **(C)** Because both the inputs used to make jewelry and the finished product are so lightweight, this type of industry can locate basically anywhere a large enough market exists to make a profit.

4. **(A)** Ancillary activities include all the necessary services to sustain and provide for a local population, such as grocery stores, haircutters, and veterinary hospitals. As the population within a region increases, as it does when an area becomes the site for a particular agglomeration, more of the services become necessary. Thus, agglomerations have tremendous impacts on the overall economy within a specific region.

5. **(C)** Sao Paolo, Brazil, is actually one of the most dominant economic areas in all of South America. Backwaters exist when other regions in a country experience great levels of economic development—like Sao Paolo—and this concentration has negative effects on other regions that cannot generate high levels of economic activity.

DEVELOPMENT, EQUALITY, AND SUSTAINABILITY AND GLOBALIZATION

1. **(C)** Although the term "sustainable development" has many definitions, this one, paraphrased from the Bruntland Report, or "Our Common Future," which was published by the World Commission on Environment and Development in 1987, is the standard that is most frequently quoted.

2. **(D)** During the past 500 years, the world economic system has undergone several periods of greater or less economic integration, or globalization.

Answers for Free-Response Questions

INDUSTRIALIZATION

1. **Main points:**
   - Deindustrialization occurs when formerly industrialized regions lose their industrial base.
   - Since the 1970s, deindustrialization has occurred throughout the older manufacturing areas of Britain, continental Europe, and North America.
   - During this time, transnational corporations have geographically reorganized. While most transnationals have maintained their headquarters in wealthy countries like the United States and Japan, they have moved their production facilities to less-developed countries in Asia and Latin America.
   - This shift of industry from the developed to the developing countries can be described in terms of a backwash effect; industrial growth for countries like Mexico and Indonesia is directly linked to industrial decline in North America and Europe.

MODELS OF DEVELOPMENT AND MEASURES OF PRODUCTIVITY AND GLOBAL ECONOMIC PATTERNS

1. **Main points:**

   - Rostow's stages-of-development model describes the economic evolution of countries from lower levels of productivity, incomes, and material consumption to higher levels of all three. According to Rostow, along the way, countries will pass through five stages of economic development.

   - Rostow's model makes several basic assumptions. It assumes that (1) all countries will have similar developmental trajectories, (2) intrinsic factors such as natural resources and culture will not affect development, (3) countries that undergo development at different times in history will undergo the same processes, (4) all countries will have the same access to development, and (5) the natural goal, path, and purpose of all economies is to increase productivity and material consumption.

   - Rostow's argument is supported by the observation that some countries actually have undergone similar courses of economic development and that, in the current world economic system, different countries appear to be at different levels in the stages-of-development model.

   - It is difficult to say whether all countries will eventually conform to Rostow's model. However, it is unlikely that, given all of the simplified assumptions listed here, Rostow's model will be universally applicable.

DEVELOPMENT, EQUALITY, AND SUSTAINABILITY AND GLOBALIZATION

1. **Main points:**

   - The countries that seem to be the driving forces behind globalization are also those that form the highly developed regions of the globe. Additionally, these are the countries that are increasingly relocating to other parts of the world to take advantage of cheap labor and relaxed environmental limitations.

   - Thus the nations that both drive and benefit from globalization are usually those that measure quite well on all indices of economic development. Conversely, the nations that do not benefit from all of globalization's amazing possibilities remain locked in economic dependency on the drivers of this process and continually exhibit poor measures of economic development.

   - It seems that technology continues to widen the gap between the developed and developing nations across the globe. Beginning with the Industrial Revolution, technology started separating different regions across the globe in terms of productivity and economic dominance on the world market. Our current era of globalization is characterized by powerful technologies that allow for instantaneous connections across the globe. The developing world is still catching up, trying to achieve technological levels characteristic of developed regions during the Industrial Revolution. As such, if the peripheral areas of the world ever achieve the benefits of globalization, it will be a long time in the future, and by then the developed world might be enjoying even greater technologies, perpetuating the technology gap that exists between core and peripheral countries.

ADDITIONAL RESOURCES

Michael Moore. 1989. *Roger & Me* [videorecording]. Dog Eat Dog Films; written, produced, and directed by Michael Moore. Burbank, CA: Warner Home Video.

An excellent documentary film that looks at the effects of deindustrialization in Flint, Michigan. It specifically explores how the relocation of General Motors firms out of the United States devastated an entire community.

Kristof, N. and Wu Dunn, C. 2010. *Half the Sky: Turning Oppression into Opportunity for Women Worldwide.* New York, and in Canada: Vintage.

Timmerman, Kelsey. 2012. *Where Am I Wearing: A Global Tour to the Countries, Factories, and People Who Make Our Clothes.* Hoboken, NJ: Wiley.

Practice Tests

ANSWER SHEET
Practice Test 1

SECTION I

1. Ⓐ Ⓑ Ⓒ Ⓓ Ⓔ
2. Ⓐ Ⓑ Ⓒ Ⓓ Ⓔ
3. Ⓐ Ⓑ Ⓒ Ⓓ Ⓔ
4. Ⓐ Ⓑ Ⓒ Ⓓ Ⓔ
5. Ⓐ Ⓑ Ⓒ Ⓓ Ⓔ
6. Ⓐ Ⓑ Ⓒ Ⓓ Ⓔ
7. Ⓐ Ⓑ Ⓒ Ⓓ Ⓔ
8. Ⓐ Ⓑ Ⓒ Ⓓ Ⓔ
9. Ⓐ Ⓑ Ⓒ Ⓓ Ⓔ
10. Ⓐ Ⓑ Ⓒ Ⓓ Ⓔ
11. Ⓐ Ⓑ Ⓒ Ⓓ Ⓔ
12. Ⓐ Ⓑ Ⓒ Ⓓ Ⓔ
13. Ⓐ Ⓑ Ⓒ Ⓓ Ⓔ
14. Ⓐ Ⓑ Ⓒ Ⓓ Ⓔ
15. Ⓐ Ⓑ Ⓒ Ⓓ Ⓔ
16. Ⓐ Ⓑ Ⓒ Ⓓ Ⓔ
17. Ⓐ Ⓑ Ⓒ Ⓓ Ⓔ
18. Ⓐ Ⓑ Ⓒ Ⓓ Ⓔ
19. Ⓐ Ⓑ Ⓒ Ⓓ Ⓔ
20. Ⓐ Ⓑ Ⓒ Ⓓ Ⓔ

21. Ⓐ Ⓑ Ⓒ Ⓓ Ⓔ
22. Ⓐ Ⓑ Ⓒ Ⓓ Ⓔ
23. Ⓐ Ⓑ Ⓒ Ⓓ Ⓔ
24. Ⓐ Ⓑ Ⓒ Ⓓ Ⓔ
25. Ⓐ Ⓑ Ⓒ Ⓓ Ⓔ
26. Ⓐ Ⓑ Ⓒ Ⓓ Ⓔ
27. Ⓐ Ⓑ Ⓒ Ⓓ Ⓔ
28. Ⓐ Ⓑ Ⓒ Ⓓ Ⓔ
29. Ⓐ Ⓑ Ⓒ Ⓓ Ⓔ
30. Ⓐ Ⓑ Ⓒ Ⓓ Ⓔ
31. Ⓐ Ⓑ Ⓒ Ⓓ Ⓔ
32. Ⓐ Ⓑ Ⓒ Ⓓ Ⓔ
33. Ⓐ Ⓑ Ⓒ Ⓓ Ⓔ
34. Ⓐ Ⓑ Ⓒ Ⓓ Ⓔ
35. Ⓐ Ⓑ Ⓒ Ⓓ Ⓔ
36. Ⓐ Ⓑ Ⓒ Ⓓ Ⓔ
37. Ⓐ Ⓑ Ⓒ Ⓓ Ⓔ
38. Ⓐ Ⓑ Ⓒ Ⓓ Ⓔ
39. Ⓐ Ⓑ Ⓒ Ⓓ Ⓔ
40. Ⓐ Ⓑ Ⓒ Ⓓ Ⓔ

41. Ⓐ Ⓑ Ⓒ Ⓓ Ⓔ
42. Ⓐ Ⓑ Ⓒ Ⓓ Ⓔ
43. Ⓐ Ⓑ Ⓒ Ⓓ Ⓔ
44. Ⓐ Ⓑ Ⓒ Ⓓ Ⓔ
45. Ⓐ Ⓑ Ⓒ Ⓓ Ⓔ
46. Ⓐ Ⓑ Ⓒ Ⓓ Ⓔ
47. Ⓐ Ⓑ Ⓒ Ⓓ Ⓔ
48. Ⓐ Ⓑ Ⓒ Ⓓ Ⓔ
49. Ⓐ Ⓑ Ⓒ Ⓓ Ⓔ
50. Ⓐ Ⓑ Ⓒ Ⓓ Ⓔ
51. Ⓐ Ⓑ Ⓒ Ⓓ Ⓔ
52. Ⓐ Ⓑ Ⓒ Ⓓ Ⓔ
53. Ⓐ Ⓑ Ⓒ Ⓓ Ⓔ
54. Ⓐ Ⓑ Ⓒ Ⓓ Ⓔ
55. Ⓐ Ⓑ Ⓒ Ⓓ Ⓔ
56. Ⓐ Ⓑ Ⓒ Ⓓ Ⓔ
57. Ⓐ Ⓑ Ⓒ Ⓓ Ⓔ
58. Ⓐ Ⓑ Ⓒ Ⓓ Ⓔ
59. Ⓐ Ⓑ Ⓒ Ⓓ Ⓔ
60. Ⓐ Ⓑ Ⓒ Ⓓ Ⓔ

SECTION II

Formulate your responses to the free-response questions on separate sheets of paper.

Practice Test 1

SECTION I

60 MINUTES, 60 QUESTIONS
PERCENTAGE OF TOTAL GRADE—50

Directions: Read each question carefully, choose the correct response, and shade the corresponding choice on the answer sheet provided.

1. Which statement best exemplifies how the residential spatial patterns of wealthy urbanites in the United States compare with those of the wealthy urbanites in most European cities?

   (A) European and US city dwellers show a very similar pattern: most of the wealthy urbanites live in the central business district (CBD).
   (B) In both European and US cities, wealthy urbanites live on large estates far removed from the urban core.
   (C) In European cities, wealthy urbanites live in the exurbs and typically telecommute, while wealthy US urbanites live in the CBD.
   (D) In European cities, although wealthy people live in the suburbs, they also keep a high presence in the inner rings; in the United States, the wealthy tend to collect in the suburbs.
   (E) Both types of cities conform to the sector model: those with wealth are concentrated in a wedge going from the urban core to the suburbs, typically along a major transportation route.

2. If a country with a rapidly growing population were to implement a strict policy of only one child per couple, how would its population pyramid change from the time of implementation to twenty years after?

(A) The pyramid's shape would change from a triangle (the true pyramid shape) to a column as the number of younger children dramatically decreased.
(B) The pyramid would not change; it would be columnar in shape for both time pyramids.
(C) The pyramid would still have a wide base twenty years later because of demographic momentum.
(D) The pyramid would show decline at the base after twenty years because of low replacement-level fertility.
(E) The pyramid would be wide at the top after twenty years as the population policy would lead to a problematic dependency ratio.

Use the following chart (data from the CIA World Factbook, 2019) to answer questions 3–5.

Country	Population living in urban areas (%)	Labor-force composition (%)
Australia	86	Agriculture: 3.6 Industry: 21.1 Services: 75.3
Zimbabwe	32.2	Agriculture: 67.5 Industry: 7.3 Services: 25.2
Bangladesh	36.6	Agriculture: 42.7 Industry: 20.5 Services: 36.9

3. On the basis of the data, how does a country's level of urbanization relate to its labor-force composition?

(A) Highly urbanized countries, like Australia, concentrate on services.
(B) Less urbanized countries, like Zimbabwe, have a highly diverse labor force.
(C) Urbanized countries tend to have less industry.
(D) Less urbanized countries tend to have less agricultural production.
(E) There is no clear relationship between urbanization and labor-force composition.

4. Given Bangladesh's data, its position in the demographic transition model is most likely

(A) stage 1
(B) stage 2
(C) stage 3
(D) stage 4
(E) stage 5

5. If the chart contained an additional column with data showing educational attainment of women, what would the data be likely to reveal?

(A) Educational attainment would be highest in Zimbabwe because service-based jobs require low levels of education.

(B) Educational attainment would be highest in Bangladesh because of increasing growth in microcredit for women.

(C) Educational attainment would be the same in Australia and Bangladesh, where industry and service occupation percentages are higher.

(D) Educational attainment would likely be highest in Australia, where service-based occupations dominate.

(E) Data on women's educational attainment would be likely to reveal that the three countries differed only minimally.

---

6. New York City and Los Angeles are 3,000 miles apart in terms of absolute distance, but they are quite connected economically and culturally. What explains this connectivity?

(A) The cities have similar transportation and communications infrastructure, which facilitates interaction.

(B) Both have diverse populations, which leads to a common cultural experience.

(C) Both have high populations, which counteract distance in the gravity model.

(D) They have similar economic bases, which leads to higher trade between the two.

(E) Both have airports that allow for large jets and frequent arrivals and departures, and so travel between the two cities is very easy.

7. Agricultural production in the developed world is difficult to classify as part of the primary economy. Why?

(A) Production no longer happens primarily in a field but rather through high-tech methods such as hydroponics.

(B) Agriculture is primarily controlled by large agribusinesses that encourage large-scale industrial production.

(C) With so little of the labor force engaged in this economic activity, the category essentially no longer exists for the developed world.

(D) Developed countries now outsource all of their agricultural production from the developing world.

(E) Agricultural production is still very much a primary economic activity in the developed world, and its methods are comparable to those of the developing world.

8. Many countries in developing regions are experiencing dramatic declines in their fertility rates. What is the **main** cause for this pattern?

    (A) Women are marrying less frequently because many developing countries struggle with gender imbalances.
    (B) Women are becoming more educated and more active in the formal economy.
    (C) Women's access to contraception is getting better.
    (D) As more men in developing countries emigrate, women are left with fewer potential marriage partners.
    (E) Women are dying from infectious diseases before they reach childbearing age.

9. Java's population density, by far the highest of Indonesia's 13,000-plus islands, is attributable to

    (A) volcanic activity, which feeds the soil.
    (B) government limitations on which islands Indonesians can inhabit.
    (C) Java's size—it is the biggest island in Indonesia.
    (D) Java's strategic geographic location.
    (E) Java's accessibility as an island.

10. The cultural landscape is an example of the human-environment relationship in that

    (A) it shows how the physical environment has been negatively altered to fulfill the needs of a particular cultural group.
    (B) it shows how humans have reshaped a local space to meet their needs.
    (C) It shows how various cultural groups have suffered or thrived on the basis of geographic conditions.
    (D) it shows how spatial variation in this relationship might be used for finding ways to curb environmental damage resulting from human activity.
    (E) the cultural landscape involves investigation and analysis of human activity, but it does not consider the physical environment.

11. The ideas of Thomas Malthus and the neo-Malthusians describe the tension that exists between

    (A) gender equality and fertility.
    (B) population growth and poverty.
    (C) social welfare (health care, education) and fertility.
    (D) economic growth and population decline.
    (E) population growth and consumption of resources, including food.

12. In squatter settlements in many Latin American cities, poor rural-to-urban migrants tend to make a living in which sector of the economy?

(A) Basic
(B) Informal
(C) Nonbasic
(D) Formal
(E) Primary

13. How is microcredit related to other common economic opportunities for women in the developing world, specifically work in an export processing zone (EPZ)?

(A) In order to fund dormitory housing, women must have microcredit to work in an export processing zone.
(B) Microcredit is the main avenue for union formation, which works to combat unfair labor conditions in factories.
(C) With microloans, women can control their own businesses and therefore set their own hours and wages.
(D) With microloans, women can pay for child care while they work factory jobs in the EPZ.
(E) In the developing world, microcredit is available only to men; women must be married to have access to it.

14. The cultural and political ideas of nationalism can work to sever the social fabric of a state. In this case, nationalism can be seen as

(A) a push factor.
(B) a pull factor.
(C) a centrifugal force.
(D) a centripetal force.
(E) an intervening opportunity.

15. _____ involves the economic and political domination of one state by another, while _____ includes official, institutionalized government rule.

(A) Regionalism . . . sectionalism
(B) Sectionalism . . . regionalism
(C) Imperialism . . . colonialism
(D) Colonialism . . . imperialism
(E) Nationalism . . . isolationism

Use the map below to answer questions 16–18.

Ukraine's Ethnic Divide
Percentage of population that speaks Russian natively

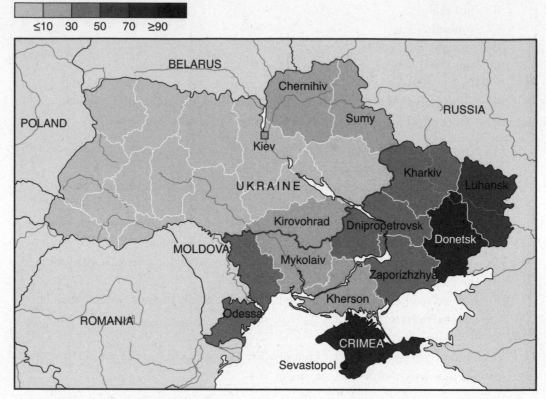

≤10   30   50   70   ≥90

Source: State Statistics from Ukraine

In 2014, Russian forces entered the Crimean Peninsula in Ukraine. A referendum was then held to determine whether the Crimean population wanted the region to remain under Ukrainian sovereignty or join Russia as a federal subject.

16. Russia's takeover of the Crimean Peninsula, given the language patterns, illustrates which concept?

   (A) Irredentism
   (B) Gerrymandering
   (C) Devolution
   (D) Terrorism
   (E) Demilitarization

17. What does the map show about differences between ethnic boundaries and political boundaries?

   (A) That ethnic boundaries do not typically differ from political boundaries
   (B) That political boundaries, which indicate sovereignty, sometimes fail to recognize cultural and ethnic diversity within a territory
   (C) That ethnic boundaries are absolute, whereas political boundaries are relative
   (D) That political boundaries are fuzzy, whereas ethnic boundaries are clearly defined
   (E) That both political and ethnic boundaries are clearly defined in the landscape

18. Why is it important to look at linguistic diversity on a national scale in addition to a global scale?

(A) It is likely that there would be little difference between global and national maps when looking at linguistic diversity.
(B) The global-scale map shows variation that would be difficult to detect on a national-scale map.
(C) The national map shows patterns that would not be apparent on a global map.
(D) Global-scale maps are more distorted than national-scale maps.
(E) Data tends to lose accuracy as a map's scale becomes larger.

---

19. Von Thunen's model of rural land use is based on which of the following premises?

(A) Value of land decreases the farther it is from the urban center.
(B) Value of land increases the farther it is from the urban center.
(C) Perishable goods are the least valuable.
(D) Railroads provide fixed transportation costs.
(E) Vegetarianism is environmentally beneficial.

20. Some political geographers argue that every war is fought essentially for space (space to rule, space in which to practice one's religion, space to harvest certain resources, etc.). This argument illustrates the power of which concept?

(A) Territoriality
(B) Proxemics
(C) Ethology
(D) Sacred space
(E) Topophilia

21. Because of its heritage of the Catalan language and culture, Catalonia, a region in northeast Spain, voted on which of the following in 2017?

(A) Electoral reapportionment
(B) Economic freedom
(C) Unitary standing
(D) Self-determination
(E) Membership in the United Nations

22. "Food security" is the term used to describe

    (A) reliable access, across scales, to food at all times.
    (B) a reliable national system for ensuring food safety.
    (C) a reliable labeling system that lets people know what they are consuming (e.g., GMOs).
    (D) a reliable international trading system that insures against fluctuation in the market.
    (E) a reliable international regulatory system that ensures similar safety standards across the globe.

23. The disproportionate siting of power plants and waste-disposal facilities in African American and Latino neighborhoods—independent of other economic and historical factors—is an example of

    (A) postmaterialism.
    (B) environmental racism.
    (C) deindustrialization.
    (D) environmental justice.
    (E) ecological inferiority.

24. The religious practices of some Native American groups combine elements from their traditional religion and from Christianity. This is an example of

    (A) a cultural complex.
    (B) a cultural confluence.
    (C) a counterculture.
    (D) a cultural diaspora.
    (E) a cultural syncretism.

25. _____ maps work well for locating and navigating between places, while _____ maps display one or more variables across a specific space.

    (A) Reference . . . thematic
    (B) Thematic . . . reference
    (C) Spatial . . . cartographic
    (D) Cartographic . . . spatial
    (E) Topologic . . . choropleth

Use the following excerpt from a *New York Times* article published on July 5, 2019 to answer questions 26–28.

Sterling Road is newly hip, thanks to the arrival of the Museum of Contemporary Art. But with its elevated profile, locals worry the old strip will lose its gritty magic.

Dismissed for decades as a postindustrial wasteland, Sterling Road, a zigzagging half-mile strip of old factories and warehouses, is getting a second life. Last summer, the North American debut of a splashy Banksy exhibition in an empty warehouse there drew a global spotlight. With the arrival of Toronto's Museum of Contemporary Art (MOCA) last fall, Sterling Road is newly hip, its appeal broadening beyond the small cadre of tuned-in artists and bohemian types who for years have had it to themselves. The street's cavernous structures have also quickly become hot real estate.

"It's like Brooklyn—great bones, great old manufacturing buildings, and a great history of artists," said Mr. Stober, whose Sterling Road restaurant, Drake Commissary, became a scene soon after it opened in a huge former pickle factory in June 2017. "The transformation's overdue."

Even bigger changes will come to Sterling Road over the next decade, with plans for a park, day-care facilities and affordable housing, according to Lynda Macdonald, the City of Toronto's director of community planning for the Toronto and East York District, which includes the area.

"Right now, people live in surrounding neighborhoods, but nobody lives on the street itself," she said. "In the coming decade, we'll have 540,000 square feet of new housing in about 900 housing units, along with 565,000 square feet of new office space."

But in a cycle familiar to big urban areas, artists who had quietly colonized the area's disused factories are feeling the squeeze. They say they are fighting for their future and the neighborhood's creative atmosphere. Soon after the Museum of Contemporary Art announced its Sterling Road location in 2016, landlords were quick to start hiking rents, and many artisans fled. Some, though, are fighting back.

"I love living in a hip neighborhood, but rents in the building next door have gone up as much as 100 percent since MOCA," said Angola Murdoch, a 16-year resident who operates the aerialist troupe LookUp Theatre from her high-ceilinged home in a converted 1890 munitions factory. "This amazing co-op woodworking shop, a painter and a metalworker—they all left." In March, Ms. Murdoch and her neighbors sued their landlord to prevent conversions of their lofts to expensive commercial spaces; they won.

With Sterling Road in the spotlight, even some newcomers realize that it will be a struggle to preserve the street's character.

"The concern is that we'll end up with a Burger King and a drugstore and a bank. If we can balance out gentrification, we can set a great example for Toronto and other cities," said Mr. Himel, who opened Henderson Brewing inside an abandoned tent factory in 2016. Sterling Road, he said, "was a wasteland then."

26. The excerpt reveals the tension in many urban areas around which of the following urban geography processes?

   (A) Gentrification
   (B) Commercialization
   (C) New urbanism
   (D) Edge cities
   (E) Primate cities

27. "Preserving the street's character" is an example of the importance of which of the following concepts?

   (A) Place
   (B) Space
   (C) Distance
   (D) Absolute location
   (E) Relative location

28. On the basis of this excerpt, those most likely to relocate to gentrified areas seem to be

   (A) families.
   (B) retired couples.
   (C) low-income or unemployed individuals.
   (D) young, creative entrepreneurs.
   (E) working-class couples.

---

29. Swahili is the language of trade for most of East Africa and the only African language in use by the African Union. As such it is a good example of a

   (A) lingua franca.
   (B) dialect.
   (C) syncretism.
   (D) Creole.
   (E) cultural hearth.

30. Slash-and-burn agriculture meets all of the following conditions EXCEPT

   (A) it occurs mainly in tropical environments.
   (B) it is associated with deforestation.
   (C) it is part of the Green Revolution.
   (D) it was practiced sustainably by indigenous peoples.
   (E) it is uncommon in the temperate latitudes.

31. The pan-Arab region of northern Africa and Southwest Asia is best classified as which region type?

(A) Functional region
(B) Formal region
(C) Administrative region
(D) Perceptual region
(E) Hierarchical region

32. Which of the following universalizing religions originated in northern India and then spread across Central and Southeast Asia, Indonesia, and Japan?

(A) Hinduism
(B) Buddhism
(C) Islam
(D) Taoism
(E) Judaism

33. The Industrial Revolution transformed Western agriculture

(A) through mechanization and the creation of new markets.
(B) with biotechnology.
(C) through technological and religious change.
(D) by eliminating agricultural pests.
(E) by eliminating plant hybridization.

34. Which of the following BEST describes the purpose of the European Union?

(A) An economic organization whose focus is on easy flow of goods, services, and people among member nations
(B) A political organization whose focus is on maintaining peaceful relations, particularly with nations to the east of Europe
(C) A diplomatic organization whose focus is on increasing democracy and capitalism in the Western world
(D) A cultural organization whose focus is on cultural integration across a variety of ethnicities and religions
(E) A social organization whose focus is on easing relations between a diverse set of ethnic and religious groups

Linguistic Groups in Nigeria

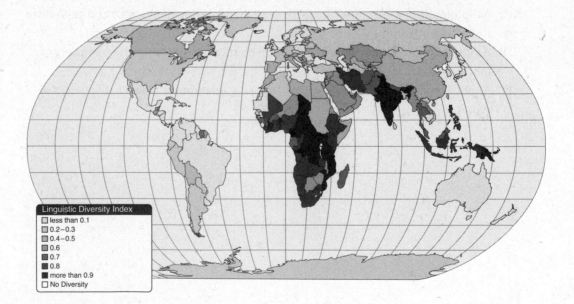

PRINCIPAL LINGUISTIC GROUPS

- Chamba
- Edo
- Efik-Ibibio
- Gwari
- Hausa and Fulani
- Ibo
- Ijaw
- Kanuri
- Nupe
- Tiv
- Yoruba
- Mixed

TIV **Tribe**

Linguistic Diversity Index
- less than 0.1
- 0.2–0.3
- 0.4–0.5
- 0.6
- 0.7
- 0.8
- more than 0.9
- No Diversity

35. In general, the countries with the least linguistic diversity tend to be located

(A) in the southern hemisphere.
(B) in the eastern hemisphere.
(C) in the world's most developed regions.
(D) in the world's least developed regions.
(E) in the Americas.

36. Which of the following are the correct scales of analysis presented by these two maps?

(A) Regional vs. national
(B) Global vs. regional
(C) Global vs. national
(D) Global vs. local
(E) National vs. province

37. Which of the following is the likely result of Nigeria's linguistic diversity?

(A) Increased political fragmentation as linguistic diversity often acts as a centripetal force
(B) Increased immigration as linguistic diversity attracts Africans from across the continent
(C) Increased global trade as linguistic diversity opens up economic opportunities
(D) Decreased political tension as diversity leads to tolerance
(E) Decreased immigration as people stick to their own ethnic enclaves

---

38. The fact that Cincinnati, OH, is able to support a major-league baseball team and an NFL football team, while its neighbor, Toledo, can support only a minor league team, is largely attributable to

(A) metropolitan-area theory.
(B) sector theory.
(C) central-place theory.
(D) urban-matrix theory.
(E) the rank-size rule.

39. One important consequence of cultural extinction is

(A) decreasing population.
(B) increased poverty in peripheral regions.
(C) loss of genetic diversity.
(D) loss of indigenous knowledge about ecosystems.
(E) increasing population.

40. States with a federal form of government must also have

    (A) territorial organization.
    (B) fascist regimes.
    (C) social-democracies.
    (D) proportional representation.
    (E) congressional oversight.

41. Which urban model best describes the typical suburban American city?

    (A) Multiple-nuclei model
    (B) Concentric-zone model
    (C) Demographic-transition model
    (D) Hoyt-sector model
    (E) Ecological model of urban land use

42. Monocultural agricultural systems tend to be most prevalent

    (A) in the least-developed countries and regions of the world.
    (B) in developing countries that are transitioning through industrialization.
    (C) in highly developed countries with labor-force concentrations in the tertiary and quaternary sectors.
    (D) in challenging climate regimes.
    (E) in tropical climate regimes where slash-and-burn is practiced.

43. To get an understanding of voting patterns within the Los Angeles metropolitan area, what would be the best scale at which to investigate these data?

    (A) Global
    (B) Country
    (C) State
    (D) County
    (E) Census tract

44. Many of the most successful and sustainable population programs focus on which of the following as a means for curbing fertility rates?

    (A) Establishing quotas on resources such as food, water, and land
    (B) Increasing access to sterilization procedures for males and females
    (C) Increasing women's access to education
    (D) Instituting penalties on families that have more than a designated number of children
    (E) Providing incentives to make it easier for women to stay at home with their children

45. Replacing imports with goods manufactured locally (import substitution) falls under which philosophy of development?

(A) Sustainable
(B) Modernization
(C) Neoliberal
(D) Dependency
(E) Maturity

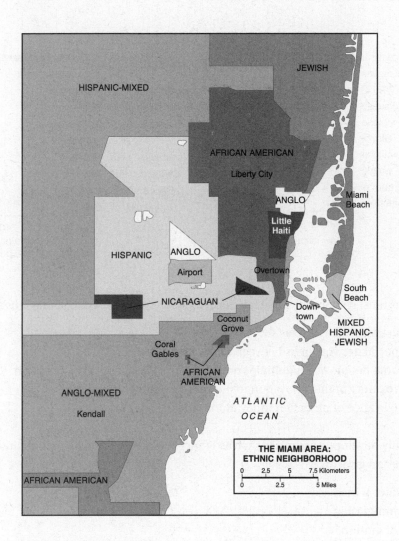

46. The pattern on this map is the likely result of which of the following processes?

(A) Stepped migration
(B) Chain migration
(C) Forced migration
(D) Redlining
(E) Blockbusting

47. The political geographic concept of the nation is defined as

(A) a group of people defined and organized by a centralized government.

(B) a smaller unit of a federal system, such as a Canadian province.

(C) a group of people represented by a singular and unique culture.

(D) a group of people with a singular culture and a singular government.

(E) being synonymous with a state or country.

Use the two charts below to answer questions 48–50.

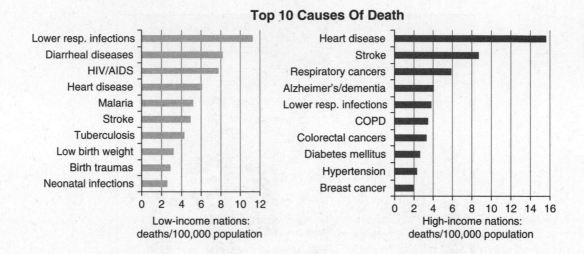

Top 10 Causes Of Death

48. What is most responsible for hierarchical diffusion, as opposed to contagious diffusion?

(A) Distance decay effects

(B) Special network links between major nodes

(C) Some people need multiple contacts before they adopt an innovation.

(D) Proximity of the innovation to the varying degrees of diffusion

(E) Relevance of particular innovations only to specific locations

49. In highly developed countries such as France and Germany, what lies behind most population growth?

(A) Natural increase

(B) Immigration

(C) Emigration

(D) Adoption

(E) Population is declining, not growing, in these countries.

50. Which of the following countries is likely to list HIV/AIDS as a significant cause of death?

(A) Finland

(B) Brazil

(C) Botswana

(D) Mongolia

(E) Turkmenistan

51. Which human geography concept or model explains the difference between causes of death in low-income vs. high-income countries?

   (A) Core-periphery model
   (B) Epidemiological transition
   (C) Demographic transition
   (D) Rostow's modernization model
   (E) Differing beliefs on vaccination

52. Development indicators are often categorized as economic or social. How should mortality rates be categorized?

   (A) As a social indicator of development
   (B) As an economic indicator of development
   (C) As both a social and an economic indicator of development
   (D) If mortality is included with economics, as an economic indicator of development
   (E) As having nothing to do with development

53. This image above illustrates which of the following processes?

   (A) Chain migration
   (B) Forced migration
   (C) Voluntary migration
   (D) Step migration
   (E) Ravenstein's third law of migration

54. Many of Africa's recent political and economic challenges can be attributed to

    (A) lack of natural resources.
    (B) climate change.
    (C) colonialism.
    (D) nepotism.
    (E) the Cold War.

55. A Pakistani family moves to a predominantly Pakistani neighborhood in London, where the family members maintain many of their cultural traditions. Which of the following terms best describes this circumstance?

    (A) A diorama
    (B) A diaspora
    (C) An ethnic enclave
    (D) Step migration
    (E) Gentrification

56. Which of the following is an example of how patterns of food production and consumption have shifted in highly developed countries, such as the United States, in recent years?

    (A) The rise of pesticides and other chemicals
    (B) The rise of genetically modified organisms (GMOs)
    (C) The rise of agricultural machinery
    (D) The rise in number and popularity of urban farmer's markets
    (E) The decline in food-related legislation

57. The growing number of retirement homes and memory-care facilities being built throughout the Western world is an indication of which of the following?

    (A) A columnar-shaped population pyramid
    (B) A population pyramid with a wide bottom
    (C) A problematic dependency ratio
    (D) A problematic family dynamic
    (E) A problematic health-care system

58. According to the gravity model, technological improvements in transportation and communications technology should

    (A) have no effect on the probability of interaction between two places.
    (B) increase the population of the two places.
    (C) decrease the amount of interaction between two places.
    (D) decrease the population in the two places.
    (E) decrease the friction of distance.

59. A geographic information system (GIS) differs from a traditional map in what ways?

(A) A traditional map typically shows only land features, whereas a GIS takes a variety of different phenomena into account.
(B) Data can be more easily manipulated or analyzed within a GIS, whereas information is static on a traditional map.
(C) A GIS produces a thematic map, whereas a traditional map tends to be a reference map.
(D) Because a GIS is on a computer, it isn't projected; a traditional map is.
(E) There are no major differences between a GIS and a traditional map.

60. Domingo owns a piece of land in the US Southwest that he irrigated heavily in order to grow tomatoes. Domingo's land is now infertile. What process is most likely to have caused Domingo's soil to become unproductive?

(A) Gentrification
(B) Deforestation
(C) Salinization
(D) Bromidification
(E) Desertification

If there is still time remaining, you may review your answers.

SECTION II

Directions for free-response questions: Read each question carefully and write your essays on standard composition paper. At the actual exam, you will be given a bound booklet containing lined pages for your free-response essays.

1. In many cities in the heartland of the United States, abandoned and dilapidated warehouses and factories can be found near and within the central business district (CBD).

   (a) Identify a heartland city that has been affected by this trend.

   (b) Describe one domestic economic process that contributes to this trend.

   (c) Describe one international economic process that contributes to this trend.

   (d) Explain how this pattern affects metropolitan areas of heartland cities.

   (e) Define at least one environmental consequence of this trend.

   (f) Describe at least one economic policy that works against the economic effects of deindustrialization.

   (g) Describe at least one environmental policy that works against the environmental effects of deindustrialization.

2. The following graphic is an urban model for Central and South American cities.

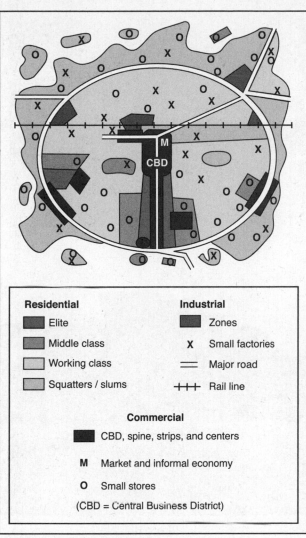

Residential
- ▮ Elite
- ▮ Middle class
- ▯ Working class
- ▯ Squatters / slums

Industrial
- ▮ Zones
- X Small factories
- ═ Major road
- +++ Rail line

Commercial
- ▮ CBD, spine, strips, and centers
- M Market and informal economy
- O Small stores

(CBD = Central Business District)

World Regional Geography Concepts
© 2009 W.H.Freeman and Company [From Yearbook of the Association of Pacific Coast Geographers 57 (1995): 28; printed with permission]

(a) Define squatter settlements.

(b) Explain why these settlements exist in Central and South American cities.

(c) Explain one factor leading to their spatial location within Central and South American cities.

(d) Describe how people living in squatter settlements make a living.

(e) Identify one challenge associated with economic development that Central and South American cities with significant squatter settlements face.

(f) Identify one challenge associated with social development that Central and South American cities with significant squatter settlements face.

(g) Describe one way in which squatter settlements in Central and South American countries might differ from squatter settlements in sub-Saharan Africa or South Asia.

Photo A

Photo B

3. The two photos above show two different types of agricultural production.

   (a) Identify and explain the type of production (intensive vs. extensive) in photo A.

   (b) Identify and explain the type of production (intensive vs. extensive) in photo B.

   (c) Explain the difference in agricultural technology between the two photos.

   (d) Describe one likely difference in the market area for each of the two products.

   (e) Describe the likely dietary differences between people living in nearby communities in photo A and people living in nearby communities in photo B.

   (f) Identify and describe one repercussion of the type of farming represented in photo B.

   (g) Identify and explain which stage of demographic transition is likely to be represented by each photo.

ANSWER KEY
Practice Test 1

SECTION I

1. **D**	21. **D**	41. **A**
2. **C**	22. **A**	42. **C**
3. **A**	23. **B**	43. **E**
4. **C**	24. **E**	44. **C**
5. **D**	25. **A**	45. **D**
6. **C**	26. **A**	46. **C**
7. **B**	27. **A**	47. **C**
8. **B**	28. **D**	48. **B**
9. **A**	29. **A**	49. **B**
10. **B**	30. **C**	50. **C**
11. **E**	31. **B**	51. **B**
12. **B**	32. **B**	52. **A**
13. **C**	33. **A**	53. **B**
14. **C**	34. **A**	54. **C**
15. **C**	35. **C**	55. **B**
16. **A**	36. **C**	56. **D**
17. **B**	37. **A**	57. **C**
18. **C**	38. **C**	58. **E**
19. **A**	39. **D**	59. **B**
20. **A**	40. **A**	60. **C**

ANSWERS AND EXPLANATIONS

Section I

MULTIPLE-CHOICE QUESTIONS

1. **(D)** While there are certainly wealthy people living in America's urban cores, most of America's urban residential wealth is concentrated in the suburbs, where families can enjoy sprawling estates, lower crime rates, and better schools, among other suburban amenities. In contrast, many European urbanites prefer to reside in the urban core to be in close proximity to shops, restaurants, and other cultural amenities.

   Metadata:

EK:	IMP-6.A.1
Unit:	6
Skill:	2.B Explain spatial relationships in a specified context or region of the world using geographic concepts, processes, models, and theories.

2. **(C)** Even with a strict policy, because of the pyramid's wide base at the time of implementation, in 20 years, *many* young people will be having just one child. It will take several generations for the population to actually stabilize, a phenomenon called *demographic momentum.*

   Metadata:

EK:	PSO-2.F.1; SPS-2.A.1
Unit:	2
Skill:	2.C Explain a likely outcome in a geographic scenario using geographic concepts, processes, models, and theories.

3. **(A)** Countries that are urbanized tend to have more highly educated populations that specialize in a variety of service-based occupations, jobs that typically concentrate in cities. Countries that are mostly rural have labor forces that obviously concentrate in the agricultural sector of the economy. As countries industrialize, they also typically urbanize, as factories and associated industries tend to concentrate in urban centers.

   Metadata:

EK:	PSO-6.A.2; SPS-7.B.1
Unit:	6, 7
Skill:	3.B Describe spatial patterns presented in maps and in quantitative and geospatial data.

4. **(C)** As countries industrialize and urbanize, birth rates tend to drop as women become more active in the labor force. A falling birth rate characterizes stage 3 countries, such as in Bangladesh, where women provide the majority of labor in the industrial sector of the economy.

   Metadata:

EK:	IMP-2.B.1
Unit:	2
Skill:	3.C Explain patterns and trends in maps and in quantitative and geospatial data to draw conclusions.

5. **(D)** Highly developed countries, such as Australia, have strong, service-based economies. These diverse economic systems require an educated population of both males and females.

   Metadata:

EK:	SPS-7.D.1
Unit:	7
Skill:	3.E Explain what maps or data imply or illustrate about geographic principles, processes, and outcomes.

6. **(C)** The gravity model predicts interaction between two places on the basis of the size of their population and the distance between them. Large populations (in the numerator) lessen the friction of distance between two places.

   Metadata:

EK:	PSO-6.C.1
Unit:	6
Skill:	2.B Explain spatial relationships in a specified context or region of the world using geographic concepts, processes, models, and theories.

7. **(B)** Primary economic activities involve direct extraction of resources from the land. Modern-day agricultural production, particularly in the developed world, is difficult to categorize as a primary economic activity because of how industrialized it has become.

   Metadata:

EK:	PSO-5.C.3
Unit:	5
Skill:	1.B Explain geographic concepts, processes, models, and theories.

8. **(B)** As women gain access to educational and economic opportunities, an additional, professional role is given priority over their roles as wives and homemakers. This additional role tends to lead to less childbearing along with greater investment in the now smaller family's educational opportunities.

    Metadata:

EK:	SPS-2.B.1
Unit:	2
Skill:	1.B Explain geographic concepts, processes, models, and theories.

9. **(A)** Because of Java's equatorial position, the soil quality should not allow for substantial agricultural production. However, because of volcanic activity on the island, the soil is nutrient-rich and more suited for higher population density than nearby Indonesian islands.

    Metadata:

EK:	PSA-2.A.1
Unit:	2
Skill:	2.B Explain spatial relationships in a specified context or region of the world using geographic concepts, processes, models, and theories.

10. **(B)** Cultural landscapes include the interaction of physical features, economic practices, religious and linguistic characteristics, and other forms of cultural expression all in one space. As a result, they nearly perfectly embody the concept of the human-environment relationship.

    Metadata:

EK:	PSO-3.B.1
Unit:	3
Skill:	1.D Describe a relevant geographic concept, process, model, or theory in a specified context.

11. **(E)** Malthus and those that subscribe to his ideas posit that when populations exceed the resource base needed to support them, the population must be "checked" through war, famine, disease, or some other means.

    Metadata:

EK:	IMP-2.B.3
Unit:	2
Skill:	1.E Explain the strengths, weaknesses, and limitations of different geographic models and theories in a specified context.

12. **(B)** Most migrants in squatter settlements have relocated to the city in search of employment in growing urban economies. Until these individuals can secure a job in the formal economy, they make a living through informal economic activities, such as the sale of goods (e.g., food, clothing, handicrafts) or services (e.g., child care, handiwork, transportation).

Metadata:

EK:	SPS-7.C.1
Unit:	7
Skill:	2.B Explain spatial relationships in a specified context or region of the world using geographic concepts, processes, models, and theories.

13. **(C)** Microlending is proving to be a revolutionary sustainable development strategy as it allows women to control how they earn and spend money. They are not at the mercy of factory owners who require them to work long hours in substandard conditions, as is often the case for female laborers in export-processing zones.

Metadata:

EK:	SPS-7.D.3
Unit:	7
Skill:	2.D Explain the significance of geographic similarities and differences among different locations and/or at different times.

14. **(C)** Centrifugal forces are those that tear apart the social fabric of a country and can potentially lead to devolution. When a state is made up of multiple ethnic groups, nationalism can become a threat if those ethnic groups compete with one another for political dominance.

EK:	SPS-4.C.1
Unit:	4
Skill:	1.B Explain geographic concepts, processes, models, and theories.

15. **(C)** A country need not have formal rule over another in order to dominate it. Imperialism can include cultural, political, and other forms of external control. Colonialism includes officially organized political and military rule.

Metadata:

EK:	PSO-4.B.2
Unit:	4
Skill:	1.A Describe geographic concepts, processes, models, and theories.

16. **(A)** Irredentism describes any political or popular movement that involves reclaiming a territory that previously belonged to the movement's members—in this case, Russia. As seen on the accompanying map, the majority of Crimea's population is of Russian descent. Russia gifted the peninsula to Ukraine in 1954. The referendum, deemed illegal by the United Nations, was an effort to reestablish the territory as Russian.

Metadata:

EK:	SPS-4.A.1
Unit:	4
Skill:	4.C Explain patterns and trends in visual sources to draw conclusions.

17. **(B)** Boundary lines can be defined in numerous ways. Sometimes political boundary lines, which indicate internationally recognized sovereignty over certain areas, do not align with ethnic, linguistic, and cultural boundaries, which tend to be fuzzier but are often more important, as they indicate a region with a common language and common customs.

Metadata:

EK:	IMP-4.B.1
Unit:	4
Skill:	4.D Compare patterns and trends in sources to draw conclusions.

18. **(C)** As shown in the map accompanying the question, some parts of Ukraine use Russian much more prevalently than Ukrainian, their official language. Having access to data at a finer scale, in this case the national scale as opposed to the global scale, allows for greater insight into spatial patterns.

Metadata:

EK:	PSO-1.D.1
Unit:	1
Skill:	5.C Compare geographic characteristics and processes at various scales.

19. **(A)** In von Thunen's model, agricultural land use changes with distance from the city center, primarily due to land and commodity values. Expensive, perishable commodities are grown closer to urban markets, while less expensive, more expansive (in terms of land area necessary for cultivation) agricultural goods are grown farther from the city center.

Metadata:

EK:	PSO-5.D.1
Unit:	5
Skill:	1.A Describe geographic concepts, processes, models, and theories.

20. **(A)** Territoriality describes the deep connection of a people, their culture, and their economic systems to their land. Because of this deep connection and all that it encompasses, disputes or wars often center on maintaining ownership or control of land.

Metadata:

EK:	PSO-4.C.2
Unit:	4
Skill:	1.B Explain geographic concepts, processes, models, and theories.

21. **(D)** Catalonia, because of its unique language and cultural identity, sought independence from Spain in a referendum in October 2017. The referendum was not recognized by the Spanish government, which considered it illegal.

Metadata:

EK:	SPS-4.B.1
Unit:	4
Skill:	1.D Describe a relevant geographic concept, process, model, or theory in a specified context.

22. **(A)** Food security is a condition related to food supply and an individual's access to it. It is a concern whose scale encompasses everything from food deserts in some of America's urban centers to entire countries that struggle against famine.

Metadata:

EK:	IMP-5.B.C
Unit:	5
Skill:	1.A Describe geographic concepts, processes, models, and theories.

23. **(B)** Environmental racism is any activity that causes people of a particular racial or ethnic group to be purposefully and disproportionately exposed to environmental hazards and pollution or to be denied access to environmental amenities such as clean water and open space.

Metadata:

EK:	SPS-6.A.1
Unit:	6
Skill:	1.B Explain geographic concepts, processes, models, and theories.

24. **(E)** A cultural syncretism occurs when two or more different cultures or cultural traits converge to form something new.

Metadata:

EK:	PSO-3.D.1
Unit:	3
Skill:	1.D Describe a relevant geographic concept, process, model, or theory in a specified context.

25. **(A)** Another way to think about cartographic representation is to divide maps into those used for reference and those that focus on a specific theme. An example of a reference map would be a road map or an atlas. An example of a thematic map would be a dot map that shows instances of petty crimes in the London metropolitan area.

Metadata:

EK:	IMP-1.A.1
Unit:	1
Skill:	1.C Compare geographic concepts, processes, models, and theories.

26. **(A)** Gentrification describes the process whereby people, typically middle- and upper-income people, move into city centers and rehabilitate much of the architecture for commercial and residential purposes. While it has its benefits, gentrification can also displace low-income populations and change the social character of certain neighborhoods.

Metadata:

EK:	SPS-6.A.4
Unit:	6
Skill:	4.C Explain patterns and trends in visual sources to draw conclusions.

27. **(A)** A space describes a location—an address or a set of coordinates—whereas the concept of "place" connotes the set of characteristics that make a location special and unique. It is essentially the difference between a house and a home: a house being the structure you live in and "home" describing the feeling you have about the structure and the people and things within it.

Metadata:

EK:	PSO-1.A.1
Unit:	1
Skill:	4.E Explain how maps, images, and landscapes illustrate or relate to geographic principles, processes, and outcomes.

28. **(D)** Gentrified urban areas typically attract young, single entrepreneurs, white-collar workers, and couples without kids. While gentrified areas offer the attraction of cultural amenities, restaurants, and bars, they do not often offer the types of amenities sought by families, such as large yards, high-quality education, and varying recreational opportunities.

Metadata:

EK:	IMP-6.E.2
Unit:	6
Skill:	4.B Describe the spatial patterns presented in visual sources.

29. **(A)** A lingua franca is a simple language that combines aspects of two or more other, more-complex languages; it is usually used for quick and efficient communication.

Metadata:

EK:	SPS-3.B.1
Unit:	3
Skill:	1.D Describe a relevant geographic concept, process, model, or theory in a specified context.

30. **(C)** Slash-and-burn agriculture historically was practiced in a sustainable manner by native people in tropical regions across the globe. However, it has now become associated with population growth, migration, and tropical deforestation in Brazil and elsewhere. It is not part of the Green Revolution, which involves increasing crop yield through new crop strains and other technological systems.

Metadata:

EK:	IMP-5.A.2
Unit:	5
Skill:	2.B Explain spatial relationships in a specified context or region of the world using geographic concepts, processes, models, and theories.

31. **(B)** A formal region is one in which the population shares one or more unique characteristics, such as language, religion, and ethnicity. The boundary that defines the pan-Arab region includes people that typically embrace Islam and are of Arab descent.

Metadata:

EK:	SPS-1.A.1
Unit:	1
Skill:	1.A Describe geographic concepts, processes, models, and theories.

32. **(B)** Buddhism, one of the three great world religions, began in the Indian subcontinent and then spread across central and eastern Asia.

Metadata:

EK:	IMP-3.B.4
Unit:	3
Skill:	1.A Describe geographic concepts, processes, models, and theories.

33. **(A)** During the Industrial Revolution, many people migrated from rural areas to large urban centers, generating a great need for agricultural goods within those centers. Furthermore, the technology characteristics of the Industrial Revolution transformed agricultural production as mechanization allowed more rapid cultivation of greater expanses of land.

Metadata:

EK:	SPS-5.C.1
Unit:	5
Skill:	1.B Explain geographic concepts, processes, models, and theories.

34. **(A)** The European Union, which stemmed from the European Economic Community, is at its heart an economic organization that integrated separate economies into one unit in order to be more competitive in the global marketplace.

Metadata:

EK:	PSO-7.A.2
Unit:	7
Skill:	1.D Describe a relevant geographic concept, process, model, or theory in a specified context.

35. **(C)** Some of the world's least-developed regions, including sub-Saharan Africa, South Asia, and Southeast Asia, are rich in linguistic diversity, as opposed to developed regions of North America, Western Europe, and Oceania, where linguistic diversity tends to be quite low.

Metadata:

EK:	IMP-3.B.2
Unit:	3
Skill:	3.D Compare patterns and trends in maps and in quantitative and geospatial data to draw conclusions.

36. **(C)** One map shows variation in linguistics across the globe—global scale. The other shows variation within a nation-state or country—national scale. Comparing data across scales allows for a richer and more nuanced understanding of a pattern or trend.

Metadata:

EK:	PSO-1.C.1
Unit:	1
Skill:	5.A Identify the scales of analysis presented by maps, quantitative and geospatial data, images, and landscapes.

37. **(A)** While Nigeria's linguistic diversity contributes to its rich history, strong family traditions, and vibrant culture, it also contributes to political tension. Operating as one cohesive nation with a multiplicity of languages and ethnicities proves challenging.

Metadata:

EK:	SPS-4.C.1
Unit:	4
Skill:	3.E Explain what maps or data imply or illustrate about geographic principles, processes, and outcomes.

38. **(C)** Central-place theory describes the organization of urban areas in a region or country. According to the theory, there should be a few large centers that provide a wide range of goods and services requiring a large market area (like a major-league baseball team). The central place is surrounded by a matrix of smaller cities, towns, and hamlets that offer progressively fewer products and enticements.

Metadata:

EK:	PSO-6.C.1
Unit:	6
Skill:	2.D Explain the significance of geographic similarities and differences among different locations and/or at different times.

39. **(D)** Many geographers, ecologists, and medical researchers have noted that indigenous knowledge about nature can be helpful in understanding ecosystems and in locating potential organisms for biomedical research. However, with decreasing cultural diversity, such knowledge is inevitably being lost.

Metadata:

EK:	SPS-3.A.4
Unit:	3
Skill:	2.C Explain a likely outcome in a geographic scenario using geographic concepts, processes, models, and theories.

40. **(A)** Federalism implies territorial organization. In territorial organization, national governments divide up their land base into smaller units, such as states and counties. These smaller units assume some level of autonomy in local governance and representation.

Metadata:

EK:	IMP-4.D.1
Unit:	4
Skill:	1.B Explain geographic concepts, processes, models, and theories.

41. **(A)** The multiple-nuclei model describes urban areas that contain numerous centers of business and cultural activity rather than a single central place. As American communities have become increasingly suburbanized, commercial and cultural opportunities have spread to the suburbs, creating multiple nuclei of activity within one urban area.

Metadata:

EK:	PSO-6.D.1
Unit:	6
Skill:	1.D Describe a relevant geographic concept, process, model, or theory in a specified context.

42. **(C)** In highly developed countries, labor is concentrated in the service sector of the economy. Machines do the bulk of agricultural work instead of people, a process most suited to monoculture crops such as corn, soybeans, cotton, and wheat.

Metadata:

EK:	PSO-5.C.1
Unit:	5
Skill:	2.B Explain spatial relationships in a specified context or region of the world using geographic concepts, processes, models, and theories.

43. **(E)** The finer the scale, the greater the ability to recognize and understand spatial patterns. If voting patterns are looked at on the state scale, Los Angeles County is likely to be categorized in one way. When the data are looked at by census tract, voting patterns in different parts of the city become observable.

Metadata:

EK:	PSO-1.D.1
Unit:	1
Skill:	5.B Explain spatial relationships across various geographic scales using geographic concepts, processes, models, and theories.

44. **(C)** As women gain access to education, employment opportunities in the formal economy open up to them. With more time invested in education and employment, less time is available to manage a family. Furthermore, as women become more educated, their awareness of family planning increases. Taken together, these factors lead to a dramatically lower fertility rate.

Metadata:

EK:	SPS-2.B.1
Unit:	2
Skill:	2.C Explain a likely outcome in a geographic scenario using geographic concepts, processes, models, and theories.

45. **(D)** The dependency theory of development describes a desire to become economically self-sufficient rather than relying on the problematic economic relationship set into play under colonization. Instead of relying on their colonizers for expensive manufactured imports, newly independent countries chose to develop a manufacturing base in order to substitute imports with domestically manufactured goods.

Metadata:

EK:	SPS-7.E.1
Unit:	7
Skill:	1.D Describe a relevant geographic concept, process, model, or theory in a specified context.

46. **(C)** Chain migration describes the social process by which immigrants from a particular area follow others, typically family members or close friends, to a new area. These new neighborhoods often take on the cultural characteristics of "home" and include restaurants, stores, and places of worship, among other things.

Metadata:

EK:	IMP-2.C.2
Unit:	2
Skill:	4.C Explain patterns and trends in visual sources to draw conclusions.

47. **(C)** A nation is not synonymous with a state or country, both of which connote political sovereignty over a particular piece of space. A nation is a tightly knit group of individuals that share a language, ethnicity, religion, and other unique cultural attributes.

Metadata:

EK:	PSO-4.A.2
Unit:	4
Skill:	1.A Describe geographic concepts, processes, models, and theories.

48. **(B)** Hierarchical diffusion describes the spread of an innovation through specific nodes that are not usually located close together in space but exhibit strong links to one another in a network.

Metadata:

EK:	IMP-3.A.1
Unit:	3
Skill:	1.C Compare geographic concepts, processes, models, and theories.

49. **(B)** France and Germany, both highly developed countries, are in stage 4 of the demographic transition model, meaning each couple replaces itself, if that. In order for the population of these countries to grow, they must allow for immigration. The opportunities available in both countries make them appealing destinations for many migrants.

Metadata:

EK:	IMP-2.A.1
Unit:	2
Skill:	2.B Explain spatial relationships in a specified context or region of the world using geographic concepts, processes, models, and theories.

50. **(C)** Botswana, located in sub-Saharan Africa, is a low-income nation that has been ravished by the HIV/AIDS epidemic. As the graph shows, low-income countries struggle much more with infectious diseases than do high-income countries.

Metadata:

EK	SPS-7.C.1
Unit:	7
Skill:	3.B Describe spatial patterns presented in maps and in quantitative and geospatial data.

51. **(B)** Epidemiological transition accounts for the replacement of infectious diseases with chronic diseases as a major cause of death. As countries become more developed, they gain access to the medical care that prevents infectious disease, as well as discover the causes of chronic diseases (sedentary occupations, processed foods, and the like).

Metadata:

EK:	IMP-2.B.2
Unit:	2
Skill:	4.E Explain how maps, images, and landscapes illustrate or relate to geographic principles, processes, and outcomes.

52. **(A)** Economic indicators of development typically describe measures of wealth, such as gross national product and gross national income. Social-welfare indicators demonstrate how a country is investing in the well-being of its population in terms of education, health, and other social services. Therefore, mortality rates are often seen as a social indicator of economic development.

Metadata:

EK:	SPS-7.C.1
Unit:	7
Skill:	1.D Describe a relevant geographic concept, process, model, or theory in a specified context.

53. **(B)** In forced migration, people leave an area for reasons other than their own choice. Factors leading to this movement include political unrest, religious persecution, ethnic cleansing, and environmental degradation. Often those fleeing their homes for these reasons seek refuge in a different country, as illustrated in this photo of a refugee camp in Pakistan.

Metadata:

EK:	IMP-2.D.1
Unit:	2
Skill:	4.A Identify the different types of information presented in visual sources.

54. **(C)** Many attribute the lack of political cohesion in much of sub-Saharan Africa to the way colonizing countries divided the space without any consideration of ethnic boundaries. Even after decolonization, much of the region's resources still flow out, with the wealth often accruing to corporations based in the countries of the former colonizers.

Metadata:

EK:	PSO-4.B.2
Unit:	4
Skill:	2.B Explain spatial relationships in a specified context or region of the world using geographic concepts, processes, models, and theories.

55. **(B)** A diaspora refers to a group of people with a common ethnic identity who are spread out over a large geographic area. Such an ethnic group often exists in an enclave that allows its members to maintain the group's cultural traditions.

Metadata:

EK:	PSO-3.D.1
Unit:	3
Skill:	1.D Describe a relevant geographic concept, process, model, or theory in a specified context.

56. **(D)** The growth in farmer's markets and urban gardening in highly developed countries, particularly the United States, indicates a growing backlash against large-scale industrial agriculture. Many consumers want to support local farms and farmers so as to be assured that their food is healthy for themselves and the environment.

Metadata:

EK:	IMP-5.B.2
Unit:	5
Skill:	1.D Describe a relevant geographic concept, process, model, or theory in a specified context.

57. **(C)** The dependency ratio refers to the percentage of people in a population who are either too old or too young to work and thus must rely on the productive labor of others to meet their needs. The growing number of retirement homes in developed countries, particularly the United States, reflects the entry of baby boomers into the later stages of life.

Metadata:

EK:	SPS-2.C.2
Unit:	2
Skill:	2.E Explain the degree to which a geographic concept, process, model, or theory effectively explains geographic effects in different contexts and regions of the world.

58. **(E)** The gravity model predicts interaction between two places. The standard model states that interaction is a function of the population of two places and the distance between them. As distance increases, the likelihood of interaction decreases. However, with technologies that overcome or shrink distance, its influence on the probability of spatial interaction decreases.

Metadata:

EK:	PSO-6.C.1
Unit:	6
Skill:	2.C Explain a likely outcome in a geographic scenario using geographic concepts, processes, models, and theories.

59. **(B)** GIS combines "geographic" (maps) and "information" (data) in one digital system that can integrate multiple layers of geospatial information. These layers can be manipulated or queried to conduct varying levels of spatial analysis. Thus, GIS is dynamic, while a traditional paper map is static.

Metadata:

EK:	IMP-1.B.2
Unit:	1
Skill:	3.D Compare patterns and trends in maps and in quantitative and geospatial data to draw conclusions.

60. **(C)** When farmers irrigate desert landscapes in an effort to increase productivity of an otherwise unproductive piece of land, the outcome typically involves salinization. The water applied to the land through irrigation evaporates quickly in the hot temperatures associated with these climates. After evaporation, a thick salty residue remains that renders the soil infertile. This process is what is meant by the term "salinization."

Metadata:

EK:	IMP-5.A.1
Unit:	5
Skill:	1.D Describe a relevant geographic concept, process, model, or theory in a specified context

Section II

FREE-RESPONSE QUESTIONS

1. Answer Explanations and Rubric:

1 point for cities in the American "heartland"	Cities include Akron, Baltimore, Buffalo, Chicago, Cincinnati, Cleveland, Columbus, Detroit, Flint, Hartford, Indianapolis, Kansas City, Milwaukee, Philadelphia, Pittsburgh, and St. Louis, among others.
1 point for any of the following domestic economic processes	■ Changing education levels: higher education leads to jobs in the service sector ■ Suburbanization: movement of the population outside of urban cores, where industrial activities are located ■ Changing technology: rise of the internet and web-based services changes employment opportunities ■ Changing technology: automation replaces human labor in factories

1 point for any of the following international economic processes	■ Globalization:
	  – Ease of transportation of raw materials and finished goods
	  – Rise of information technology allows for easy flow of information and money
	■ The rise of special economic zones (SEZs) and cheaper international labor

1 point for any of the identified effects on metropolitan areas	■ Loss of tax base as industries leave the city ■ Less income to maintain infrastructure and other social services (schools, police, etc.) in urban areas ■ Loss of tourism revenue
1 point for any of the identified environmental consequences	■ Brownfields (contaminated factory sites) ■ Elevated greenhouse gas emissions (trend coincides with suburbanization, which requires more automobile transportation)
1 point for any of the identified economic policies	■ Tax incentives that encourage gentrification (rehabilitation of abandoned buildings) of residential and commercial spaces ■ Investment in public transportation that makes it easier to travel to and within downtown areas
1 point for any of the identified environmental policies	■ Government funding for brownfield cleanup

2. Answer Explanations and Rubric:

1 point for an appropriate definition of a squatter settlement	■ A residential development characterized by poverty and little provision of social services, where housing is not typically owned or rented by its occupants ("squatters")
1 point for a correct explanation for the presence of squatter settlements	■ As countries become increasingly industrialized, they experience a rural-to-urban migration. People often migrate before securing a job and therefore don't have the means to buy housing.
1 point for a correct explanation of the spatial location of a squatter settlement	■ Squatters are typically found on the edges of cities (see model) where land is not suitable for development (e.g., hilly) or in other parts of the city where land is less desirable (e.g., along railroad tracks).
1 point for a correct description of the occupations of most people living in squatter settlements	■ Many who live in squatter settlements are without a job or a home. People so situated typically make money in the informal economy. Some of these occupations are legal—selling food, handicrafts, child-care services—while others would be considered illegal in most places (drug dealing, prostitution, etc.).
1 point for the correct identification of a challenge in terms of economic development	■ Since many who work in the informal economy earn income that is not taxed, governments do not have an accurate account of GDP, nor do they have the tax income for government services. ■ In many Central and South American countries, a significant portion of the population lives in these settlements. Local and national governments struggle to provide basic services (education, waste management, water, electricity, etc.) to these areas. ■ These settlements can limit or affect tourism.
1 point for the correct identification of a challenge in terms of social development	■ Disparity in certain social services, such as education, can have compounding effects on kids that grow up in squatter settlements. ■ Crime tends to be higher in squatter settlements than in other parts of the city.

1 point for the correct identification of a difference in settlements between Central and South America and either sub-Saharan Africa or South Asia	■ As Central and South America began industrializing before sub-Saharan African and South Asian cities, squatter settlements are more established. Students must mention any one of the following possible differences that result because of this fact: – Greater provision of services in Central and South American settlements (e.g., electricity, waste removal, water, education) – More solid structures in Central and South American settlements (as they have been around longer) – Some policies in place that allow home ownership or transfer of property in Central and South American settlements

3.   Answer Explanations and Rubric:

1 point for correct identification AND explanation of production type	■ Photo A shows intensive agriculture as labor is completed by humans and animals; as a consequence, production is likely to take place over a smaller area in comparison to photo B.
1 point for correct identification AND explanation of production type	■ Photo B shows extensive agricultural production as labor is completed by machinery; crops can be grown over a wide, expansive space.
1 point for correct explanation of difference in agricultural technology	■ Photo B, with the tractor, represents industrialized agriculture. The second agricultural revolution brought technology to agricultural production; tractors and other machines made it more efficient, larger in scale, and less dependent on human labor. In photo A, production has not yet been mechanized.
1 point for inclusion of any of the listed descriptions	■ Market area is likely to be much larger for product in photo B. ■ Market area is likely to be international for product in photo B and local for product in photo A. ■ Product from photo A is likely to go directly to local markets and consumers, while product from photo B probably goes to an industrial facility for processing before being distributed to corporations that will then distribute it to retail facilities to be purchased by consumers.

1 point for an explanation of dietary differences	■ Mechanized agriculture typically produces goods that need processing (e.g., field corn); thus the diet of community B is likely to be much more processed, and most of the food consumed is likely to be purchased rather than grown. In community A, the diet is likely to be less processed and composed of food grown through subsistence agriculture.
1 point for any of the listed repercussions	■ Represents a more processed diet, which can lead to a variety of health issues ■ Dependent on fossil fuels (for machinery) and transport of goods, which leads to greater greenhouse gas emissions ■ Typically requires a variety of chemicals that then enter the soil and groundwater ■ Monocultural production systems often strip soils of nutrients (thus requiring more chemicals). ■ Efficiency in production allows greater food security (more food produced on less land). ■ Efficiency in production allows diversification of the economy (people can pursue occupations other than farming).
1 point for correct identification of stages for both photos	■ Photo A: either stage 2 or 3. Stage 2 and 3 countries typically have a large portion of the labor force engaged in the primary sector of the economy as farming is not industrialized and requires more labor. ■ Photo B: stage 4. With industrialized agriculture, people can work in the tertiary sector (service-based) of the economy, which is typical in highly developed stage 4 countries.

ANSWER SHEET
Practice Test 2

SECTION I

1. Ⓐ Ⓑ Ⓒ Ⓓ Ⓔ
2. Ⓐ Ⓑ Ⓒ Ⓓ Ⓔ
3. Ⓐ Ⓑ Ⓒ Ⓓ Ⓔ
4. Ⓐ Ⓑ Ⓒ Ⓓ Ⓔ
5. Ⓐ Ⓑ Ⓒ Ⓓ Ⓔ
6. Ⓐ Ⓑ Ⓒ Ⓓ Ⓔ
7. Ⓐ Ⓑ Ⓒ Ⓓ Ⓔ
8. Ⓐ Ⓑ Ⓒ Ⓓ Ⓔ
9. Ⓐ Ⓑ Ⓒ Ⓓ Ⓔ
10. Ⓐ Ⓑ Ⓒ Ⓓ Ⓔ
11. Ⓐ Ⓑ Ⓒ Ⓓ Ⓔ
12. Ⓐ Ⓑ Ⓒ Ⓓ Ⓔ
13. Ⓐ Ⓑ Ⓒ Ⓓ Ⓔ
14. Ⓐ Ⓑ Ⓒ Ⓓ Ⓔ
15. Ⓐ Ⓑ Ⓒ Ⓓ Ⓔ
16. Ⓐ Ⓑ Ⓒ Ⓓ Ⓔ
17. Ⓐ Ⓑ Ⓒ Ⓓ Ⓔ
18. Ⓐ Ⓑ Ⓒ Ⓓ Ⓔ
19. Ⓐ Ⓑ Ⓒ Ⓓ Ⓔ
20. Ⓐ Ⓑ Ⓒ Ⓓ Ⓔ

21. Ⓐ Ⓑ Ⓒ Ⓓ Ⓔ
22. Ⓐ Ⓑ Ⓒ Ⓓ Ⓔ
23. Ⓐ Ⓑ Ⓒ Ⓓ Ⓔ
24. Ⓐ Ⓑ Ⓒ Ⓓ Ⓔ
25. Ⓐ Ⓑ Ⓒ Ⓓ Ⓔ
26. Ⓐ Ⓑ Ⓒ Ⓓ Ⓔ
27. Ⓐ Ⓑ Ⓒ Ⓓ Ⓔ
28. Ⓐ Ⓑ Ⓒ Ⓓ Ⓔ
29. Ⓐ Ⓑ Ⓒ Ⓓ Ⓔ
30. Ⓐ Ⓑ Ⓒ Ⓓ Ⓔ
31. Ⓐ Ⓑ Ⓒ Ⓓ Ⓔ
32. Ⓐ Ⓑ Ⓒ Ⓓ Ⓔ
33. Ⓐ Ⓑ Ⓒ Ⓓ Ⓔ
34. Ⓐ Ⓑ Ⓒ Ⓓ Ⓔ
35. Ⓐ Ⓑ Ⓒ Ⓓ Ⓔ
36. Ⓐ Ⓑ Ⓒ Ⓓ Ⓔ
37. Ⓐ Ⓑ Ⓒ Ⓓ Ⓔ
38. Ⓐ Ⓑ Ⓒ Ⓓ Ⓔ
39. Ⓐ Ⓑ Ⓒ Ⓓ Ⓔ
40. Ⓐ Ⓑ Ⓒ Ⓓ Ⓔ

41. Ⓐ Ⓑ Ⓒ Ⓓ Ⓔ
42. Ⓐ Ⓑ Ⓒ Ⓓ Ⓔ
43. Ⓐ Ⓑ Ⓒ Ⓓ Ⓔ
44. Ⓐ Ⓑ Ⓒ Ⓓ Ⓔ
45. Ⓐ Ⓑ Ⓒ Ⓓ Ⓔ
46. Ⓐ Ⓑ Ⓒ Ⓓ Ⓔ
47. Ⓐ Ⓑ Ⓒ Ⓓ Ⓔ
48. Ⓐ Ⓑ Ⓒ Ⓓ Ⓔ
49. Ⓐ Ⓑ Ⓒ Ⓓ Ⓔ
50. Ⓐ Ⓑ Ⓒ Ⓓ Ⓔ
51. Ⓐ Ⓑ Ⓒ Ⓓ Ⓔ
52. Ⓐ Ⓑ Ⓒ Ⓓ Ⓔ
53. Ⓐ Ⓑ Ⓒ Ⓓ Ⓔ
54. Ⓐ Ⓑ Ⓒ Ⓓ Ⓔ
55. Ⓐ Ⓑ Ⓒ Ⓓ Ⓔ
56. Ⓐ Ⓑ Ⓒ Ⓓ Ⓔ
57. Ⓐ Ⓑ Ⓒ Ⓓ Ⓔ
58. Ⓐ Ⓑ Ⓒ Ⓓ Ⓔ
59. Ⓐ Ⓑ Ⓒ Ⓓ Ⓔ
60. Ⓐ Ⓑ Ⓒ Ⓓ Ⓔ

SECTION II

Formulate your responses to the free-response questions on separate sheets of paper.

Practice Test 2

SECTION I

60 MINUTES, 60 QUESTIONS

PERCENTAGE OF TOTAL GRADE—50

> **Directions:** Read each question carefully, choose the correct response, and shade the corresponding choice on the answer sheet provided.

1. Which of the following best describes the difference between hierarchical and contagious diffusion?

   (A) In hierarchical diffusion, transmission occurs due to the relationships between places, whereas in contagious diffusion, transmission occurs due to the proximity between places.
   (B) Hierarchical diffusion is rare, whereas contagious diffusion is common.
   (C) Hierarchical diffusion occurs when people stay in place, whereas contagious diffusion occurs when they move.
   (D) Hierarchical diffusion is a form of expansion diffusion, whereas contagious diffusion is a form of relocation diffusion.
   (E) In hierarchical diffusion, transmission occurs due to the proximity between places, whereas in contagious diffusion, transmission occurs due to the relationships between places.

2. When recent college graduates and young professionals move to large cities with cultural amenities and job opportunities, to what are they responding?

   (A) Suburban amenities
   (B) Push factors
   (C) Pull factors
   (D) Mobility opportunities
   (E) Centripetal forces

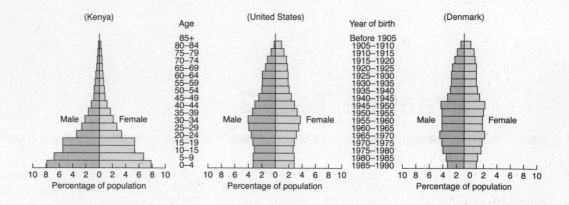

(Kenya)

(United States)

(Denmark)

Age
85+
80–84
75–79
70–74
65–69
60–64
55–59
50–54
45–49
40–44
35–39
30–34
25–29
20–24
15–19
10–15
5–9
0–4

Year of birth
Before 1905
1905–1910
1910–1915
1915–1920
1920–1925
1925–1930
1930–1935
1935–1940
1940–1945
1945–1950
1950–1955
1955–1960
1960–1965
1965–1970
1970–1975
1975–1980
1980–1985
1985–1990

Male   Female

Male   Female

Male   Female

10 8 6 4 2 0 2 4 6 8 10
Percentage of population

10 8 6 4 2 0 2 4 6 8 10
Percentage of population

10 8 6 4 2 0 2 4 6 8 10
Percentage of population

3. Which of the countries represented in the figure above is experiencing the slowest rate of population growth?

    (A) Denmark
    (B) United States
    (C) Kenya
    (D) Denmark and the United States are roughly equivalent.
    (E) The country with the greatest proportion of young people

4. The largely urbanized landscape that stretches along the US eastern seaboard from Washington, D.C., to Boston is an example of what?

    (A) A megacity
    (B) An edge city
    (C) A megalopolis
    (D) An urban agglomeration economy
    (E) A primate city

5. Which of the following describes religions that combine elements of Christianity with other traditional beliefs or practices?

    (A) Independent innovation of cultural traits
    (B) False idols
    (C) Cultural lags
    (D) Creolized religion
    (E) Cultural appropriation

6. Which of the following does not describe the Green Revolution?

    (A) Development and improvement of agricultural chemicals
    (B) Technological improvements such as hybrid seeds
    (C) Transfer of knowledge and technology from developed to developing countries
    (D) Application of more environmentally sustainable farming techniques
    (E) Intensification of farming practices in developing regions

7. A _____ is a group of people with a common cultural identity, whereas a
    _____ is a country with established and recognized borders.

    (A) territory . . . state
    (B) nation . . . territory
    (C) state . . . territory
    (D) nation . . . state
    (E) territory . . . nation

8. Which of the following is not included as a term in the demographic accounting
    equation?

    (A) Natural increase
    (B) Natural decrease
    (C) Total fertility rate
    (D) Immigration
    (E) Emigration

9. Which of the following refers to a simplified language used to communicate among
    people who speak different native tongues?

    (A) Dialect
    (B) Pidgin
    (C) Lingua franca
    (D) Creole
    (E) Accent

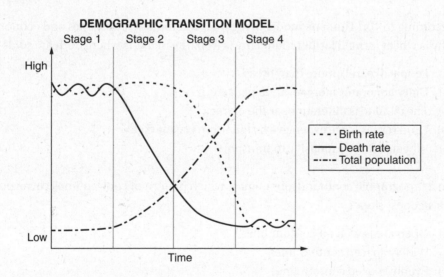

10. Which of the following describes a country in stage 3 of the demographic transition
    model shown above?

    (A) It has a high and consistent death rate.
    (B) It has a high and consistent birth rate.
    (C) It has a stable population.
    (D) It is undergoing rapid but declining population growth.
    (E) It is undergoing slow but increasing population growth.

11. Which of the following best describes a primate city?

   (A) The major seat of political power within a country
   (B) A city that is disproportionately larger than other cities in a country
   (C) The largest edge city in a country
   (D) The top city in the rank-size hierarchy
   (E) A suburban metropolis

12. According to the gravity model, technological improvements in transportation and communication between two places should result in which of the following?

   (A) They should have no effect on the two places' interactions.
   (B) They should cause an increase in the populations of both places.
   (C) They should cause a decrease in the amount of interaction between the two places.
   (D) They should cause a decrease in the populations of both places.
   (E) They should cause a decline in the friction of distance between the two places.

13. Which of the following terms best describes the makeshift neighborhoods, constructed of inexpensive scrap materials, found in all of the world's largest peripheral cities?

   (A) Edge cities
   (B) Squatter settlements
   (C) Swidden lands
   (D) Gentrified districts
   (E) Informal economic sectors

14. According to von Thunen's model of regional agricultural landscapes and economies, why is wheat farmed farther than dairy is from the markets where both are sold?

   (A) People like milk more than bread.
   (B) Dairy generates more revenue per acre.
   (C) The climate is different near the market.
   (D) Land rent is more expensive farther from the market.
   (E) Wheat is more expensive to transport.

15. On a topographic contour (isoline) map, which pattern of contour lines corresponds to the steepest slope?

   (A) Open areas with no contour lines
   (B) Widely spaced contour lines
   (C) Evenly spaced contour lines
   (D) Closely spaced contour lines
   (E) Only elevation, not slope, can be deciphered from contour lines

16. Which is a good example of a functional region?

    (A) The Bible Belt states
    (B) The area served by a local bus line
    (C) The state of California
    (D) An individual's perception of his/her daily activity space
    (E) An area where one dominant language prevails

17. On a Mercator projection map, where will you find the landmasses most exaggerated in relative size?

    (A) Near the poles
    (B) Near the prime meridian
    (C) Near the equator
    (D) Near the major oceans
    (E) Near the tropics

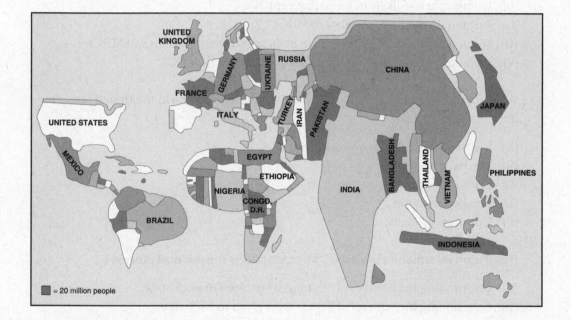

18. Which of the following is the proper term for the above map depicting the size of each country in proportion to its population?

    (A) An image map
    (B) A choropleth map
    (C) A projected map
    (D) An isoline map
    (E) A cartogram

19. When looking at labor-force distribution on a global scale, which of the following patterns is generally true?

(A) In highly developed (rich) countries, the labor force is evenly distributed across all economic sectors (primary, secondary, tertiary, and quaternary).

(B) In the least-developed countries, most laborers are in service-based industries (tertiary sector).

(C) Since the world's wealthiest countries tend to have vast natural resources, their labor forces are mainly in the primary sector (agriculture and mining).

(D) A poor country's labor force is concentrated in the primary sector, while a richer country's labor force skews toward the tertiary and quaternary sectors.

(E) Both highly developed and least-developed countries.

20. Which is generally true regarding the conditions of women in more-developed countries (MDCs) compared to those in less-developed countries (LDCs)?

(A) Women are more likely to be employed in agriculture in MDCs.

(B) Women are less likely to be literate in MDCs.

(C) Women do not perform labor in MDCs.

(D) Women are more likely to participate in government positions in MDCs.

(E) Women are more likely to have multiple children in MDCs.

21. Which of the following did not contribute to suburbanization in the United States in the 1950s and 1960s?

(A) Greater availability of mortgage loans

(B) The baby boom

(C) Economic growth

(D) Road construction

(E) Urban gentrification

22. Population information from the US Census affects the political process by

(A) determining the number of electoral votes given to each state.

(B) determining the number of Senate seats given to each state.

(C) determining the number of electoral votes given to each political party.

(D) determining voting district boundaries within states.

(E) determining when each state holds its primary election.

23. Mexico's maquiladoras are examples of which of the following?

(A) Offshore financial centers

(B) Brick-and-mortar businesses

(C) Ancillary activities

(D) Informal economic activities

(E) Export processing zones

Labor Force by Occupation

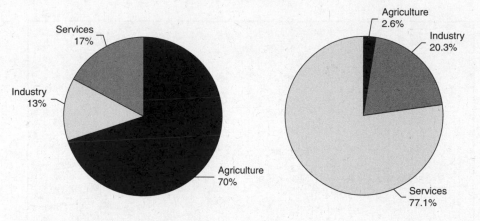

Services
17%

Industry
13%

Agriculture
70%

Agriculture
2.6%

Industry
20.3%

Services
77.1%

24. The pie charts above most likely represent the labor force by occupation in which of the following pairs of countries?

(A) Mexico; Denmark
(B) Saudi Arabia; South Africa
(C) Cameroon; Denmark
(D) New Zealand; Russia
(E) Chile; Oman

25. Rostow's stages-of-economic-development model predicts that each country's economy will progress along which of the following trajectories?

(A) High consumption to ecological sustainability
(B) Low output to high output
(C) Low income to high income and high mass consumption
(D) Low employment in tertiary activities to high employment in primary activities
(E) Low dependency to high dependency on other economies

26. Which of the following best defines the term "gerrymandering"?

(A) An early technique for setting maritime boundaries
(B) The large number of federal regulations controlling pollution by corporations
(C) Redistricting to ensure a majority-minority population within a district
(D) Legal "covenants" prohibiting people of color from residing in certain neighborhoods
(E) Drawing voting district boundaries to favor a political party or candidate

Questions 27 and 28 apply to the figure below.

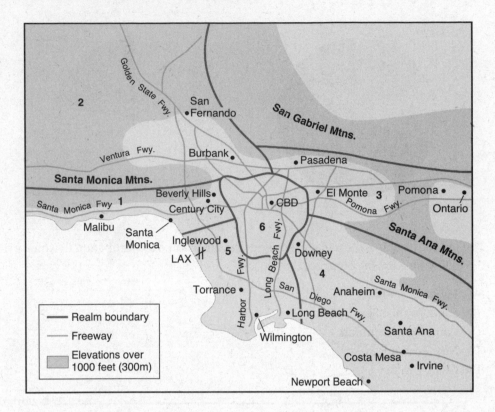

27. The map suggests which of the following about Los Angeles?

   (A) It is organized around a dominant central business district.
   (B) It is a primate city.
   (C) It is a multinucleated urban region.
   (D) It is organized into concentric rings.
   (E) It is a new urbanist city.

28. Based on the map, which of the following best describes Los Angeles?

   (A) It is a compact medieval city.
   (B) It has no squatter settlements.
   (C) It is a rink-sized city.
   (D) It contains much suburban sprawl.
   (E) Its highest rents are located in the outer ring.

29. Which of the following terms describes national governments that are organized into geographically defined subunits, such as states or provinces?

   (A) Federalist
   (B) Territorial
   (C) Consolidated
   (D) Electoral
   (E) Republican

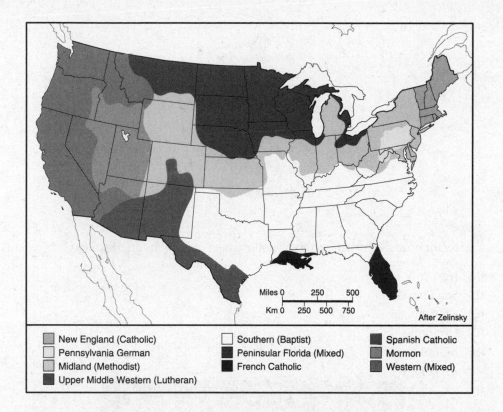

New England (Catholic)
Pennsylvania German
Midland (Methodist)
Upper Middle Western (Lutheran)

Southern (Baptist)
Peninsular Florida (Mixed)
French Catholic

Spanish Catholic
Mormon
Western (Mixed)

After Zelinsky

30. Which of the following best explains the above map of dominant religions by region in the United States?

(A) Climate determines people's religious affiliations.
(B) Different religions tend to take hold in rural rather than urban areas.
(C) Different US states have different official religions.
(D) Religious affiliation is related to historic patterns of immigration.
(E) Republicans tend to be Protestant, whereas Democrats tend to be Catholic.

31. Canada is an example of which of the following?

(A) A primate state
(B) A multinational state
(C) A prorupted state
(D) A centrifugal state
(E) A supranational state

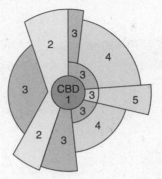

32. The sector model of urban geography above best fits which American city?

(A) New York
(B) San Francisco
(C) Chicago
(D) Honolulu
(E) Miami Beach

33. In the aerial photograph above, the grid of rural north–south trending roads is offset, with a short jog to the left about halfway up the image for anyone heading north. Which of the following best explains this?

(A) The original surveyors must have made an error in laying out the roads.
(B) Due to the curvature of the earth, north–south lines converge as they near the poles.
(C) Frontier geographies often contain such unusual local features.
(D) These grid corrections are an example of relative-versus-absolute distance.
(E) These grid corrections are an example of space-time convergence.

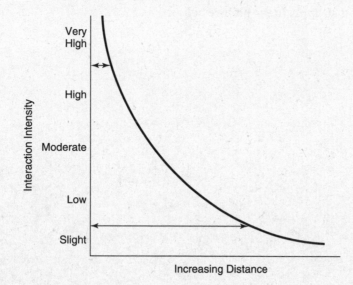

34. Which of the following is correct regarding the above distance decay function?

   (A) The friction of distance has little effect on connectivity.
   (B) Connectivity is related to the position of a central place.
   (C) Distance imposes a significant barrier to spatial interactions.
   (D) Relative distance acts independently of absolute distance.
   (E) Absolute decay and relative decay are different.

Questions 35 and 36 apply to the picture below.

35. Edge cities, such as Camarillo, California (pictured above), which is located about 50 miles east of Los Angeles, have which of the following qualities?

    (A) Low taxes
    (B) Export processing zones
    (C) Commuter rail
    (D) Sprawling suburbs with varied amenities
    (E) Light traffic and clean air

36. Which of the following statements is false regarding edge cities?

    (A) The first true edge cities had appeared by the late nineteenth century.
    (B) Edge cities tend to be located near national borders.
    (C) Edge cities often have high-quality public transportation systems.
    (D) Edge cities serve many of the same functions as traditional cities, but many lack downtown cores.
    (E) Edge cities are often considered examples of sustainable urban design.

37. Opened in 1915, San Francisco's City Hall, pictured above, combines classical forms with modern industrial features. It is an example of which of the following?

(A) Beaux Arts architecture
(B) Gentrification
(C) Postmodern architecture
(D) Node planning
(E) Urban revitalization

38. The area of the Middle East that extends from Egypt through Israel, Lebanon, Syria, and Iraq is known as which of the following?

(A) The demilitarized zone
(B) The green triangle
(C) The Sahara Desert
(D) Judea and Samaria
(E) The Fertile Crescent

39. Which of the following is incorrect of political boundaries?

(A) They establish the limits of sovereignty.
(B) They often indicate cultural, national, or economic divisions.
(C) In the ocean, they are defined by rules established through the United Nations Convention on the Law of the Sea.
(D) They are rarely contested.
(E) They usually contain well-defined nation-states.

Children Who Speak a Language Other Than English at Home (Percentage)—2015

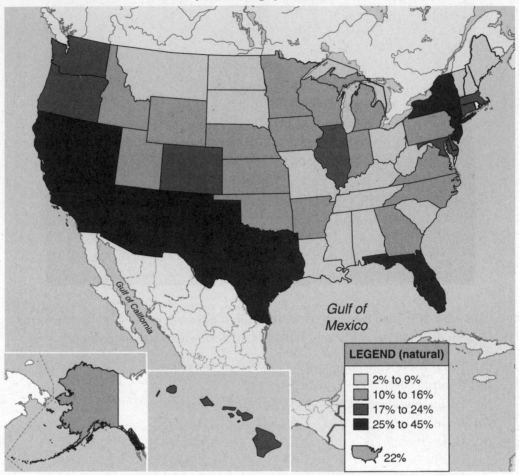

40. Which of the following statements probably best explains the data displayed in the map above?

(A) Children who speak languages other than English at home are randomly distributed throughout the United States.

(B) Children who speak languages other than English at home live mostly in rural areas.

(C) States with smaller populations tend to have a greater proportion of children who speak languages other than English at home.

(D) Most children speak languages other than English at home.

(E) Children who speak languages other than English at home tend to live in states with large Latino populations.

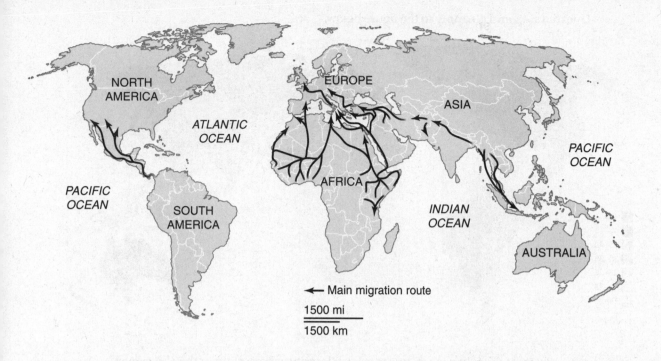

← Main migration route

1500 mi

1500 km

41. The above map shows the world's most heavily trafficked international migration routes. Based on the data displayed in it, which of the following statements is probably true?

(A) Migrants almost always move from south to north.
(B) Migrants usually end up in conflict zones.
(C) Most cross-border migrants travel from less-developed to more-developed countries.
(D) Most migrants pass easily through international borders.
(E) Most migrants end up back in their place of origin.

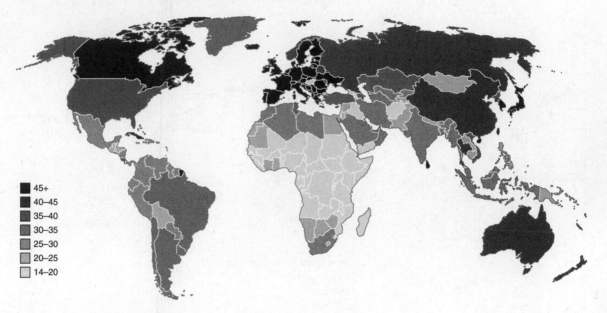

■	45+
■	40–45
■	35–40
■	30–35
■	25–30
■	20–25
□	14–20

42. Which of the following statements best describes the pattern in the above map showing the average ages of people by country?

(A) Wealthier countries tend to have younger populations.
(B) Wealthier countries tend to have older populations.
(C) Poorer countries tend to have larger populations.
(D) Landlocked countries tend to have older populations.
(E) People living in protracted countries tend to live protracted lives.

43. Countries with large proportions of young people tend to be located

(A) in landlocked areas.
(B) in the Arctic.
(C) in fertile agricultural regions.
(D) in developing regions of Africa, Asia, and Latin America.
(E) in the metropolitan core.

44. Which of the following is the best predictor of the birth rate in a given country?

(A) Gross domestic product
(B) Demographic transition models
(C) Pro-natalist policies
(D) Immigration policies
(E) Women's level of health, education, and empowerment

2010 TOP 20 CITIES BY POPULATION

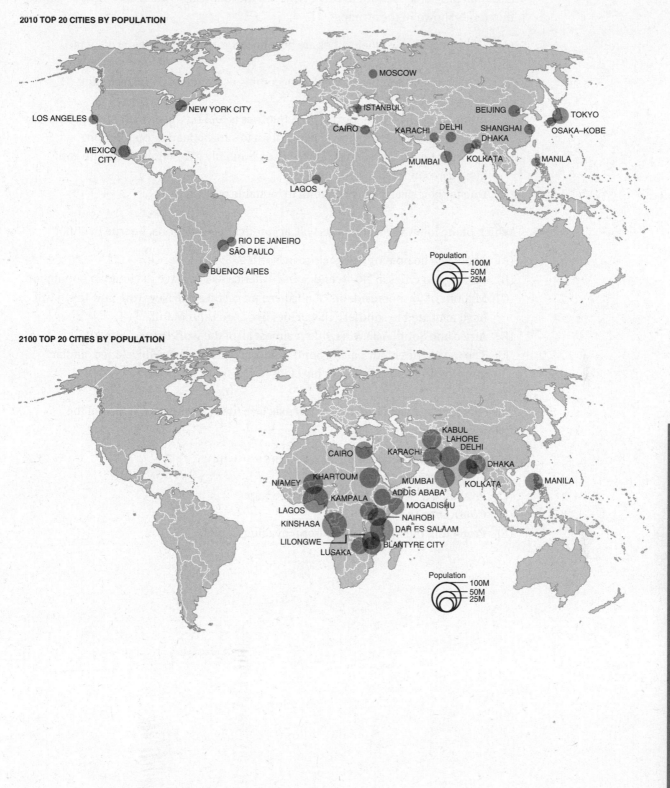

2100 TOP 20 CITIES BY POPULATION

45. The two maps above show the world's top twenty megacities, by population, in 2010 and then projected forward to 2100. Which of the following conclusions best describes the trends shown in these maps?

(A) The world's largest megacities are becoming more geographically dispersed over time.
(B) The world's largest megacities are becoming concentrated in developing countries.
(C) Immigration from less-developed countries is fueling the growth of megacities in more-developed countries, causing them to surge ahead in population.
(D) Growth of megacities in the 21st century is mainly a function of climate and distance to markets.
(E) Due to their size, megacities tend to be stable over time.

46. Which of the following statements will, according to projections, be true by 2100?

(A) There will be no more megacities in North or South America.
(B) The biggest cities in North and South America will all have declined in population.
(C) Millions of former residents of northern megacities like New York and Tokyo will have relocated to southern megacities like Lagos and Manila.
(D) Africa and South Asia will contain almost all of the world's megacities.
(E) The megacities of Africa and South Asia will probably have developed similar patterns and appearances as big cities

47. The emerging megacities in Africa and Asia face many challenges. Which of the following are not among these?

(A) Building and maintaining adequate infrastructure
(B) Fighting crime
(C) Providing a clean and healthy environment
(D) Controlling the cost of living
(E) Protecting their populations from invading armies

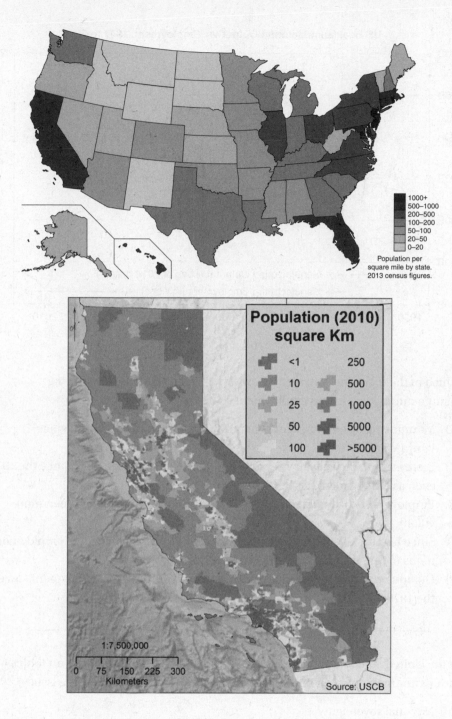

Population per square mile by state. 2013 census figures.

48. The two maps above depict population density at two geographic scales: the whole United States and the state of California. Based on this information, which of the following statements is incorrect?

(A) The scale of a map can fundamentally alter the information and insights it conveys.

(B) The population of California is distributed highly unevenly, with large numbers of people living in relatively small areas.

(C) The states with the greatest population densities tend to contain large cities.

(D) As the state with the largest population, most of California's land area is urbanized.

(E) Maps that include larger areas tend to display fewer details.

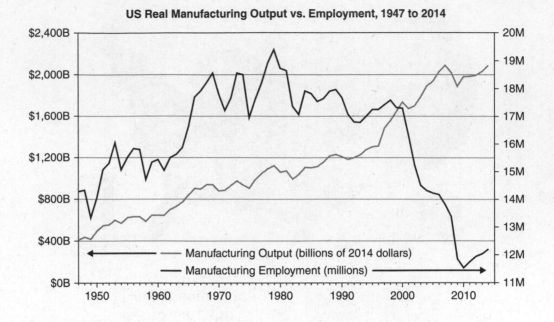

US Real Manufacturing Output vs. Employment, 1947 to 2014

Manufacturing Output (billions of 2014 dollars)

Manufacturing Employment (millions)

49. Which of the following statements best explains the above figure showing manufacturing output and employment since 1947?

(A) Manufacturing employment generally has kept pace with output since World War II.

(B) Increases in efficiency have enabled output to continue growing since the 1970s, even as employment has declined.

(C) Employment has undergone gradual changes, while output has been more volatile.

(D) Since highly skilled manufacturing jobs cost firms relatively little, employment tends to remain consistent or grow over time.

(E) Deindustrialization has led to declines in both output and employment since the 1970s.

50. In the United States, national political debates about federalism focus on topics like the environment and health care and often raise which of the following issues?

(A) National sovereignty

(B) Self-determination

(C) States' rights

(D) The heartland theory

(E) Buffer states

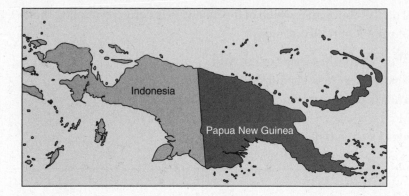

51. Which of the following terms best describes the boundary separating Papua New Guinea and Indonesia on the island of New Guinea, as pictured above?

   (A) Antecedent boundary
   (B) Fictitious boundary
   (C) Prorupted boundary
   (D) Superimposed boundary
   (E) Demilitarized boundary

Questions 52 and 53 apply to the figure below.

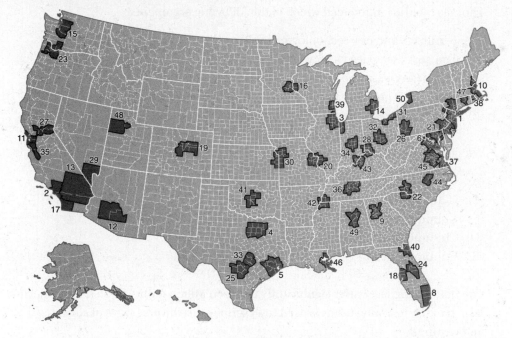

52. The map above shows the 50 largest metropolitan statistical areas in the United States by population. Which of the following best defines these areas?

   (A) An area containing several smaller urban areas that act together as a whole
   (B) A geographical center of activity within a larger urban region
   (C) A leading city, often with a population disproportionately larger than other urban areas in the same country
   (D) A global center of culture, government, or finance
   (E) A city placed in a remote area to draw population for economic or symbolic reasons

53. Based on the above map, which of the following is correct regarding the 50 largest US metropolitan statistical areas?

    (A) They are mostly located in the West.
    (B) They appear to be randomly distributed.
    (C) They have similar land-area sizes and shapes.
    (D) Most are located east of the Mississippi River.
    (E) Every state has one.

    _____

54. Which of the following terms refers to "the fair treatment and meaningful involvement of all people regardless of race, color, national origin, or income with respect to the development, implementation, and enforcement of environmental laws, regulations, and policies"?

    (A) Self-determination
    (B) Environmental justice
    (C) Acculturation
    (D) Relative distance
    (E) A sense of place

55. Thomas Malthus introduced which of the following arguments?

    (A) Fertilizers and new technologies could greatly increase crop output.
    (B) The best way to feed a growing population is to conserve resources.
    (C) Americans waste more than 30 percent of the food they produce.
    (D) Population growth will outstrip food production, leading to "negative checks."
    (E) Geography is an interdisciplinary field that draws from four distinct traditions.

56. Which of the following describes a cultural tradition's center and place of origin?

    (A) Cultural center
    (B) Cultural complex
    (C) Cultural trait
    (D) Cultural hearth
    (E) Cultural genesis

57. The Italian language varies significantly between Milan, Rome, Naples, and Palermo. Which of the following terms is used to describe these diverse ways of speaking and writing?

    (A) Pidgins
    (B) Lingua franca
    (C) Language groups
    (D) Dialects
    (E) Idioms

Questions 58–60 apply to the figure below.

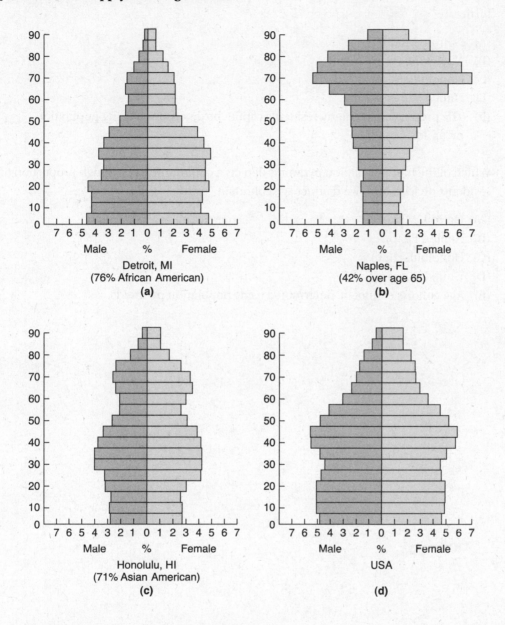

Detroit, MI
(76% African American)
(a)

Naples, FL
(42% over age 65)
(b)

Honolulu, HI
(71% Asian American)
(c)

USA
(d)

58. Which of the four population pyramids shown above has the greatest proportion of elderly residents?

(A) Detroit, MI (a)
(B) Naples, FL (b)
(C) Honolulu, HI (c)
(D) United States (d)
(E) Dependency ratios cannot be determined using population pyramids.

59. Which of the four population pyramids depicts the community with the highest birth rate?

(A) Detroit, MI (a)
(B) Naples, FL (b)
(C) Honolulu, HI (c)
(D) United States (d)
(E) The proportion of elderly residents cannot be determined using population pyramids.

60. Which of the four population pyramids depicts a community with a high proportion of residents divided into two distinct age cohorts?

(A) Detroit, MI (a)
(B) Naples, FL (b)
(C) Honolulu, HI (c)
(D) United States (d)
(E) Age cohorts cannot be determined using population pyramids.

STOP

If there is still time remaining, you may review your answers.

SECTION II

Directions for free-response questions: Read each question carefully and write your essays on standard composition paper. At the actual exam, you will be given a bound booklet containing lined pages for your free-response essays.

1. The world's population is currently around 7.5 billion people. This is the largest it has ever been, and it means that around 1 in 14 people who have ever lived are alive today. Answer (a)–(g), on the subject of human population growth.

   (a) Describe global human population growth since 1700.

   (b) How does the rate of growth since 1700 compare to the long-term rate of growth over the previous ten thousand years?

   (c) Explain the demographic transition model.

   (d) Explain the role of women in determining the rate of population growth in a country.

   (e) Describe some of the results and consequences of China's one-child policy, which was in place from 1980 to 2016.

   (f) Several countries have aging populations. Name three, and explain the reasons and consequences of this.

   (g) Thomas Malthus predicted that population would outstrip food production, resulting in wars, famines, and other "negative checks." Yet, although malnutrition is common in many countries, true famines have rarely, if ever, occurred in countries with democratic institutions and free presses. Explain what may account for this distinction between food production and food availability.

2. With a 2019 population of 14.4 million people, Dhaka, Bangladesh, pictured above, is a classic primate city.

    (a) Define a primate city.

    (b) Name two other primate cities, along with their home countries.

    (c) Explain why primate cities exist in many countries, including their possible origins and some of the functions they serve.

    (d) Canada, Germany, and India are examples of countries without primate cities. Explain why a country may lack a primate city.

    (e) Some primate cities are also world cities. Define a world city, and explain the difference between the two concepts.

    (f) Some primate cities are also megacities. Define a megacity, and explain the difference between the two concepts.

    (g) The United States lacks a primate city, but it has several world cities. Name two of these, and explain what makes them world cities.

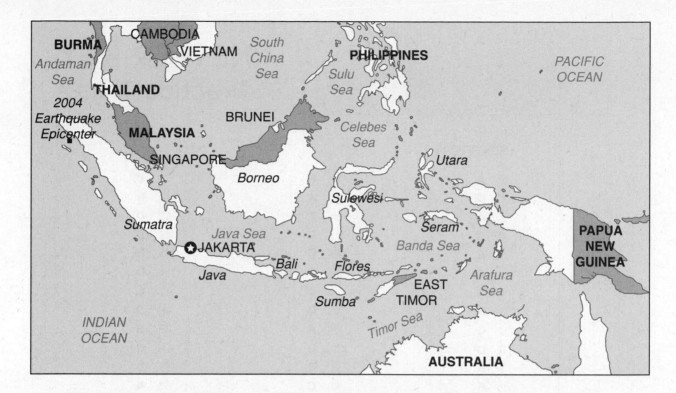

3. With more than 17,000 islands and 260 million people spread over an area of more than 3 million square miles, Indonesia is one of the world's largest and most diverse countries. As such, it has been the subject of considerable research by geographers and other scholars trying to understand what holds this vast archipelago together. Respond to the following prompts, using Indonesia as an example.

(a) Define the concept of the multinational state.

(b) Define centrifugal and centripetal forces.

(c) Explain how language, culture, and religion can act as either centrifugal or centripetal forces in maintaining the Indonesian state.

(d) Explain the role of colonialism and postcolonialism in assembling and maintaining a far-flung country like Indonesia.

(e) Benedict Anderson defined "imagined communities" as social constructs brought into being or maintained by people who perceive themselves to be a part of the group. Explain how this concept may apply to Indonesia.

(f) Indonesia has experienced several separatist movements in recent years. Explain why Indonesia may be especially vulnerable to these efforts.

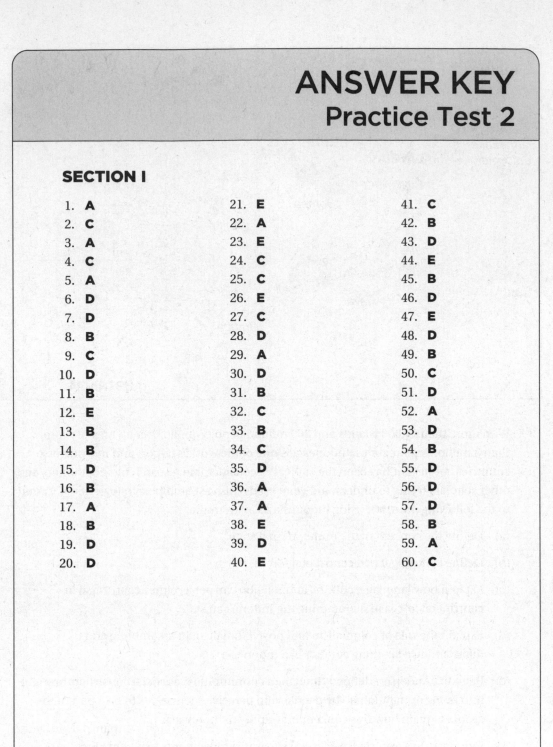

ANSWER KEY
Practice Test 2

SECTION I

1.	**A**	21.	**E**	41.	**C**
2.	**C**	22.	**A**	42.	**B**
3.	**A**	23.	**E**	43.	**D**
4.	**C**	24.	**C**	44.	**E**
5.	**A**	25.	**C**	45.	**B**
6.	**D**	26.	**E**	46.	**D**
7.	**D**	27.	**C**	47.	**E**
8.	**B**	28.	**D**	48.	**D**
9.	**C**	29.	**A**	49.	**B**
10.	**D**	30.	**D**	50.	**C**
11.	**B**	31.	**B**	51.	**D**
12.	**E**	32.	**C**	52.	**A**
13.	**B**	33.	**B**	53.	**D**
14.	**B**	34.	**C**	54.	**B**
15.	**D**	35.	**D**	55.	**D**
16.	**B**	36.	**A**	56.	**D**
17.	**A**	37.	**A**	57.	**D**
18.	**B**	38.	**E**	58.	**B**
19.	**D**	39.	**D**	59.	**A**
20.	**D**	40.	**E**	60.	**C**

ANSWERS AND EXPLANATIONS

Section I

MULTIPLE-CHOICE QUESTIONS

1. **(A)** Hierarchical diffusion occurs when a cultural trait is transmitted from one place to another because the two places have some common characteristic or relationship. It is different from contagious diffusion, in which a cultural trait is passed from one person or place to another simply because of their proximity.

   Metadata:

EK:	IMP-3.A.1
Unit:	3
Skill:	1.D Describe a relevant geographic concept, process, model, or theory in a specified context.

2. **(C)** Pull factors, such as job and recreational opportunities, attract people to move to new places. Push factors, such as violence and economic stagnation, encourage people to leave their old homes and search for new places to live with better prospects.

   Metadata:

EK:	IMP-2.C.1
Unit:	2
Skill:	2.B Explain spatial relationships in a specified context or region of the world using geographic concepts, processes, models, and theories.

3. **(A)** The population pyramids provided show that Denmark, which has the smallest proportion of young people and largest proportion of older people, is experiencing the slowest population growth.

   Metadata:

EK:	PSO-2.F.1
Unit:	2
Skill:	2.C Explain a likely outcome in a geographic scenario using geographic concepts, processes, models, and theories.

4. **(C)** A megalopolis is an entire region that has become highly urbanized. In North America, megalopolises include the Boston–New York–Philadelphia–Baltimore–Washington, D.C. urban mass along the eastern seaboard and the Los Angeles–Orange County–San Diego–Tijuana megalopolis in southern California and Baja, Mexico.

   Metadata:

EK:	PSO-6.A.4
Unit:	6
Skill:	5.B Explain spatial relationships across various geographic scales using geographic concepts, processes, models, and theories.

5. **(A)** Creolized religions combine aspects of world religions with local beliefs and traditional practices to create something new.

Metadata:

EK:	SPS-3.B.1
Unit:	3
Skill:	5.B Explain spatial relationships across various geographic scales using geographic concepts, processes, models, and theories.

6. **(D)** The Green Revolution was an effort to increase agricultural yields in the short term through the use of selective breeding, technology, and fertilizers. It did not seek to be more environmentally sustainable in the long term.

Metadata:

EK:	SPS-5.D.2
Unit:	5
Skill:	2.D Explain the significance of geographic similarities and differences among different locations and/or at different times.

7. **(D)** A nation is a group of people with a common cultural identity, whereas a state is a country with established and recognized borders.

Metadata:

EK:	PSO-4.A.2
Unit:	4
Skill:	1.D Describe a relevant geographic concept, process, model, or theory in a specified context.

8. **(B)** Natural decrease is not a term in the demographic accounting equation, but a country can have a negative rate of natural increase.

Metadata:

EK:	IMP-2.A.2
Unit:	2
Skill:	3.C Explain patterns and trends in maps and in quantitative and geospatial data to draw conclusions.

9. **(C)** A lingua franca is a simplified language often used to communicate when traveling or engaging in trade.

Metadata:

EK:	SPS-3.A.1
Unit:	3
Skill:	2.B Explain spatial relationships in a specified context or region of the world using geographic concepts, processes, models, and theories.

10. **(D)** In stage 3, death rates decline dramatically, followed by birth rates. The decline in death rates results, for a time, in rapid growth, but this growth slows considerably as birth rates drop. Longer lives and lower fertility are common in more-developed countries.

Metadata:

EK:	IMP-2.B.1
Unit:	2
Skill:	3.B Describe spatial patterns presented in maps and in quantitative and geospatial data.

11. **(B)** A primate city is one that is disproportionately larger—at least twice as large and often several times bigger—than other cities in a country. Many countries have primate cities. Paris and London are examples, as are Mexico City and Tokyo. The United States, China, India, and Germany are examples of countries without true primate cities.

Metadata:

EK:	PSO-6.C.1
Unit:	6
Skill:	2.C Explain a likely outcome in a geographic scenario using geographic concepts, processes, models, and theories.

12. **(E)** The gravity model says that greater connectivity between two places will reduce the friction of distance between them. They will seem closer together even though they remain the same distance apart.

Metadata:

EK:	PSO-6.C.1
Unit:	6
Skill:	5.B Explain spatial relationships across various geographic scales using geographic concepts, processes, models, and theories.

13. **(B)** Squatter settlements are unplanned neighborhoods that have developed around the world in fast-growing cities in developing countries. These neighborhoods often lack basic infrastructure and services and are plagued by crime, but their resilient residents are known for finding creative ways to overcome these challenges and hardships.

Metadata:

EK:	SPS-6.A.2
Unit:	6
Skill:	4.E Explain how maps, images, and landscapes illustrate or relate to geographic principles, processes, and outcomes.

14. **(B)** Dairy generates greater profits from a smaller land area, making it better suited for the higher rents closer to markets. Dairy is also heavier than wheat per unit sold, which increases transportation costs.

Metadata:

EK:	PSO-5.D.1
Unit:	5
Skill:	5.B Explain spatial relationships across various geographic scales using geographic concepts, processes, models, and theories.

15. **(D)** Since topographic contour lines follow points of equal elevation, the closer the contour lines are, the steeper the slope they describe.

Metadata:

EK:	IMP-1.A
Unit:	1
Skill:	3.A Identify the different types of data presented in maps and in quantitative and geospatial data.

16. **(B)** Unlike other region types, which are typically defined according to some sort of uniform characteristic within an area, a functional region is defined by a particular function, usually involving spatial interaction, which takes place within a fairly defined boundary. The only option among the possible responses that describes a functional region of interaction is that of the area a local bus line serves.

Metadata:

EK:	SPS-1.A.2
Unit:	1
Skill:	1.A Describe geographic concepts, processes, models, and theories.

17. **(A)** Because it is impossible to project a rounded surface onto a flat one with full accuracy, all map projections distort either area, distance, shape, or direction. The Mercator projection distorts area such that landmasses near the poles appear far bigger than they actually are relative to landmasses at lower latitudes.

Metadata:

EK:	IMP-1.A.3
Unit:	1
Skill:	3.A Identify the different types of data presented in maps and in quantitative and geospatial data.

18. **(B)** A choropleth map uses area to depict relative values, such as population size.

Metadata:

EK:	IMP-1.A.3
Unit:	1
Skill:	3.A Identify the different types of data presented in maps and in quantitative and geospatial data.

19. **(D)** Poor countries have labor forces concentrated in the primary sector, while a richer country's labor force skews toward the tertiary and quaternary sectors.

Metadata:

EK:	SPS-7.B.1
Unit:	7
Skill:	2.B Explain spatial relationships in a specified context or region of the world using geographic concepts, processes, models, and theories.

20. **(D)** Women in MDCs tend to have more education and political empowerment, and thus they are more likely to participate in government.

Metadata:

EK:	SPS-7.D.1
Unit:	7
Skill:	3.D Compare patterns and trends in maps and in quantitative and geospatial data to draw conclusions.

21. **(E)** Several factors—including more accessible loans, higher fertility rates, economic growth, and road building—enabled large-scale suburbanization in the decades after World War II. Urban gentrification is mainly associated with the period since 1990, when cities that declined during the postwar era started to receive increasing investment and, as a result, property values rose.

Metadata:

EK:	PSO-6.A.4
Unit:	6
Skill:	3.D Compare patterns and trends in maps and in quantitative and geospatial data to draw conclusions.

22. **(A)** US Census information is used in many ways. For example, it helps set the amount of funding that goes to different areas on the basis of each area's population and socio-economic needs. The census also determines the number of electoral votes given to each state in upcoming presidential elections.

Metadata:

EK:	IMP-4.B
Unit:	4
Skill:	2.A Describe spatial patterns, networks, and relationships.

23. **(E)** Mexico's maquiladoras are export-processing zones where goods are manufactured and distributed to consumers mostly in the United States and Canada.

Metadata:

EK:	PSO-7.A.6
Unit:	7
Skill:	5.B Explain spatial relationships across various geographic scales using geographic concepts, processes, models, and theories.

24. **(C)** Cameroon is a less-developed country with a large agricultural labor force, whereas Denmark is a highly developed country with a large service sector.

Metadata:

EK:	SPS-7.E.1
Unit:	7
Skill:	1.E Explain the strengths, weaknesses, and limitations of different geographic models and theories in a specified context.

25. **(C)** Rostow's model predicts that countries will develop over time, proceeding from low income to high income and high mass consumption. It describes the trajectories of some countries, but the complexity of history and geography require that we consider it only as a general concept and not a predictive model applicable to specific circumstances.

Metadata:

EK:	SPS-7.E.1
Unit:	7
Skill:	1.E Explain the strengths, weaknesses, and limitations of different geographic models and theories in a specified context.

26. **(E)** Gerrymandering involves drawing voting district boundaries to favor a political party or candidate. The United States Supreme Court generally has upheld the idea that partisan gerrymandering is allowed, to a point, under the Constitution even though computer mapping has increased the ability of parties to draw biased boundaries and doing so has negative consequences for political representation.

Metadata:

EK:	IMP-4.B.5
Unit:	4
Skill:	5.A Identify the scales of analysis presented by maps, quantitative and geospatial data, images, and landscapes.

27. **(C)** Los Angeles is a classic multinucleated urban region in that it contains multiple nodes of commerce and government of similar social, political, and financial status. Los Angeles has a large downtown, but its central business district is not nearly as dominant, for example, as Chicago's Loop, which is the center for business and finance in the Midwest.

Metadata:

EK:	PSO-6.D.1
Unit:	6
Skill:	1.E Explain the strengths, weaknesses, and limitations of different geographic models and theories in a specified context.

28. **(D)** Los Angeles is a vast, mostly low-rise city that contains extensive areas of suburban sprawl. This does not, however, mean that Los Angeles has low population density. Many people are surprised to learn that, according to the US Census, the Los Angeles metropolitan area is the most densely populated in the United States.

Metadata:

EK:	PSO-6.A.4
Unit:	6
Skill:	1.E Explain the strengths, weaknesses, and limitations of different geographic models and theories in a specified context.

29. **(A)** Federalist systems of government are organized hierarchically into geographically defined subunits to which the national government delegates some power.

Metadata:

EK:	IMP-4.D.1
Unit:	4
Skill:	2.A Describe spatial patterns, networks, and relationships.

30. **(D)** The dominant religious affiliation in any region is usually a product of historic patterns of immigration. In the US Intermountain West, for example, the arrival of Mormons during the 19th century still profoundly shapes religious institutions and beliefs in the region.

Metadata:

EK:	MP-3.B.3
Unit:	3
Skill:	4.E Explain how maps, images, and landscapes illustrate or relate to geographic principles, processes, and outcomes.

31. **(B)** Canada is a multinational state because the province of Quebec contains a distinct history, language, and identity and has a long-established separatist movement.

Metadata:

EK:	PSO-4.A.2
Unit:	4
Skill:	5.A Identify the scales of analysis presented by maps, quantitative and geospatial data, images, and landscapes.

32. **(C)** Chicago is a distinctive American city that has both concentric rings of increasingly recent development as one moves outward from the center and rail lines that act as spokes extending into these newer communities.

Metadata:

EK:	PSO-6.D.1
Unit:	6
Skill:	1.E Explain the strengths, weaknesses, and limitations of different geographic models and theories in a specified context.

33. **(B)** Due to the curvature of the earth, north–south lines converge as they near the poles. This means that in regions with rectangular grid layouts and infrastructure, occasional "corrections" must be built into the grid.

Metadata:

EK:	IMP-1.A
Unit:	1
Skill:	3.B Describe spatial patterns presented in maps and in quantitative and geospatial data.

34. **(C)** This distance decay function suggests that distance poses a significant barrier to spatial interactions. The intensity of this barrier can vary due to many factors, such as infrastructure, migration history, and digital connectivity.

Metadata:

EK:	PSO-1.A.1
Unit:	1
Skill:	3.B Describe spatial patterns presented in maps and in quantitative and geospatial data.

35. **(D)** Edge cities, such as Camarillo, CA, have sprawling suburbs with varied amenities and are often located 25 to 100 miles from a major metropolitan core.

Metadata:

EK:	PSO-6.A
Unit:	6
Skill:	1.E Explain the strengths, weaknesses, and limitations of different geographic models and theories in a specified context.

36. **(A)** Edge cities began to appear in the 1980s, not in the late 19th century.

Metadata:

EK:	PSO-6.A
Unit:	6
Skill:	1.E Explain the strengths, weaknesses, and limitations of different geographic models and theories in a specified context.

37. **(A)** San Francisco City Hall is a classic example of Beaux Arts architecture, which marries older, classical forms with newer, industrial ones. Beaux Arts planners designed wide thoroughfares, spacious parks, and civic monuments that stressed progress, freedom, and national unity.

Metadata:

EK:	MP-6.B
Unit:	6
Skill:	3.C Explain patterns and trends in maps and in quantitative and geospatial data to draw conclusions.

38. **(E)** The Fertile Crescent, which spanned the Middle East from Egypt to Iraq, was one of the world's earliest centers of sedentary farming.

Metadata:

EK:	SPS-5.A.1
Unit:	5
Skill:	2.B Explain spatial relationships in a specified context or region of the world using geographic concepts, processes, models, and theories.

39. **(D)** National boundaries are frequently contested due to their complex histories and geographies.

Metadata:

EK:	IMP-4.B
Unit:	4
Skill:	5.D Explain the degree to which a geographic concept, process, model, or theory effectively explains geographic effects across various geographic scales.

40. **(E)** Children who speak languages other than English at home tend to live in states with large Latino populations, as Spanish is the most frequently spoken language other than English in homes across this country.

Metadata:

EK:	IMP-3.B.2
Unit:	4
Skill:	4.E Explain how maps, images, and landscapes illustrate or relate to geographic principles, processes, and outcomes.

41. **(C)** The map above shows that most cross-border migrants travel from less-developed to more-developed countries, as push factors drive them from their homes and pull factors attract them to places with greater security, opportunities, and quality of life.

Metadata:

EK:	IMP-2.C
Unit:	2
Skill:	2.B Explain spatial relationships in a specified context or region of the world, using geographic concepts, processes, models, and theories.

42. **(B)** Wealthier countries in Europe, North America, and parts of East Asia, which have low death and fertility rates, tend to have older populations.

Metadata:

EK:	IMP-2.B
Unit:	2
Skill:	3.B Describe spatial patterns presented in maps and in quantitative and geospatial data.

43. **(D)** Countries with large proportions of young people tend to be located in the fast-growing developing regions of Africa, Asia, and Latin America.

Metadata:

EK:	IMP-2.B
Unit:	2
Skill:	3.B Describe spatial patterns presented in maps and in quantitative and geospatial data.

44. **(E)** Women's level of health, education, and empowerment is the best predictor of the birth rate in a country. As women gain more of all three, they tend to pursue careers, have children later in life, and have fewer of them.

Metadata:

EK:	SPS-2.B.1
Unit:	2
Skill:	3.B Describe spatial patterns presented in maps and in quantitative and geospatial data.

45. **(B)** Over the 21st century, the world's largest megacities are projected to become more concentrated in rapidly growing developing countries.

Metadata:

EK:	PSO-6.A.3
Unit:	6
Skill:	2.D Explain the significance of geographic similarities and differences among different locations and/or at different times.

46. **(D)** By 2100, Africa and South Asia are predicted to have almost all of the world's biggest megacities. Megacities will still exist elsewhere, but they will be eclipsed in population by these booming urban regions containing tens of millions of people.

Metadata:

EK:	PSO-6.A.3
Unit:	6
Skill:	2.D Explain the significance of geographic similarities and differences among different locations and/or at different times.

47. **(E)** Although many megacities in Africa and Asia have high rates of violent crime, they are in no danger of being invaded by outside forces.

Metadata:

EK:	PSO-6.A.3
Unit:	6
Skill:	2.D Explain the significance of geographic similarities and differences among different locations and/or at different times.

48. **(D)** California is the US state with the largest population, having surpassed New York in 1962. Yet, California's population is concentrated in a few densely populated major metropolitan areas, with vast, thinly populated expanses in between. In addition to its human population, California also has the country's biggest agricultural economy and more area in protected nature reserves than any state except Alaska.

Metadata:

EK:	IMP-6.A
Unit:	3
Skill:	3.D Compare patterns and trends in maps and in quantitative and geospatial data to draw conclusions.

49. **(B)** Increases in efficiency, including investments in technologies such as factory robots, have enabled output to continue growing since the 1970s even as employment has declined. Deindustrialization in the United States is thus more a story of a loss of jobs than it is a loss of industries or total productivity.

Metadata:

EK:	PSO-7.A.7
Unit:	7
Skill:	5.B Explain spatial relationships across various geographic scales using geographic concepts, processes, models, and theories.

50. **(C)** Federalism debates often raise the question of states' rights. Under the US Constitution, federal laws passed by Congress supersede those passed by the states. The federal government's authority to legislate is limited, however, to specific "enumerated" areas, whereas the states, which have a "general police power," may pass laws pertaining to almost any issue or topic.

Metadata:

EK:	IMP-4.C.1
Unit:	4
Skill:	2.A Describe spatial patterns, networks, and relationships.

51. **(D)** A superimposed boundary is one that does not reflect the complex social, cultural, or ecological landscape. It often comes in the form of a linear or other geometric boundary.

Metadata:

EK:	IMP-4.A.1
Unit:	4
Skill:	1.D Describe a relevant geographic concept, process, model, or theory in a specified context.

52. **(A)** A metropolitan statistical area is a US Census Bureau designation that pertains to a region containing several smaller urban areas that act together as a whole.

Metadata:

EK:	PSO-6.A.1
Unit:	2
Skill:	2.C Explain a likely outcome in a geographic scenario using geographic concepts, processes, models, and theories.

53. **(D)** Most of the 50 largest metropolitan statistical areas are located east of the Mississippi River, but some of the biggest—such as those centered on Houston, Dallas–Ft. Worth, Phoenix, San Francisco, and Los Angeles—are located west of the Mississippi.

Metadata:

EK:	SPS-6.A.1
Unit	2
Skill:	2.C Explain a likely outcome in a geographic scenario using geographic concepts, processes, models, and theories.

54. **(B)** Environmental justice is "the fair treatment and meaningful involvement of all people regardless of race, color, national origin, or income with respect to the development, implementation, and enforcement of environmental laws, regulations, and policies."

Metadata:

EK:	SPS-6.A.1
Unit:	6
Skill:	4.E Explain how maps, images, and landscapes illustrate or relate to geographic principles, processes, and outcomes.

55. **(D)** Malthus argued that population growth, which increases geometrically, will outstrip food production, which increases only linearly, leading to famine, war, and other "negative checks" on the human population. He is believed to have underestimated the potential of technology to produce more food and the role of distribution systems, markets, and democratic decision-making to supply the food needs of diverse people.

Metadata:

EK:	IMP-2.B.3
Unit:	2
Skill:	2.B Explain spatial relationships in a specified context or region of the world, using geographic concepts, processes, models, and theories.

56. **(D)** A cultural hearth is the place where a cultural tradition begins and from which it emanates outward.

Metadata:

EK:	IMP-3.B.1
Unit:	3
Skill:	4.E Explain how maps, images, and landscapes illustrate or relate to geographic principles, processes, and outcomes.

57. **(D)** A dialect is a local variation of a more widespread language.

Metadata:

EK:	IMP-3.B.1
Unit:	3
Skill:	4.E Explain how maps, images, and landscapes illustrate or relate to geographic principles, processes, and outcomes.

58. **(B)** Naples, FL, a community with a large number of retirees, has the highest proportion of elderly residents.

Metadata:

EK:	SPS-2.C.2
Unit:	2
Skill:	2.C Explain a likely outcome in a geographic scenario using geographic concepts, processes, models, and theories.

59. **(A)** Detroit, MI, has the largest proportion of young people and therefore the highest birth rate.

Metadata:

EK:	SPS-2.C.2
Unit:	2
Skill:	2.C Explain a likely outcome in a geographic scenario using geographic concepts, processes, models, and theories.

60. **(C)** Honolulu, HI, has large proportions of residents in the 20–50 and 60–80 age ranges.

Metadata:

EK:	SPS-2.C.2
Unit:	2
Skill:	2.C Explain a likely outcome in a geographic scenario using geographic concepts, processes, models, or theories.

Section II

FREE-RESPONSE QUESTIONS

1. Answer Explanations and Rubric:

1 point for description of global population growth since 1700	■ Rapid global population growth since 1700, from around 600 million to 7.5 billion
1 point for description of long-term human population size over the past 10,000 years	■ Human population growth was extremely slow over the 10,000 years that preceded 1700. Brief periods of growth in certain regions were often followed by periods of decline. Rapid global population growth did not begin until around 1700.
1 point for explaining the demographic transition model	■ In the demographic transition model, countries go through a series of stages from low levels of economic production and low rates of population growth to higher levels of economic production and high levels of population growth to very high levels of economic production, very high per-capita incomes, and low levels of population growth. In the middle phase, birth rates dramatically outpace death rates, while in the first and third phases, birth rates and death rates are more closely aligned with one another.
1 point for explaining the role of women in rates of population growth	■ When women gain educational and economic opportunities, their roles within their societies shift. In addition to being wives and mothers, they participate in the paid labor force at greater rates and take on other obligations outside their homes. These additional roles tend to lead to lower fertility levels.
1 point for explaining China's one-child policy	■ China's one-child policy, in effect for 36 years, slowed the country's rate of population growth but violated the human rights of millions of people, imposed harsh penalties on some, was unevenly enforced, produced a population with an unnaturally large proportion of men, and had other unintended consequences.

1 point for naming three countries with aging populations and the reasons for their demographic trends	■ Three countries with aging populations are Denmark, Germany, and Japan. A combination of high development and women's empowerment, which leads to low growth rates, and a baby boom in the mid-twentieth century, has created a situation in which these countries have large numbers of aging and elderly residents.
	■ [Dozens of other countries have declining populations. Most are highly developed countries located in North America, Europe, and East Asia.]
1 point for describing why Thomas Malthus's thesis fails to adequately explain the occurrence of famines and the distinction between food production and food availability	■ Malthus focused on people's ability to produce enough food to supply a growing population. He did not, however, sufficiently account for technological change, distribution networks, markets, armed conflict, and political power. All of these forces shape not only the amount of food produced but also the ability of political institutions and economic systems to import, export, transport, deliver, and efficiently allocate food. These factors turn out to be just as important in ensuring that people have enough to eat as the total amount of food produced.

2.   Answer Explanations and Rubric:

1 point for defining a primate city	■ A primate city is a city that is disproportionately large—usually at least twice the size of other cities—and also has more influence compared with other cities in the same country.
1 point for naming two primate cities and their home countries	■ London, United Kingdom, and Mexico City, Mexico, are good examples of primate cities.
	■ [There are dozens of other countries around the world with primate cities. Other good examples include Paris, Buenos Aires, Santiago, Tokyo, Manila, Seoul, Bangkok, Istanbul, Cairo, and Lagos.]
1 point for explaining why so many countries have primate cities	■ Many primate cities became important seats of political power or commerce long ago and continued to attract migrants throughout their histories. Some primate cities were colonial capitals. Others have such distinct advantages as deep-water ports and natural defenses. Today, primate cities are often both political capitals and centers of culture, education, and finance. As they grow and offer new amenities and opportunities, they continue to attract new residents.

1 point for explaining why some countries may not have primate cities	■ Some countries, like Canada, cover so much land area that they have far-flung cities that serve as capitals for entire regions. Other countries, like Germany, were cobbled together centuries ago from multiple independent city-states, each with its own large and historic capital city. Still other countries, like India, have big cities in several regions, each with its own history, cultural legacy, and industries. In the United States, New York is a center of culture and finance, Boston of education, Washington, D.C., of government, Los Angeles of entertainment, and San Francisco of technology.
1 point for defining world cities and explaining the difference between primate and world cities	■ Whereas primate cities are the biggest cities in their countries, world cities are global centers of economic, cultural, or political activity. World cities are categorized into tiers according to the extent of their influence, but all world cities enjoy a significant amount of influence and prosperity. Top-tier world cities include the economic and cultural powerhouses of New York, London, and Tokyo. Second-tier world cities, such as Moscow, Paris, and Washington, D.C., include seats of government.
1 point for defining megacities and explaining the difference between primate cities and megacities	■ Megacities are giant urban regions, often containing 10 million or more people. Although a few megacities are located in highly developed countries, most are now found in less-developed countries of Latin America, Africa, and Asia.
1 point for naming two US world cities and explaining why they qualify as such	■ In the United States, New York and Los Angeles are both world cities due to their profound global economic and cultural influences. Other US world cities include Chicago and Houston.

3.   Answer Explanations and Rubric:

1 point for defining a multinational state	■ Multinational states are countries that contain numerous nations or nationalities within their bounds. Indonesia, with its remarkable cultural, linguistic, religious, and ethnic diversity, is a perfect example.
1 point for defining centripetal and centrifugal forces	■ Centrifugal forces reduce civic cohesiveness and pull countries apart. They include regionalism, ethnic strife, and territorial disputes. Centripetal forces increase cohesiveness and bind countries together. They include strong national institutions, a sense of common history, social trust, and reliance on strong central government.

1 point for explaining how language, culture, and religion can act as centripetal or centrifugal forces	■ States with single dominant languages, cultures, or religions have strong centripetal forces and approach the theoretical ideal of the nation-state: a sovereign governing unit containing a single, coherent group of people. States with great linguistic, cultural, or religious diversity contain people with different life experiences, family backgrounds, and personal attitudes, values, and beliefs. This makes for strong centrifugal forces that can result in conflict or violence or, in extreme circumstances, even civil war.
1 point for explaining the role of colonialism and postcolonialism in assembling and maintaining far-flung countries, like Indonesia	■ Prior to European colonialism, most of today's multinational countries, like Indonesia, were made up of diverse ethnic groups, tribes, city-states, or other forms of social and political organization. Colonial powers redrew these maps, usually for the purposes of extracting resources, facilitating commerce, or controlling local people and thus produced many of the multinational postcolonial countries we know today, including the United States and India.
1 point for explaining how Benedict Anderson's concept of "imagined communities" applies to national identity	■ The idea of imagined communities is central to national identity because most sovereign states are not in fact nations with clear, unified, and longstanding cultural, linguistic, or religious identities. Cultivating a sense of loyalty to state entities is not natural and thus requires ongoing effort.
1 point for explaining how Anderson's idea may apply to Indonesia	■ Indonesia is a country with extreme cultural and linguistic diversity, vast geographic distances, and a challenging environment, all of which contribute to powerful centrifugal forces. Anderson's notion of imagined communities applies to Indonesia because it helps us understand how centralized colonial and postcolonial governments used political, military, educational, and economic institutions, as well as symbolism, narrative, physical force, and other tools, to produce a sense of national identity.
1 point for explaining why Indonesia may be particularly vulnerable to separatist movements	■ As one of the world's most diverse and expansive countries, Indonesia has many groups of people who consider themselves unique, separate, and nearly sovereign. Separatist movements in places like East Timor show that in certain parts of the archipelago, some people still yearn for independence.

Index